Ceramics for Energy Conversion, Storage, and Distribution Systems

Ceramics for Energy Conversion, Storage, and Distribution Systems

Ceramic Transactions, Volume 255

A Collection of Papers Presented at CMCEE-11, June 14–19, 2015, Vancouver, BC, Canada

Edited by
Thomas Pfeifer
Josef Matyáš
Palani Balaya
Dileep Singh
John Wei

Volume Editors
Mrityunjay Singh
Tatsuki Ohji
Alexander Michaelis

The American Ceramic Society

WILEY

Published by John Wiley & Sons, Inc., Hoboken, New Jersey.
Published simultaneously in Canada.

Library of Congress Cataloging-in-Publication Data is available.

ISBN: 978-1-119-23448-7
ISSN: 1042-1122

10 9 8 7 6 5 4 3 2 1

Contents

CERAMIC-RELATED MATERIALS, DEVICES, AND PROCESSING FOR HEAT-TO-ELECTRICITY DIRECT CONVERSION

MATERIAL SCIENCE AND TECHNOLOGIES FOR ADVANCED NUCLEAR FISSION AND FUSION ENERGY

ADVANCED BATTERIES AND SUPERCAPACITORS FOR ENERGY STORAGE APPLICATIONS

Preface

The global challenges we face require innovative thinking and sustainable technology to meet increased demands for energy, clean water, and infrastructure. Research of materials—specifically, ceramic materials—continues to provide solutions to everyday challenges such as environmental protection, energy supply and generation, and healthcare. The 11th International Symposium on Ceramic Materials and Components for Energy and Environmental Applications (11th CMCEE), held June 14-19, 2015 at the Hyatt Regency Vancouver in Vancouver, B.C., Canada, identified key challenges and opportunities for ceramic technologies to create sustainable development.

This Ceramic Transactions volume contains papers submitted from the following six symposia held in Track 1: Ceramics for Energy Conversion, Storage, and Distribution Systems:

- High-Temperature Fuel Cells and Electrolysis
- Ceramic-Related Materials, Devices, and Processing for Heat-to-Electricity Direct Conversion
- Material Science and Technologies for Advanced Nuclear Fission and Fusion Energy
- Advanced Batteries and Supercapacitors for Energy Storage Applications
- Materials for Solar Thermal Energy Conversion and Storage
- High-Temperature Superconductors: Materials, Technologies, and Systems

After a peer-review process, 25 papers were accepted for inclusion in this proceedings volume. The editors wish to extend their gratitude and appreciation to all the symposium co-organizers for their help and support, to all the authors for their cooperation and contributions, to all the participants and session chairs for their time and efforts, and to all the reviewers for their valuable comments and suggestions. We also acknowledge the skillful organization and leadership of the meeting chairs, Mrityunjay Singh, Tatsuki Ohji, and Alexander Michaelis.

We hope that this proceedings will serve as a useful resource for the researchers and technologists in the field of energy conversion, storage, and distribution systems.

THOMAS PFEIFER, Fraunhofer Institute for Ceramic Technologies and Systems—IKTS, German
JOSEF MATYÁŠ, Pacific Northwest National Laboratory, USA
PALANI BALAYA, National University of Singapore, Singapore
DILEEP SINGH, Argonne National Laboratory, USA
JOHN WEI, University of Toronto, Canada

High-Temperature Fuel Cells and Electrolysis

EFFECT OF ADDITIVES ON SELF-HEALING OF PLASMA SPRAYED CERAMIC COATINGS

N. Sata, A. Ansar and K. A. Friedrich
German Aerospace Center (DLR), Pfaffenwaldring 38-40, 70569 Stuttgart, Germany

ABSTRACT

$MgAl_2O_4$ coatings, used for electrical insulation along the metallic seals in high temperature fuel cell stacks, tend to fail prematurely due to generated stresses during on-off thermal cycles. To enhance stress bearing capability and the durability of these coatings, self-healing was investigated by introducing additives consisting of SiC as primary additive along with a secondary additive material. Secondary additives were one of the following compounds: BaO, CaO, ZnO, Y_2O_3, Al_2O_3, La_2O_3, TiO_2, GeO_2, $Ce_{0.9}Gd_{0.1}O_{1.95}$ (GDC) or Ta_2O_5. Crack healing can be attained due to reaction between SiC, secondary additive and oxygen that was transported to the additive due to a crack. Such reaction would be associated to a phase formation in the additive material linked with mass and volume increase assisting in closure of the advancing crack. Using TGA/DSC reaction temperatures and mass gains of SiC with or without secondary additives were identified. SiC+Y_2O_3 and SiC+ZnO were opted as promising additive materials. 20 wt% (SiC+Y_2O_3) containing $MgAl_2O_4$ coatings were produced by plasma spraying. In these developed coatings, healing was demonstrated after heat treatment at 1050°C in air for 10 hour. Defect healing in spinel coating with SiC+ZnO is under investigation.

INTRODUCTION

Defects in the sealing in high temperature solid oxide fuel cells (SOFC) has been reported as the foremost cause of failure in the fuel cell stacks[1,2]. The sealing, traditionally made of glass or glass-ceramic composites, ensures flow of fuel gas and air in designated compartments of SOFC stack. A leakage across the seal leads to catastrophic loss in cell potential and power output. Glass-based seals exhibit limited reliability which suffers further when the stack should undergo thermal transients such as during intermittent operation. In our earlier work[3], an alternative approach was proposed in which Ag-based filler material is used for sealing of two consecutive cells. Despite mismatch of coefficient of thermal expansion (CTE) between filler alloy and neighboring components of stack, the high ductility and creep of the filer alloy compensate for stresses. However, as the filler alloy is electronically conductive, short circuiting between the cells is avoided by introducing an Mg-spinel ($MgAl_2O_4$) insulating coating in between. The spinel deposit is produced by plasma spraying. The schematic of sealing the approach is given in Figure 1. In spite of enhanced reliability compared to glass-based seals, the coating-braze based seals suffer from defects and cracks in the coating. These defects, associated to the manufacturing process or arise due to thermal cycling, give site for further crack nucleation and propagation, decrease the elastic modulus, yield strength and fracture energy of the coatings. At elevated temperatures crack initiation and propagation mechanism changes and failure may occur at the featureless zones, as suggested by Lowrie and Rawlings[4]. Overall, the increase in temperature from room temperature to 800°C caused a 23–30% reduction in flexural strength of bulk 8YSZ. Ansar et al[5] have reported that the elastic modulus of plasma sprayed 8YSZ reduces from 35±2 GPa at room temperature (instead of 120 GPa for bulk material) to 16±1 GPa at 800°C. The decrease in elastic properties of such a coating was almost twice to bulk material and this was associated to the intrinsic elastic modulus of the YSZ but also to the structure of the splat boundaries.

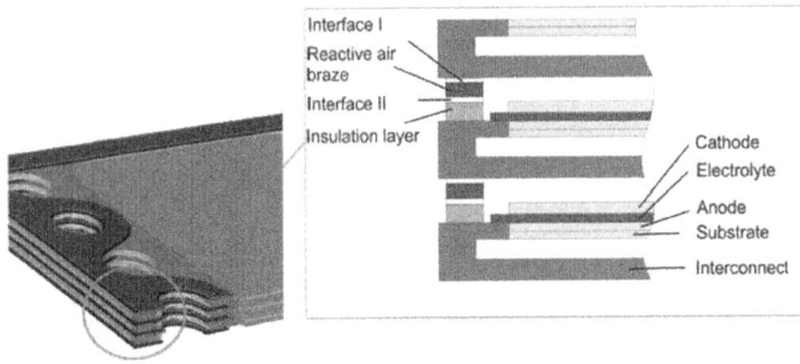

Figure 1: Schematic of the SOFC stack sealing based on active braze and insolating coating.

Catastrophic failure in these insulation layers is a major limiting factor constraining the use of fuel cell in automotive applications. One approach to address this shortcoming consists of incorporating crack healing capabilities, which can offer improved reliability and service time of ceramic components. Though reported already in 1970's[6, 7], this approach has limited work attributed to it to this day. Most studied self-healing ceramics are oxide ceramic matrix composites (CMCs) having 15 to 20 wt% of well distributed SiC particles such as Al_2O_3/SiC[8, 9] and mullite/SiC[10, 11]. It was established that in SiC containing ceramic composites the healing occurs due to oxidation reaction, associated with volume expansion of new phases into the defects[12]:

$$SiC + 2O_2 \rightarrow SiO_2 + CO_2\ (CO) \qquad (1)$$

Other studies suggested healing can be improved by promoting vitreous phase formation by using $SiC+Y_2O_3$[13]:

$$2SiC + Y_2O_3 + 4O_2 \rightarrow Y_2Si_2O_7 + 2CO_2\ (CO) \qquad (2)$$

As a result, cracks up to 100 µm were fully healed and the bending strength was increased by several folds.
Self-healing was also developed in other systems including ZrO_2/SiC by thermal decomposition transformation to $ZrSiO_4$ and carbon black mixture[14]. Additionally, crack healing ability was investigated in non-oxide ceramics such as Si_3N_4 by introducing SiC[15,16]. In all materials presence of oxidizing environment is mandatory for a healing mechanism to occur and in most cases the minimum temperature at which full crack healing takes place has been reported at above 1000°C[8-16].

Incorporation of a material that can heal the defects in insulating ceramic layer in SOFC at elevated temperatures can significantly improve the stress bearing capability and fracture strength of these coatings. Insulating spinel coatings are primarily exposed to oxidizing environment which makes it realistic to achieve a crack healing mechanism. However the potential of healing capability can be best utilized if the reaction can be achieved at operating temperature of fuel cells which is around 800°C. The potential of such an approach is investigated in this paper. A range of potential self-healing additives consisting of primarily SiC along with secondary additives, BaO, CaO, ZnO, Y_2O_3, Al_2O_3, La_2O_3, TiO_2, GeO_2, $Ce_{0.9}Gd_{0.1}O_{1.95}$ (GDC) or Ta_2O_5 were tested and reported. Self-healing was demonstrated in Mg-spinel plasma sprayed insulating coatings using $SiC+Y_2O_3$ as reference additive particles at 1050°C. It was suggested that addition of other tested additive compounds with $SiC+Y_2O_3$ can effectively reduce reaction temperature well below 1000°C and in-operando healing can be potentially attained.

EXPERIMENTAL PROCEDURE

Self-healing additives

SiC and the secondary additive powders studied in this work are listed in Table 1. SiC with two different particle sizes were investigated submicronic sized with d50 of 0.65 µm and nano sized with d50 of 50-60 nm. SiC and additive materials were mixed in molar ratio using an agate mortar and pestle. In order to understand the reaction mechanism of the mixed particles and to select the healing additives, Thermal gravimetric analysis (TGA)/ Differential scanning calorimetry (DSC), X-ray diffraction and Raman spectrometry have been performed. Reaction temperatures were defined by the commencement of weight gain (associated with volume expansion) on TGA data of the additive mixtures.

Table 1: List of the powder samples.

material	particle size	supplier
SiC (submicron)	650 nm	Iolitec
SiC (nano)	50-60 nm	Iolitec
Al_2O_3	1-2µm	Sigma-Aldrich
CaO	< 160 nm	Sigma-Aldrich
TiO_2	1-2 µm	Merck
ZnO	< 100 nm	Sigma-Aldrich
GeO_2	1-2 µm	Sigma-Aldrich
Y_2O_3	30-50 nm	Sigma-Aldrich
BaO	1-2 µm	Sigma-Aldrich
La_2O_3	1-2 µm	Fluka
GDC (GDC10)	1-3 µm	fuelcellmaterials
Ta_2O_5	< 20µm	Sigma-Aldrich

The TGA measurements were conducted in parallel with the DSC measurements using the STA 449C Jupiter® from Netzsch (D). The samples were tested in Pt-Rh crucibles from room temperature to 1200 °C (except for SiC+Ta_2O_5~1100 °C) with a constant heating rate of 5°K/min. Synthetic air (O_2:N_2=20:80) and argon (Argon 5.0) at a flow rate of 30 ml/minwere used as reaction and protection gases, respectively..

The mixed powders were analyzed by X-ray diffraction before and after the TGA/DSC measurements. The measurements were conducted using D8 Discover GADDS, equipped with a VÅNTEC-2000 area detector from Bruker AXS. A tuned monochromatic and collimated X-ray beam (Cu-Kα) was used.

For selected additive mixtures, Raman spectroscopy was conducted with a confocal Raman microscope (LabRam 800, Horiba Jobin Ybon). A green laser line (wavelength= 532 nm) was used for excitation. Mixed powders have been annealed in ambient air for 1, 5 and 10 hrs for Raman measurements.

Coating Fabrication and Characterization

$MgAl_2O_4$ and $MgAl_2O_4$ with SiC+Y_2O_3 were sprayed using air plasma spraying (APS) with a Triplex Pro 210 gun (Oerlikon Metco, Switzerland). 70±6 µm thick coatings were produced on Fe-Cr substrates from Thyssen group, Germany. In coated samples cracks were induced and were then annealed for different temperatures and times to investigate healing by microscopy. The coating samples were also used for electrical conductivity measurements. For conductivity measurements, coatings were air brazed using filler alloy with a counter plate of 50x50 mm² Fe-Cr plate. The procedure is explained elsewhere[3]. Four probe dc measurements were conducted as function of temperature. In addition, free-standing 1 mm thick coatings were prepared for dilatometry. For this purpose, coatings were sprayed on 7 mm x 44 mm plain carbon steel substrates followed by substrate dissolution by acid treatment in dilute sulfuric acid solution (1N) and applying dc voltage between the substrate and a counter metal electrode of 1.0-1.1 V. Self-healing in $MgAl_2O_4$ with SiC+ZnO are under investigation.

RESULTS AND DISCUSSION

Self-healing additives

Figure 2 (a) illustrates the TGA/DSC data of SiC with both particle sizes used in this work. In the DSC curve of nano-SiC, a broad depression is observed up to 900 °C. Such a behavior of nano SiC has been reported as oxygen adsorption on the surface[17,18]. This was corroborated by the absence of this depression in the DSC curve in Ar atmosphere as shown in the figure 2(a). Corresponding weight gain due to oxygen adsorption on nano-SiC powder was less than 1%. Sub-micron SiC powder did not show this effect because of its lower specific surface area. Regardless of SiC particle size, an exothermic reaction above 900°C and an increase in its weight can be attributed to SiC oxidation. The oxidation is an exothermic reaction (heat value=943 kJ) which is described by equation (1).

Both micro and nano sized SiC was mixed Y_2O_3 and the TGA/DSC curves are given in Figure 2(b). In the case of SiC+Y_2O_3 powder mixture, a decrease in the mass up to 600°C is attributed to the evaporation of adsorbed water molecules from the Y_2O_3 surface owning to high surface area of nano-sized Y_2O_3. Such desorption in nano-sized Y_2O_3 was as previously also reported by Kuroda et al.[19]. Despite of the difference in DSC curves due to difference in SiC grain size at lower temperature zone (due to oxygen adsorption with nano-SiC), the data relevant to reaction zone appear independent of SiC particle size. As it is safe to say that there was negligible observable effect of particle size of SiC powders and as it does not influence TGA profile of the mixture significantly, the SiC particle size will not be discussed hereafter.

In Fig. 3, TGA/DSC curves of SiC+Al_2O_3, SiC+TiO, SiC+GDC and SiC+Ta_2O_5 are given. Both TGA and DSC are qualitatively similar to those of SiC. The results demonstrate that these additives have limited influence on promoting SiC oxidation or reaction formation.

The DSC data for SiC+CaO and SiC+BaO show an endothermic peak at the temperature where remarkable weight loss is observed and several exothermic peaks at higher temperatures (Fig. 4 (a)). The endothermic peak can be attributed to desorption of water molecules which have been adsorbed on the powder surface or contained in the powder as hydrates. The results demonstrate that those two samples have high reactivity at high temperatures and very hygroscopic in air. A similar characteristic, a small endothermic peak and remarkable weight loss, is also observed for SiC+La_2O_3 (Fig. 4 (a)). The TGA/DSC profile of SiC+GeO_2 shows a similar behavior as those in Fig. 3, however, a small endothermic peak is observed at 1110 °C which is followed by a remarkable exothermal reaction of about 16 mW/mg above 1180°C. The mixed powder SiC+GeO_2 turned to black in color and crystalline fragments after the measurements. Though the weight gain is faster than pure SiC but is decreased above 1180 °C as is shown in the inset of Fig. 4(b). It is suggested that the SiC starts to be oxidized to gain weight above 1000 °C and then reacted with GeO_2 to form a new phase. This was validated by data from XRD.

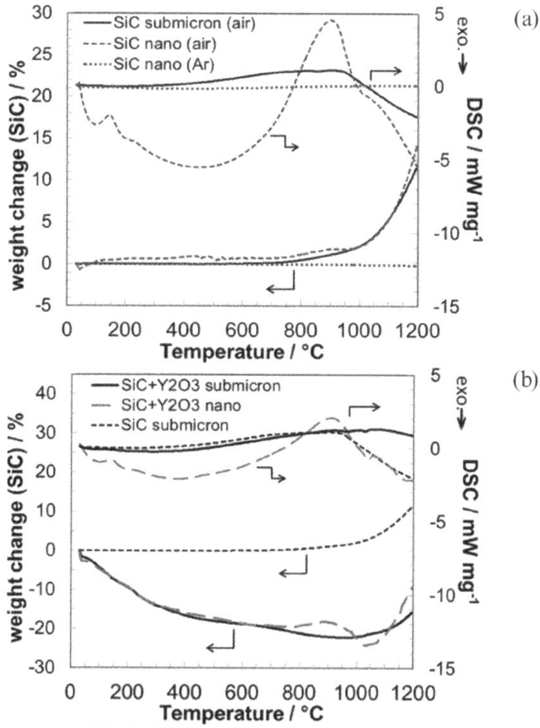

Figure 2: Comparison of TGA/DSC profiles of submicron and nano SiC powders (a) and $SiC+Y_2O_3$ mixtures with submicronic or nano sized SiC (b). Note here that the weight gain values are given with respect to the SiC weight, while the DSC values are as obtained.

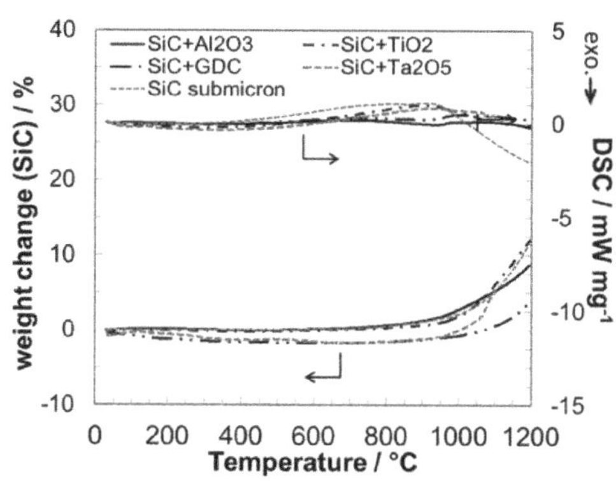

Figure 3: TGA/DSC data of submicron SiC and additives, Al_2O_3, TiO_2, GDC and Ta_2O_5, which show little influence on SiC oxidation. Note here that the weight gain values are given with respect to the SiC weight, while the DSC values are as obtained.

In the case of SiC+Y_2O_3 and SiC+ZnO in Fig. 5, small exothermic peaks are observed around 1050 °C and around 950 °C, respectively, at which or slightly lower temperature weight gain started. As already described above, the weight loss in SiC+Y_2O_3 can be attributed to the evaporation of absorbed water molecules from the Y_2O_3 surface, which is due to the small particle size of this powder. It is noteworthy that the weight gain of these mixtures is larger than that of pure SiC. As the only weight gain comes from oxidation of SiC, the result suggests that oxidation kinetics can be enhanced by these additives.

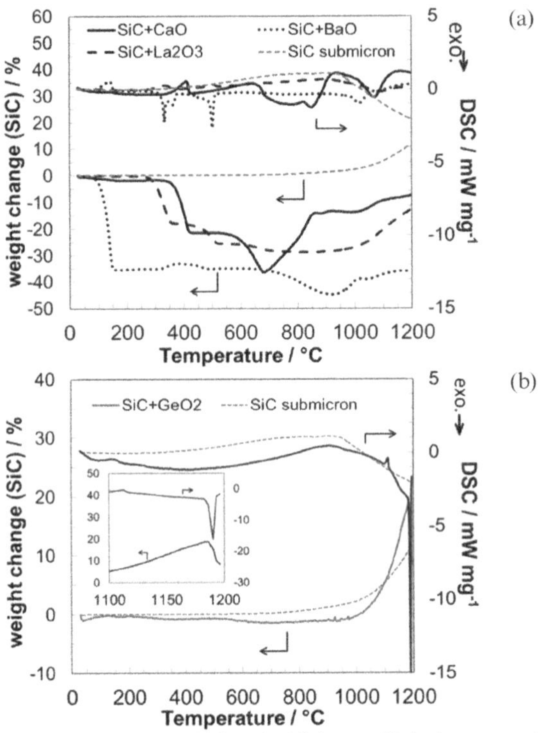

Figure 4: TGA/DSC data of submicron SiC and additives, which show remarkable differences in weight and/or DSC from those of pure SiC. (a) CaO, BaO, La_2O_3, (b) GeO_2.

Fig. 6 (a) shows XRD patterns of SiC powder before and after TGA measurements. Although the intensities of the XRD patterns before and after TGA/DSC are not numerically comparable in the measurement condition, they can be used for qualitative comparison. Remarkable decrease in intensity is noted but no additional peak appeared after TGA in Fig. 6. The powder color change from dark gray to light gray and the weight gain at 1050 °C in TGA demonstrate SiC oxidation, i.e. formation of new oxide phase, such as SiO_2. However no SiO_2 phase, either glassy phase or crystalline phase, was observed in XRD. Since diffraction of glassy phase is very broad and very weak in general, it is suggested most of the oxidized phase was in glassy phase. XRD results of SiC and additive mixtures which showed similar behavior in TGA/DSC, SiC+Al_2O_3, SiC+GDC and SiC+Ta_2O_5, confirmed that no new phase was formed in the mixtures. In the case of SiC+TiO_2, however, it was found that the starting powder of TiO_2 was mainly anatase, which turned to rutile phase after TGA/DSC. It is known that anatase may turn to rutile phase at about 800 °C [20] and the rutile is the most stable TiO_2 phase.

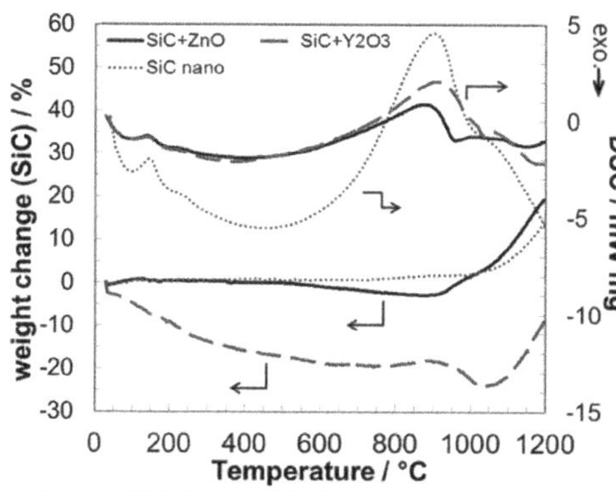

Figure 5: TGA/DSC data of SiC and additives: Y_2O_3 and ZnO

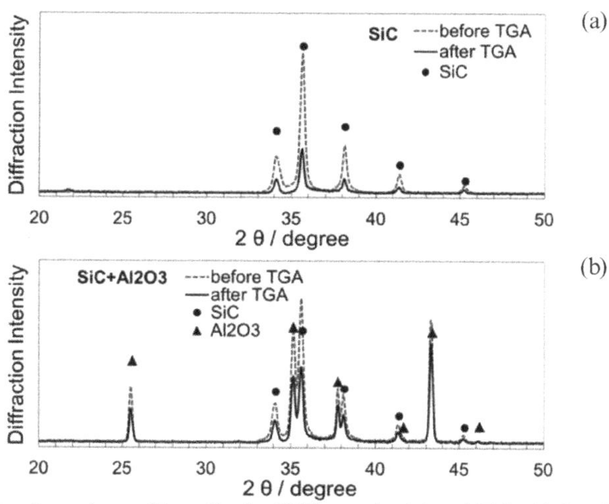

Figure 6: XRD patterns (zeroed base line) of SiC powder (a) and SiC+Al_2O_3 mixture (b) before and after TGA/DSC measurements.

The powder mixtures of SiC+CaO, SiC+BaO and SiC+La_2O_3 show weight losses between 100 and 700 °C, which may be attributed to water or carbon dioxide desorption adsorbed on the secondary additive materials. As shown in Fig. 7 (a), it is confirmed that CaO in the mixture had $Ca(OH)_2$ and $CaCO_3$ phases which may have been formed in ambient atmosphere before the measurement. The SiC+CaO mixture shows a significant change in XRD pattern. The hydroxyl and carbonated compounds disappeared and calcium silicate phase, $Ca_3(SiO_3)_3$ and $CaSiO_3$, were found after the TGA measurement. Note here that only $Ca_3(SiO_3)_3$ is indicated in the figure as a main silicate phase, while $CaSiO_3$ phase is much smaller in intensity. SiC+BaO and SiC+La_2O_3

showed similar behavior and many silicate phases, namely $BaSiO_3$, Ba_2SiO_4 and $Ba_2Si_3O_8$, were found for SiC+BaO mixture, while La_2O_3 and small amount of $La_{9.33}(SiO_4)_6O_2$ were detected for SiC+La_2O_3 mixture after TGA/DSC measurements.

The results for SiC+GeO_2 indicate that GeSi phase was formed (Fig. 7 (b)), which is consistent with the TGA/DSC result. Since the measurement temperature was higher than the stable melting point (1116 °C) and the metastable melting point (1086 °C) of GeO_2 [21], GeO_2 may have reacted with SiC to form the GeSi phase rather to promote the SiC oxidation above this temperature. Though the brazing and SOFC operating temperatures are lower than those melting points, it would be risky to use GeO_2 as a promoting additive.

The XRD data in Fig. 8 for SiC+Y_2O_3 and SiC+ZnO demonstrate that metal silicate phases, Y_2SiO_5 and Zn_2SiO_4, are formed in the mixtures. For both mixtures, remarkable decrease of SiC diffraction intensity is also noticed. Considering highermolar ratio of these metal silicates (metal:Si=2:1) than the composition of the mixed powder (metal:Si=1:1), it is possible that SiC oxidized in SiO_2 phase at the same time. This could probably be formed in a glassy phase, since no crystalline SiO_2 is observed by XRD.

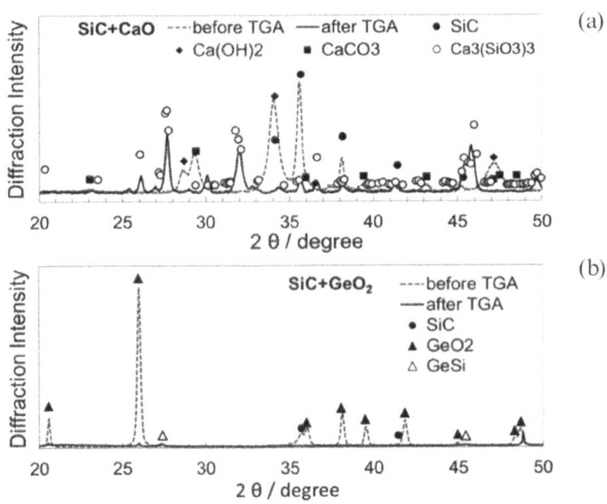

Figure 7: XRD patterns of the mixtures before and after TGA/DSC (zeroed base line): (a) SiC+CaO, (b) SiC+GeO_2.

The interaction between SiC and the additive was further investigated by the microscopic Raman spectroscopy. In Figs. 9, Raman spectra of nano-SiC and nano-SiC+ZnO mixture are given. Si-C vibration modes are observed between 700 and 1000 cm^{-1}. These are corresponding to the TO and LO phonons of Si-C, whose intensity is significantly weaker and broader than that of bulk crystalline SiC because of the small grain size[22,23]. The broad structure above 1000 cm^{-1} is corresponding to the stretching modes of the C-C bonds[24]. Such C-C vibrations are of amorphous or graphite carbons on the surface[17], which can be present on nano-sized SiC powder. Raman spectrum of nano-SiC and ZnO mixture is very similar to that of pure nano-SiC. This is due to significantly low raman scattering intensity of ZnO.

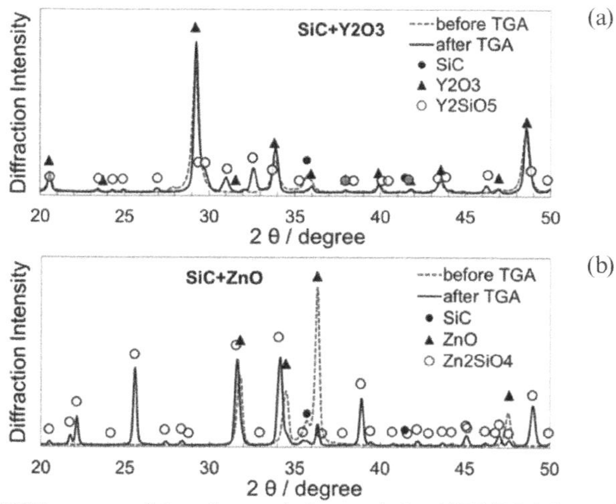

Figure 8: XRD patterns of the mixtures before and after TGA/DSC (zeroed base line): (a) SiC+Y_2O_3, (b) SiC+ZnO.

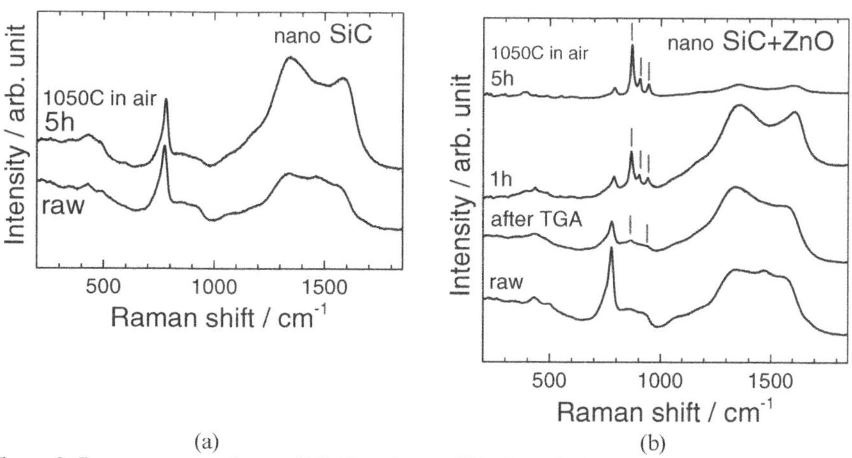

Figure 9: Raman spectra of nano SiC (a) and nano SiC+ZnO (b) heat treated at 1050°C in air for different times and after TGA measurement. The peaks appeared at 868, 903 and 942 cm^{-1} are the Raman modes of Zn_2SiO_4.

After annealing in ambient air for 5 hours, intensity of the C-C vibration modes of nano SiC powder was increased, while little change of the Si-C modes was noticed. On the other hand, the SiC-ZnO mixture showed two major changes after TGA measurement. One is intensity decrease of Si-C modes and the other is intensity increase of C-C modes. The latter is similar to that of nano-SiC powder, however, the former is possibly to the influence of ZnO, as is observed in TGA and XRD results in which promotion of SiC oxidation is suggested. When SiC-ZnO mixture was annealed in ambient air, Si-C modes is further decreased by annealing time and new Raman peaks were observed. The new peaks, indicated by vertical lines in Fig.9(a), are of Zn_2SiO_4 phase [25], which is consistent with the XRD results. Although the peak intensity is low, the silicate phase is also detected after TGA as indicated in Fig.9(a). The stretching modes of the Si-C and C-C bonds are significantly weaker when the mixture is annealed for 5 hours in contrast to the case of nano-SiC. This suggests that higher kinetics of the healing reaction can be attained in the case of the mixture.

From these data Y_2O_3 and ZnO appears as potential secondary additives: BaO should be excluded because of its high reactivity and lack of stability. CaO and La_2O_3 are also not suitable secondary additives for the self-healing coating, since they are also instable and they do not show any remarkable promotion for the SiC oxidation. The melting point of GeO_2 is very close to the brazing temperature. Therefore, it is risky to be reduced to form SiGe phase like in the TGA measurement. No other secondary additive promoted the oxidation reaction or reactive phase formation. The addition of Y_2O_3, despite of its weight loss due to the evaporation of absorbed water molecules in the Y_2O_3 surface, can promote the weight gain – thought the reaction temperature was almost identical to SiC alone. It is evident that ZnO is the most promising candidate which initiated and promoted the SiC oxidation at temperatures below 1000°C. Therefore, ZnO and Y_2O_3 would be considered further as the most promising secondary additives for the coating. Although Al_2O_3 alone did not exhibit reaction with SiC or promotion of the SiC oxidation, it is considered as a good catalyst for vitreous phase formation and therefore Al_2O_3 has been also selected as a candidate additive for the further investigation.

Self-Healing Coatings

Figure 10 illustrates the structure of self-healing air plasma sprayed coating made of $MgAl_2O_4$ as gray matrix, and SiC+Y_2O_3 as bright splats of additive material, In the higher magnification micrograph of a additive splat (Fig. 10 (b)), it can be observed that the dark particles of SiC are embedded in the brighter Y_2O_3. The self-healing coatings were fractured and heat treated at 1050°C in air for 1, 5 and 10 hours. As it can be seen from the micrographs of the as-sprayed fractured coating and the ones heat treated for one hour in Figs. 11 (a) and (b), crack healing was already observed after 1 hour. However limited numbers of cracks with median length of less than 1 µm were healed. This can also be attributed to a "thermal healing". Hence, despite indication from the SiC+Y_2O_3 data from that a reactive phase formation may already occur after one hour, it was not conclusive when tested in a plasma sprayed coating for 1 hours. On the other hand, reactive vitreous phase formation was evident after 5 and 10 hours of heat treatment and larger defects could be healed (Fig. 12 (a to d). Within a matrix of parent ceramic, it can be expected that the additive material would have access to lower concentration of oxygen and slower transformation kinetics.

The resistance data of pure spinel and spinel + 20wt% (SiC+Y_2O_3) coatings as a function of temperature is given in Fig. 14, revealing that the self-healing additive did not deteriorate the insulating properties of the coatings. The coefficient of thermal expansion (CTE) data, as summarized in Table 2, established that CTE of the spinel coating decreased as a function of SiC content which can be explained by the lower CTE of SiC. In the presence of composite additive (SiC+Y_2O_3), the CTE values were in between that of pure spinel and spinel+20wt%SiC which can be owing to the comparable CTE of Y_2O_3 to that of spinel. The coating, heat treated at 1050°C in air for 10 hours, had a comparable CTE to as-sprayed counterpart.

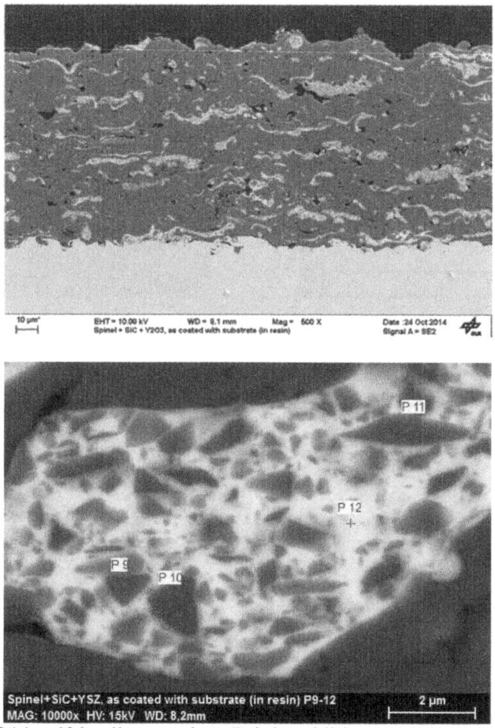

(a)

(b)

Figure 10: SEM of (a) self-healing coating: Bright additive particles are distributed in dark spinel matrix. (b) a splat of additive material. Dark particles are SiC and bright zone is Y_2O_3.

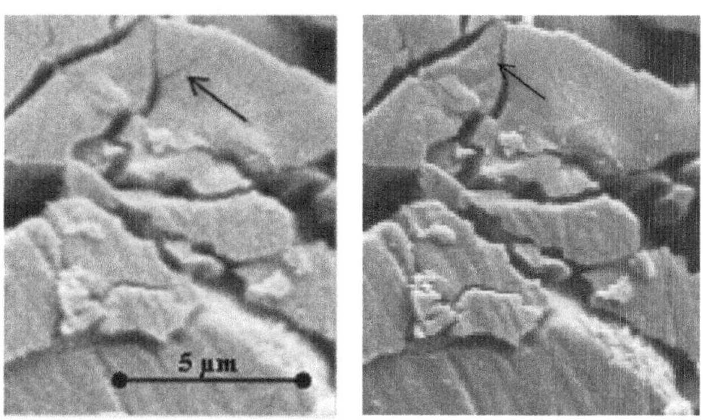

(a) (b)

Figure 11: (a) SEM micrographs of self-healing ceramic coating of in as-sprayed condition, and (b) of same zone and magnification after heat treatment at 1050°C for 1 hour in air.

Figure 12: (a and c) Micrographs of fractured self-healing ceramic coating in as-sprayed condition, and (b and d) same zone and magnification after heat treatment at 1050°C for 5 hours (b) and 10 hour (d) in air.

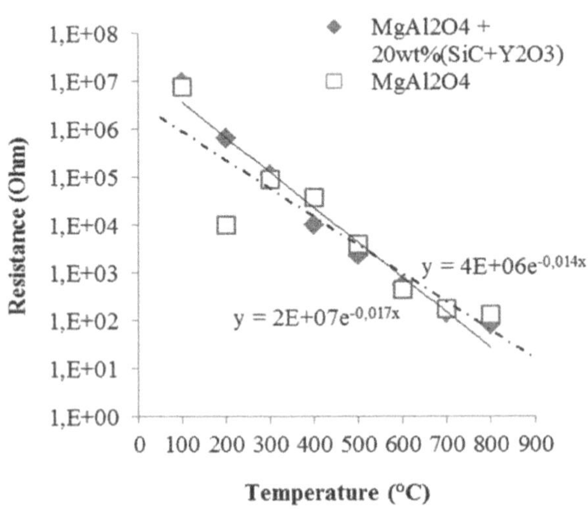

Figure 13: Electrical resistance values for $MgAl_2O_4$ and $MgAl_2O_4$+20wt% ($SiC+Y_2O_3$) coatings.

Table 2: CTE values of spinel and spinel+additive coatings

Coating Material	Technical α (10^{-6} /K^{-1})	
	At 800°C	At 850°C
$MgAl_2O_4$	8.31	8.39
$MgAl_2O_4$+ 5wt%SiC	8.07	8.15
$MgAl_2O_4$+ 20wt%SiC	7.94	8.01
$MgAl_2O_4$+ 20wt%(SiC+Y_2O_3) - As-sprayed	8.08	8.16
$MgAl_2O_4$+ 20wt%(SiC+Y_2O_3) - After heat treatment	8.06	8.13

CONCLUSIONS

The current work attempted to incorporate self-healing capability in $MgAl_2O_4$ ceramic coatings by introducing healing additives. Moreover, as the coating was employed in solid oxide fuel cells, it was targeted to select suitable additives that can engender healing at cell operating temperature, typically 800°C. As healing additive, micro-sized or nano-sized SiC was investigated with addition of one of the following BaO, CaO, ZnO, Y_2O_3, Al_2O_3, La_2O_3, TiO_2, GeO_2, $Ce_{0.9}Gd_{0.1}O_{1.95}$ (GDC) or Ta_2O_5. It was established that decreasing the size of SiC particles has only marginal effect on promoting its oxidation or reaction with secondary additives. Moreover, independent of the size of the particles, with only SiC as additive, any healing reaction may be observable after heat treatment at 1050°C for more than 10 hours; SiC was still not fully converted in SiO_2 in exposed areas. Addition of Y_2O_3 was established to promoted reaction kinetics. Though the temperature of reaction temperature remained comparable to SiC, i.e. above 1000°C, the reaction time was reduced to 1 hour. However, once embedded in a matrix of coating, healing was demonstrated at 1050°C but for dwell time of 5 hours and more. 10 hours treatment was opted for healing. Functional properties of the self-healing insulating coating, CTE and resistivity, remained comparable to spinel coatings. For sub-1000°C healing, promising healing results were attained when ZnO instead of Y_2O_3 was used as secondary additive. TGA/DSC data revealed that temperature of healing reaction could be reduced to 950°C. This, however, need to be established with further trials which are on-going. The coatings of $MgAl_2O_4$ with SiC+ZnO will also be investigated.

REFERENCES

[1]S. Taniguchi, M. Kadowaki, T. Yasuo, Y. Akiyama, Y. Miyake and K. Nishio, "Improvement of Thermal Cycle Characteristics of a Planar-Type Solid Oxide Fuel Cell by Using Ceramic Fiber as Sealing Material", *J. Power Sources,* 2000, 90 (2), p 163–169.
[2]C.-K. Lin, T.-T. Chen, Y.-P. Chyou, L.-K. Chiang, "Thermal Stress Analysis of a Planar SOFC Stack", *J. Power Sources*, 2007, 164 (1), p 238-251
[3]J. Arnold, A. Ansar U. Maier, R. Henne, *International Thermal Spray Conference (ITSC),* (org.) German Welding Society (DVS), Maastricht, Netherlands 2008, CD-version
[4]F.L. Lowrie, R.D. Rawlings, "Room and High Temperature Failure Mechanisms in Solid Oxide Fuel Cell Electrolytes", *J. Eur. Ceram. Soc*,. 2000, 20 (6), p 751–760.
[5]A. Ansar, G. Antou, and A. Denoirjean, *Proceedings of ITSC 2007,* (org.) ASM international, CD-version.
[6]J.J. Petrovic, L.A. Jacobson, "Controlled Surface Flaws in Hot-Pressed SiC", *J. Am. Ceram. Soc.*, 1976, 59 , p 34-37
[7]T.K. Gupta, "Crack Healing and Strengthening of Thermally Shocked Alumina", *J. Am. Ceram. Soc.*, 1976, 59 [5-6], p 259-262.
[8]Z. Chlup, P. Flasar, A.Kotoji, I. Dlouhy, "Fracture Behaviour of Al_2O_3/SiC Nanocomposite Ceramics after Crack Healing Treatment", *J. Eur. Ceram. Soc.*, 2008, 28 (5),p 1073-1077.
[9]K. Ando, K. Furusawa, K. Takahashi, M.C. Chu, S. Sato, "Crack-Healing Behavior of Structural Ceramics under Constant and Cyclic Stress at Elevated Temperature", *J. Ceram. Soc. Jpn.*, 2002, 110 (8), p 741-747.
[10]M.C. Chu, S. Sato, Y. Kobayashi, K. Ando, "Damage Healing and Strengthening Behaviour in Intelligent Mullite/SiC Ceramics", *Fatigue Fract. Eng. Mater. Struct.*, 1995, 18 (9), p 1019-1029
[11] S.K. Lee, M. Ono, W. Nakao, K. Takahashi, K. Ando, "Crack-Healing Behaviour of Mullite/SiC/Y_2O_3 Composites and Its Application to the Structural Integrity of Machined Components", *J. Eur. Ceram. Soc.*, 2005, 25 (15), p 3495-3502.
[12]S.R. Choi, V. Tikare, "Crack Healing Behavior of Hot Pressed Silicon Nitride Due to Oxidation", *Scripta Metall. Mater.*, 1992, 26 (8), p 1263-1268.
[13] K. Ando, M.C. Chu, S. Matsushita, S. Sato, "Effect of Crack-Healing and Proof-Testing Procedures on Fatigue Strength and Reliability of Si_3N_4/SiC Composites", *J. Eur. Ceram. Soc.*, 2003, 23 (6), p 977-984.
[14]B.Y. Ma, J.K. Yu, Q. Zhu, Y. Sun, "Thermodynamic Analysis and Preparation of β-SiC/ZrO_2 Composites ", *Int. J. Min. Metall. Mater.*, 2009, 16 (5), p 581-585.
[15] K. Takahashi, Y.S. Jung, Y. Nagoshi, K. Ando, "Crack-Healing Behavior of Si_3N_4/SiC Composite under Stress and Low Oxygen Pressure", *Mater. Sci. Eng. A*, 2010, 527 (15), p 3343-3345.
[16] M. Kasiarova, J. Dusza, M. Hnatko, P. Sajgalik, "Microstructure and Fracture-Mechanical Properties of Carbon Derived Si_3N_4 + SiC Nanomaterials", *Mater. Sci. Eng. C*, 2006,26 (5-7), p 862-866.
[17]K. Shimoda, T. Koyanagi, Surface Properties and Dispersion Behaviors of SiC Nanopowders, *Colloids and Surfaces A: Physicochem. Eng. Aspects*, 2014, 463, p 93–100.
[18]J. Zhan, H. Wang, F. Zhu, S. Song. "Analysis on the Governing Reactions in Coal Oxidation at Temperatures up to 400°C", *Inter. J. Clean Coal and Energy*, 2014, 3, p 19-28.
[19]Y. Kuroda, H. Hamano, T. Mori, Y. Yoshikawa, M. Nagao, "Specific Adsorption Behavior of Water on a Y_2O_3 Surface", *Langmuir*, 2000, 16, p 6937-6947.
[20]E. M. Levin, C. R. Robbins and H. F. McMurdie, "Phase diagrams for cermists", third edition, 1974, The American Ceramic Society, inc.
[21]N. A. Toropov, V. P. Barzakovskii, I. A. Bondar′, and Yu. P. Udalov, "Handbook of phase diagrams of silicate systems, vol. II, metal-oxygen compounds in silicate systems", second revised

edition, Leningrad, 1970, translated from Russian, translated by J. Schmorak and R. Kondor, Israel Program for Scientific Translations Ltd. , Jerusalem, 1972.
[22]Junho Jeong, Gwiy-Sang Chung and Shigehiro Nishino, Raman Scattering Investigation of Polycrystalline 3C-SiC Film Deposited on SiO_2 by Using APCVD with Hexamethyldisilane, *Journal of the Korean Physical Society*, 2008, 52, p. 43-47.
[23]A. Prez-Rodriguez, Y. Pacaud, L. Calvo-Barrio, C. Serre, W. Skorupa, and J.R. Morante, Analysis of Ion Beam Induced Damage and Amorphization of 6H-SiC by Raman Scattering, *Journal of Electronic Materials*, 1996, 25(3), 541-547.
[24]M. Havel, Ph. Colomban, "Rayleigh and Raman Images of the Bulk/Surface Nanostructure of SiC Based Fibres", *Composites B*, 2004, 35, p 139–147.
[25]Chung-Cherng Lin and Pouyan Shen, The Role of Ti^{4+} on the Structure and Transformations of Gel-Produced Zn_2SiO_4, Journal of Solid State Chemistry, 1994, 112, p 381-386.

DEVELOPMENT OF CERAMIC FUNCTIONAL LAYERS FOR SOLID OXIDE CELLS

Günter Schiller, Rémi Costa, K. Andreas Friedrich
German Aerospace Center (DLR), Institute of Engineering Thermodynamics, Pfaffenwaldring 38-40, 70569 Stuttgart, Germany

ABSTRACT

Ceramic multi-layer systems are the key components of solid oxide cells (SOC) to be used either for power generation or energy storage. Over the last decades continuous improvement in materials, architecture and manufacturing processes have been achieved to improve performance, durability and lifetime. The advanced concept of a metal-supported SOC where the functional ceramic layers are deposited onto a mechanically stable porous metal support is the most advanced approach for mobile application as auxiliary power units (APU) having strict requirements in terms of low volume, limited weight and improved ability for rapid start-up and thermal cycling. The paper provides an overview of the metal-supported cell development at DLR where functional layers are consecutively deposited by plasma spray technology onto a metal substrate. The current status of this development including materials aspects, stack technology and electrochemical performance is reported. When operating temperature is governed mainly by the electrocatalytic properties of the functional electrode materials and thickness of the electrolyte, thin-film electrolyte opens up the opportunity of low temperature operating SOC. Recently, further research efforts have started to develop a metal-supported cell with thin-film electrolyte. Hermetic thin films have been applied through PVD technology resulting in bilayer structured coatings. Recent results of this development are also presented.

INTRODUCTION

Solid oxide cells for both fuel cell (SOFC) and electrolyzer operation (SOEC) have been developed worldwide for many years because of their promise for highly efficient mutual conversion of chemical and electrical energy and their fuel flexibility. As a consequence, continuous improvement in materials, cell design and manufacturing processing has been achieved in terms of performance and durability, but successful commercialization is still in an initial stage. Different cell types - planar and tubular – with several designs with respect to the supporting component, such as anode-supported cells (ASC), electrolyte-supported cells (ESC) and cathode-supported cells (CSC) have been developed mainly using wet powder processing and sintering techniques as manufacturing technology. In recent years, alternative metal-supported cell designs (MSC) have attracted high interest driven by their low material cost, ruggedness and manufacturing benefits[1,2]. At DLR, an advanced concept of a metal-supported SOC based on thermal spray technology has been developed for mobile application as auxiliary power unit (APU) having strict requirements in terms of low volume, limited weight and improved ability for rapid start-up and thermal cycling. The cell development has been transferred to stack level. In contrast to the ASC, ESC and CSC designs where the mechanical support is a brittle ceramic or cermet, the MSC concept utilizes ceramic layers only as thick as necessary for electrochemical functions. The mechanical support is made from inexpensive and robust porous metal and the electrochemically active layers are applied directly to this metal support. The MSC concept provides significant cost benefits due to reduction of expensive material input and manufacturing advantages. Furthermore, rapid thermal cycling, redox tolerance and other operational advantages make it a very promising candidate for commercialization.

When operating temperature is governed mainly by the electrocatalytic properties of the functional electrode materials and thickness of the electrolyte, thin-film electrolyte opens up the opportunity of low temperature SOC operation. Recently, further research efforts have started at

DLR to develop an innovative metal-supported cell with thin-film electrolyte applied through PVD technology, using wet powder processing techniques such as tape casting and screen printing for the manufacture of the electrodes. This paper provides an overview of the metal-supported cell development at DLR with both MSC concepts as well as the technical achievements to date.

DEVELOPMENT OF SOC FUNCTIONAL LAYERS BY APPLYING THERMAL SPRAY TECHNOLOGY

DLR spray concept for SOC

The principle of the planar DLR concept of a metal-supported SOFC which is based on plasma deposition technology for the manufacture of the entire cell is shown in Fig.1. The mechanical strength of the cell as well as excellent electrical and thermal conductivity is provided by an open porous metal substrate which also serves as a fuel gas distributor. All functional layers of the cell - anode, electrolyte and cathode - are consecutively deposited onto this substrate by a multi-step vacuum plasma spray (VPS) process in a single procedure[3]. The move from a ceramic to a metallic substrate support reduces considerably the problem of crack formation resulting in larger cell size and simplified stack design. A precondition for such a cell and stack design with porous metal support is the availability of a long-term stable and corrosion-resistant material with an adequate thermal expansion behaviour related to the ceramic cell components.

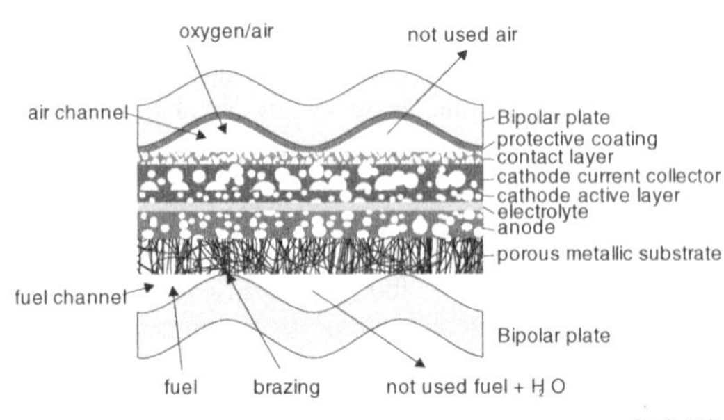

Figure 1. Principle of planar SOC design according to DLR spray concept

In cooperation with an industrial partner a stack design for mobile application has been developed on the basis of plasma sprayed cells and ferritic steel interconnects. This design is based on thin stamped metal sheets which are laser welded to form a metallic cassette arrangement. The integration of the porous substrate is done by welding and brazing of the substrate into the cassette. Cell layers are then plasma- sprayed on the porous substrate-interconnector unit. During this coating process the gas tight electrolyte layer is also sprayed over the weld seam which connects the substrate and the frame. With this process step a seal on the side of the cell is established which is more reliable than the glass seal alone. The brazing and welding of the substrate also provides an excellent electrical contact to the anode side. The concept with integrated metal substrate is shown in Fig. 2.

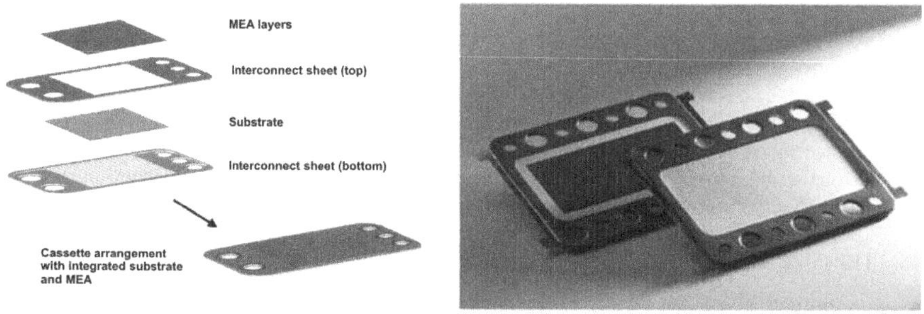

Figure 2. Cassette arrangement with plasma sprayed cells for stack assembly

MANUFACTURING OF FUNCTIONAL SOC LAYERS BY PLASMA SPRAY DEPOSITION

The metal support on which the functional electrode and electrolyte layers are deposited is a core component of plasma-sprayed cells. It has to exhibit a sufficient open porosity of at least 40%, an adapted coefficient of thermal expansion to that of the electrolyte to enable rapid thermal cycling, a low oxidation rate and low cost. In early investigation at DLR ferritic stainless steel substrates with high chromium (> 20 wt.-%) and low manganese content (~ 0.02 - 0.05 wt.-%) were identified to be well suited for this application. In particular corrosion stability was found to be adequate. For this reason a powder metallurgically (P/M) manufactured porous ODS (oxide dispersive strengthened) FeCr-alloy which forms a protective oxide scale during cell operation developed by Plansee SE, Austria, is the preferred choice to be used as metal support. Figure 3 shows a typical morphology of the Plansee manufactured substrates illustrating the porous structure characteristics. The pore size distribution is in the range of about 5 - 50 µm and the porosity could be determined to approx. 48 vol.% by means the quantitative image analysis (QPA) measurement technique.

Furthermore, a diffusion barrier layer with a thickness of 10-15 µm consisting of LSCr and deposited by plasma spraying has been introduced to prevent the diffusion of iron into the SOFC anode and the diffusion of nickel into the substrate[4].

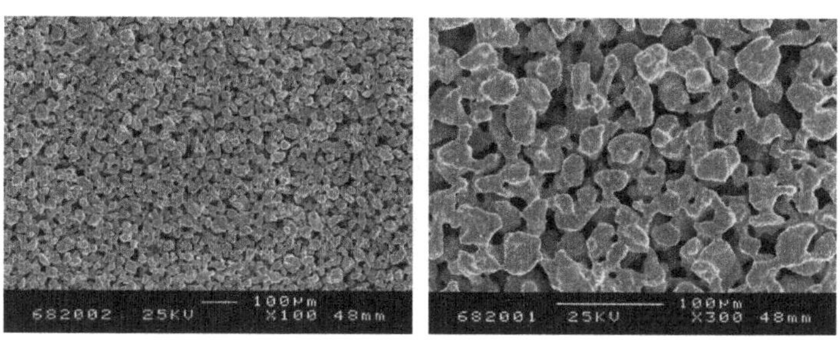

Figure 3. SEM image of typical morphology of the metallic PM Fe-26Cr-(Mo,Ti,Mn,Y_2O_3) porous substrate from Plansee

For the deposition of the ceramic layers special plasma spraying equipment has been developed and used at DLR. Fast plasma jet and spray material velocities which are needed for dense electrolyte layers were realized by using a direct current (DC) torch with Laval-like contoured anodes. The torch was integrated into a low pressure chamber[5]. To fabricate the electrode layers a three-cathode plasma gun TriplexPro 200 (Oerlikon Metco, Switzerland) was applied under atmospheric conditions (APS). Besides standard-sized powder particles (20-60 µm) also suspensions with sub-micron particles were injected. An in-house developed injection system for suspension plasma spraying (SPS) and solution precursor plasma spraying (SPPS) was adapted to get denser electrolyte coatings as well as electrode coatings with higher triple phase boundaries (TPB).

Although there has been extensive study on alternative solid electrolyte materials, e.g. scandium doped zirconia (ScSZ), gadolinia doped ceria (GDC), lanthanum gallate-based perovskite (LSGM)[6], the most commonly used electrolyte material for SOFC is still 8 mol% yttria-stabilized zirconia (8YSZ) due to its high stability and low electronic conductivity at elevated temperatures[7]. However to reduce the resistivity related to oxygen ion diffusion in 8YSZ, either higher operating temperature or thinner electrolyte layers are options. The production of thin electrolyte layers for operating temperatures below 800 °C is a major challenge for plasma-sprayed SOFC. Using a D-optimal design of experiments (DOE), influence of different 8YSZ feedstock powders and plasma spray parameters were investigated on deposition efficiency, gas-tightness and electrochemical behaviour of vacuum plasma sprayed (VPS) electrolytes [8]. It was found that the influence of particle velocity on open porosity of the deposit is four times higher than the particle temperature. Owing to this finding thickness and area-specific leak rate could be reduced from 65 µm to 35 µm and from 15.2 to 1.7 Pa m/s, respectively, which consequently resulted in lower ohmic resistance and higher open circuit voltage (OCV) for cells. Despite this tremendous progress in plasma spraying of electrolytes further improvement is needed. For instance, the ionic conductivity of as-sprayed layers is threefold lower than that of sintered ones which is attributed to the splat boundaries which hinder ion diffusion across the deposit[9]. In order to enhance ionic conductivity as well as gas-tightness of the electrolyte layer, experiments have been started with suspension plasma spraying (SPS) using nanostructured powders.

The requirements of the anode to be fulfilled are a high electronic conductivity to minimize electric losses and to supply electrons and to permit access of the fuel due to its high porosity. The electrochemical performance of an anode layer is mainly governed by active reaction sites (triple phase boundary: TPB) where oxide ions from the electrolyte, fuel gas and electrode are in close contact. Usually, a cermet of Ni and YSZ is used as anode material. With plasma spraying NiO and YSZ powder feedstock is used; NiO is then reduced during operation by hydrogen to nickel. A further increase of the anode porosity to around 25-30 vol.% is obtained during the reduction process. An optimal ratio between 8YSZ and nickel after reduction is considered to be around 70/30 vol.%. Another challenge for plasma-sprayed SOC anodes is to achieve a surface roughness as low as possible. The smoother the anode the thinner a gas-tight plasma sprayed electrolyte can be deposited on top of it. Generally, the roughness in plasma spraying is limited by the splat characteristics and, thus, by the powder particle size.

As cathode material mainly perovskite-type oxides such as $La_{0.6}Sr_{0.4}MnO_3$ (LSM) or $La_{0.6}Sr_{0.4}Co_{0.2}Fe_{0.8}O_3$ (LSCF) are used[10]. Compared to the anode the porosity of the cathode is limited with conventional plasma spraying to less than 20% due to the lack of the "pore forming" reduction process taking place at the cathode. Higher cathode porosity can be obtained by developing mixed nano-micro-structured cathodes. The nanoparticles offer a high effective surface area, however, nanoparticles tend to agglomerate and sinter at cell operating temperatures. Micrometer-sized particles were therefore introduced to impede sintering of the nanoparticles. A 20 to 80 wt.% ratio of nanometer- to micrometer-sized particles was adapted in which the the nanometer-sized particles were in the range of 60 to 90 nm and micometer-sized particles were 0.8 to 2 µm. The particles were agglomerated to a size of 5-45 µm using spray-drying prior to plasma spraying. Special care is needed during plasma spraying to avoid decomposition of these materials at high temperature and reducing atmosphere. A porosity beyond 20% was achieved in this way. A metallographic cross section of an entirely plasma sprayed cell deposited onto a porous metallic substrate is shown in Fig. 4.

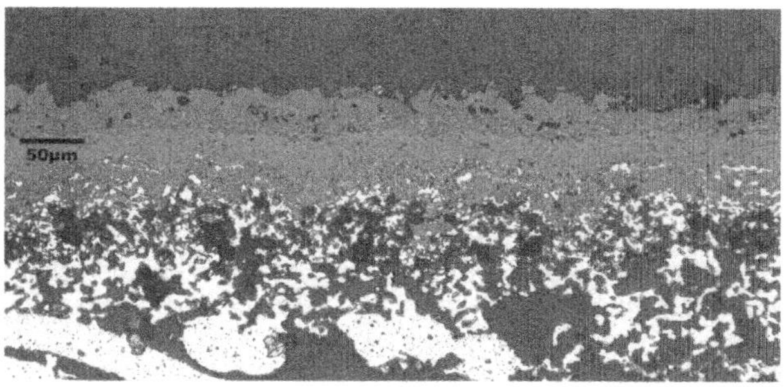

Figure 4. Metallographic cross-section of an entirely plasma sprayed cell on metallic support

Besides the manufacture of active layers as described, plasma sprayed coatings can also be applied for corrosion protection, for insolation between cell components and for sealing in SOC stacks which has been described in more detail in[4,11].

ELECTROCHEMICAL PERFORMANCE OF METAL SUPPORTED SOC

Electrochemical test results of small cells with an area of 12.5 cm^2 using simulated reformate gas as fuel ($H_2/N_2/H_2O$:48.5/48.5/3.0 vol.%) and air as oxidizing gas showed at an operating temperature of 800 °C a power density of up to 800 mW/cm^2. Up-scaling to 85 cm^2 and 100 cm^2 effective area cells was accomplished and the cells revealed a power density of 400 mW/cm^2 at a fuel utilization of 32% at 800 °C. The test of short-stacks and stacks up to 10 cells (250 W) gave a power density > 300 mW/cm^2 [2]. Further improvement in stack power is needed and can be achieved with improved stack design to decrease enhanced ohmic losses.

Plasma-sprayed cells show a superior resistance to strong temperature and redox cycling tests compared to ASC cells. Strong anode reduction during SOFC operation was conducted by switching off the fuel gas abruptly and purging with nitrogen for 5 min. to remove residual hydrogen. Subsequently, 50 ml/min pure oxygen was fed for 1 hour to oxidize the nickel in the anode completely. 20 of such redox cycles together with 15 thermal cycles could be performed with a 10-cell stack. The degradation rate after 20 cycles was 9.1% due to micro-cracks caused by volume changes during redox cycling. A long-term test of a 4-cell stack over 5000 h showed a degradation rate of 10.7%/1000 h (4%/1000 h during the first 2500 h) which is a promising value due to the fact that no protective coatings had been used. For industrial applications the degradation rate must be further reduced[2].

The metal supported cells were also tested in electrolysis mode. During electrolysis operation at an operating temperature of 850 °C a cell voltage of 1.28 V was achieved at a current density of -1.0 A cm^{-2}; at 800 °C the cell voltage was 1.40 V at the same operating conditions. The impedance spectra revealed a significantly enhanced polarization resistance during electrolysis operation compared to fuel cell operation which was mainly attributed to the hydrogen electrode. During a long-term test run of a single cell over 2000 hours a degradation rate of 3.2% per 1000 hours was observed for operation with steam content of 43% at 800 °C and a current density of -0.3 Acm^{-2} [12].

BEYOND THE PLASMA SPRAY CONCEPT

Though showing improved mechanical properties, the developed MSC concept still face critical challenges correlated with the nature of the structural and electro-catalytic materials employed so far. Indeed, considering long term application – 5000 hours and more – the porous ferritic stainless steel substrate suffer irreversible corrosion at operating conditions which makes the MSC concept less attractive for stationary applications. With the target to operate the SOFC system with syngas, a major drawback remains the sensitivity of traditional anodes regarding sulfur poisoning. This issue is encountered in all present cell designs, if the anode is based on the traditional cermet system, namely nickel/yttria-stabilized zirconia, which moreover suffer irreversible degradation over redox cycles[1]. Among alternative anodic electro-catalysts, compounds based on strontium titanate have shown high potential[13] and demonstrated capabilities to withstand redox cycles[14].

Aiming at replacing nickel, as electro-catalyst from the anode, while increasing lifetime of the metal supported cell, a new cell design has been recently conceived and is being developed in the frame of an European funded project EVOLVE (Grant Agreement 303429). The innovation of the EVOLVE cell concept remains in its anode compartment without having nickel as structural component (Figure 5). The anode compartment is based on a metal alloy forming a protective alumina layer enhancing stability during re-oxidation cycles and an electronic conducting perovskite for the current collection. This combination is expected to create a cermet type hybrid current collector, with a mechanical stability similar to the MSCs and a microstructure similar with the one from anodic substrate of ASCs.

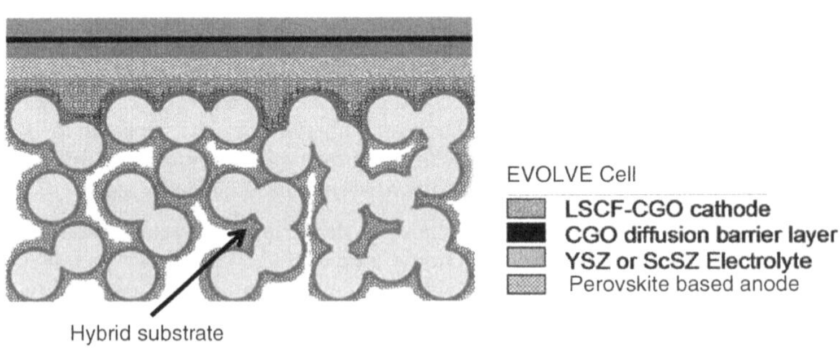

Figure 5. Schematic of the EVOLVE type cell

In this concept, a NiCrAl foam (Alantum GmbH, Germany) with open pore size of about 450 µm (Figure 6.a) is used as metallic substrate to enhance the stability of the cell. A perovskite namely $La_{0.1}Sr_{0.9}TiO_{3-\alpha}$ – LST (CerPoTech, Norway) – was used for the current collection when combined with the metal foam and as anodic electro-catalyst for anode in combination with $Ce_{0.9}Gd_{0.1}O_{2-\alpha}$ - GDC (Treibacher, Austria) - in a composite, as replacement of the Ni - YSZ cermet. Coarsened LST ceramic was infiltrated into the NiCrAl foam further and fired at 1000 °C (Figure 6(b)) allowing a pore size reduction down to the submicronic range making a compromise between the requirement for good gas permeation while allowing the deposition of a thin functional anode layer without macroscopic defects[15]. The NiCrAl metal foam as the structural framework ensured the mechanical strength of the substrates. The functional LST-GDC anode layer is then coated on top, using a fine LST powder in order to maximize the solid-gas exchange surface area which seems to control the performance of perovskite based anode towards hydrogen oxidation though the exact mechanism is not yet fully elucidated[16]. The use of a fine powder in the functional anode layer allows further reducing the pore size down to less than 100 nm at the top surface, crucial for the manufacturing of a thin electrolyte layer (Figure 6(c)). YSZ layer is coated on top.

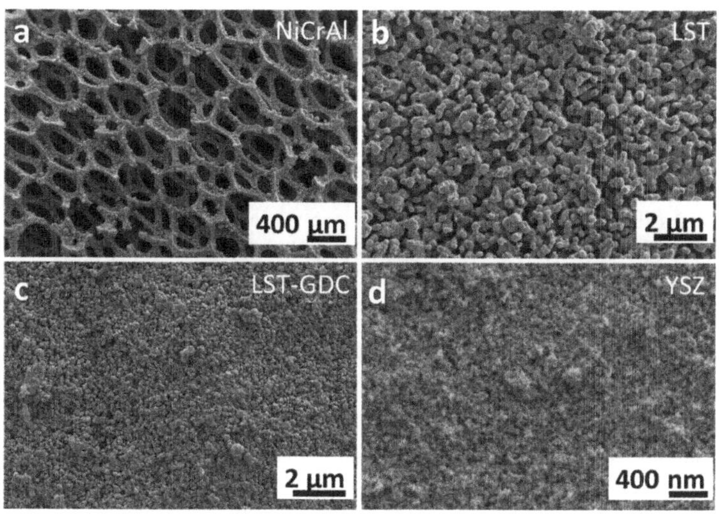

Figure 6. SEM of NiCrAl foam and NiCrAl-LST substrates: (a) top view of NiCrAl foam, (b) top view of NiCrAl foam fired with impregnated LST ceramic, (c) top view of the anode functional layer LST-GDC, (d) top view of the YSZ layer after dip coating (non sintered)[15].

It is observed in Figure 7 (a) and (c) that the YSZ layer and an EB-PVD GDC layer have been homogeneously deposited on top of the LST-GDC anode layer. The thickness of the calcined LST-GDC anode layer was about 20 µm whereas total thickness of the crack-free YSZ and GDC double layer was approximately 3 µm. The top view of fine structured GDC layer shown in Figure 7(b) reveals that the GDC layer is dense and the grain size is below 200 nm. The elemental mapping for the SEM in Figure 7(c) is shown in Figure 7(d), indicating at the LST-

GDC/YSZ/GDC interfaces no element interdiffusion between the YSZ and GDC phases. Due to the homogeneous YSZ layer and the dense GDC layer, the half cells realized excellent gas-tightness with air leakage rate as low as 0.4 $Pa.m.s^{-1}$ despite the extremely thin electrolyte. This low leak rate, combined with the thin film electrolyte is expected both to increase the OCV value of such a cell, while reducing ohmic drop linked to the thickness of the electrolyte when compared to the plasma sprayed ones. Work is ongoing to deposit the cathode and perform electrochemical testing of the cells and the results will be published in the future.

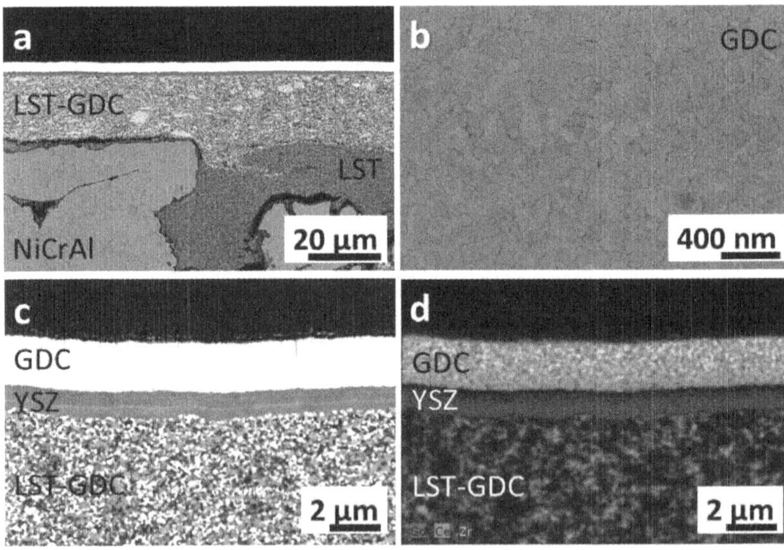

Figure 7. SEM of half cells with dip-coated YSZ layer and EB-PVD GDC layer: (a) overview cross section, (b) top view of GDC layer, (c) cross section at LST-GDC/YSZ/GDC interfaces, (d) elemental mapping at LST-GDC/YSZ/GDC interfaces[15].

CONCLUSION

Metal-Supported-SOCs represent a promising concept to overcome the commercialization hurdles for SOFC and SOEC systems. The plasma-sprayed cell concept was successfully scaled up to 84 cm^2 and demonstrated suitable power densities and durability. Also, innovative concepts with thin-layer electrolyte open new paths to enhanced performance and system cost reduction.

ACKNOWLEDGEMENTS

Dr. F. Han (DLR), and Dr. R. Semerad (Ceraco GmbH) are greatly acknowledged for providing samples with thin film electrolytes. A part of this work was funded by the European Union's Seventh Framework Programme (FP7/2007-2013) for the Fuel Cells and Hydrogen Joint Technology Initiative under grant agreement n°303429.

REFERENCES

[1]M.C. Tucker; "Progress in metal-supported solid oxide fuel cells", J. Power Sources, 195, pp. 4570-4582, 2010

[2]D. Soysal, J. Arnold, P. Szabo, R. Henne, S.A. Ansar, "Thermal plasma spraying applied on solid oxide fuel cells", J. Therm. Spray Technol., 22(5), pp. 588-598, 2013

[3]G. Schiller, R. Henne, M. Lang, M. Müller, "Development of solid oxide fuel cells by applying DC and RF plasma deposition technologies", Fuel Cells, 4(1-2) , pp. 56-61 , 2004

[4]T. Franco, R. Ruckdäschel, M. Lang, G. Schiller, P. Szabo, "Diffusion and protecting barrier layers in a substrate-supported SOFC concept", Proceedings 7th European Solid Oxide Fuel Cell Forum, 3-7 July 2006, Lucerne, Switzerland, P0802, 2006

[5]R. Henne, "Solid oxide fuel cells: A challenge for plasma deposition processes", J. Therm. Spray Technol., 16(3), pp. 381-403, 2007

[6]L. Jia, C. Dossou-Yovo, C. Gahlert, F. Gitzhofer, "Induction plasma spraying of samarium doped ceria as electrolyte for solid oxide fuel cells", Thermal Spray 2004: Advances in Technology and Application, ASM International, 10-12 May, Osaka, Japan, pp. 85-89

[7]D. Stolten, Hydrogen and Fuel Cells, Wiley-VCH Verlag, Weinheim, 2010

[8]S.A. Ansar, Z. Ilhan, J. Arnold, "Plasma sprayed metal supported SOFCs having enhanced performance and durability", Int. Thermal Spray Conference, 4-6 May 2009, Las Vegas, USA

[9]C. Christenn, S.A. Ansar, "Sprayed and constrained-sintered zirconia-based electrolytes", Thermal Spray 2009: Expanding Thermal Spray Performance to New Markets and Applications, ASM International, pp. 131-135, 2009

[10]Y. Shen, V.A.B. Almeida, F. Gitzhofer, "Preparation of nanocomposite GDC/LSCF cathode material for IT-SOFC by induction plasma spraying", J. Therm. Spray Technol. 20(1-2), pp. 145-153, 2011

[11]J. Arnold, S.A. Ansar, U. Maier, R. Henne, "Insulating and sealing of SOFC devices by plasma sprayed ceramic layers", Int. Thermal Spray Conference, Thermal Spray 2008: Crossing Borders (DVS-ASM), 2-4 June 2008, Maastricht, Netherlands

[12]G. Schiller, A. Ansar, M. Lang, O. Patz, "High temperature water electrolysis using metal supported solid oxide electrolyzer cells (SOEC)", J. Appl. Electrochem., 39, pp. 293-301, 2009

[13]M.C. Verbraeken, T. Ramos, K. Agersted, Q. Ma, C.D. Savaniu, B.R. Sudireddy, J.T.S. Irvine, P. Holtappels, F. Tietz, "Modified strontium titanates: from defect chemistry to SOFC anodes", RSC adv., 5, pp. 1168-1180, 2015

[14]Q. L. Ma, F. Tietz, A. Leonide, E. Ivers-Tiffee, "Anode-supported planar SOFC with high performance and redox stability", Electrochemistry Communications, 12, pp. 1326-1328, 2010

[15]F. Han, R. Semerad, R. Costa, "Development of metal foam supported SOFCs", Proceedings of the 11th European SOFC & SOE Forum 2014, 1-4 July 2014, Lucerne, Switzerland, A1506

[16]V. Yurkiv, G. Constantin, A. Hornes, A. Gondolini, E. Mercadelli, A. Sanson, L. Dessemond, R. Costa, J. Power Sources, 287, pp. 58-67, 2015

BICU(TI)VOX AS A LOW/INTERMEDIATE TEMPERATURE SOFC ELECTROLYTE: ANOTHER LOOK

Paul Fuierer[1], Kevin Ring[1], Joerg Exner[2], Ralf Moos[2]
[1]Department of Materials and Metallurgical Engineering, New Mexico Institute of Mining and Technology, 801 Leroy Place, Socorro, NM, 87801 USA
[2]Department of Functional Materials, University of Bayreuth, Universitätsstraße 30, Bayreuth 95440, Germany.

ABSTRACT

Bismuth vanadate ($Bi_2VO_{5.5-\delta}$) ceramics, stabilized by partial substitution of V^{5+} with transition metal ions, exhibit high ionic conductivity at temperatures below 500°C. For use as SOFC electrolyte, the ceramic has been dismissed by some due to reports of mixed conduction at higher temperature and reducing or low pO_2 atmosphere. The two dimensional conductivity of the oxide vacancies within the layered crystal structure also poses challenges. In the current study, an unique processing scheme was used to produce high density and textured polycrystalline $Bi_2(V_{0.9}Cu_{0.1})O_{5.5-\delta}$ ceramics. Platelet seeds, produced by molten salt synthesis (MSS), were dispersed within a matrix of fine powders to incite templated grain growth (tgg) during hot forging (HF). Bulk ceramics show high oxide ion conductivity ($10^{-1}\ \Omega cm^{-1}$) at 500°C, and modest texture gives modest anisotropy. Long term stability of the conductivity at this temperature over a range of pO_2 was demonstrated, however at lower temperatures (400°C), larger-grained and textured samples exhibit instability associated with oxygen vacancy ordering. Small grain size in conventionally sintered samples appears to mitigate this instability, as does Ti co-doping. Measured thermal expansion (16-20 ppm/°C), also showing small anisotropy in textured samples, may be compatible with metals in SOFC design.

INTRODUCTION

Interest in "Green" technologies have stimulated renewed research activity in oxide ion conducting ceramics for fuel cell electrolytes, gas sensors, and gas separation membranes. The established material is yttria stabilized zirconia ($Zr_{1-x}Y_xO_{2-x/2}$ "YSZ"), ideal in many ways, including high strength and toughness, chemical stability in both oxidizing and reducing atmospheres, and isotropic and well characterized properties.[1] However, depending on the application and electrolyte thickness, YSZ must be heated to between 700°C and 1000°C to provide adequate conductivity. These high operating temperatures create problems for the accompanying materials and balance of plant components.[2] Thermal cycling, material interface reactions, expansion-contraction mismatches, and gas sealing all pose challenges that limit applications. As a result, interest has grown in the development of alternative ceramics that can serve as solid state electrolytes at lower temperatures. Low or intermediate temperature SOFCs (≤650°C) can significantly reduce system cost due to wider material choices for interconnects and compressive non-glass/ceramic seals, as well as reduce balance of plant (BOP) costs. Below 600°C, both radiative heat transfer (Stefan-Boltzmann) and sintering rates become insignificant, thus reducing insulation costs and primary performance degradation mechanisms, respectively. At even lower temperatures (≤350°C), cheap stamped stainless steel interconnects, elastomeric/polymeric seals, and BOP components are possible. Lower operating temperatures would also allow more rapid start-up and reduce the negative effect of repeated thermal cycling.[3]

Bismuth vanadate, written as $Bi_2VO_{5-\delta}$ or $Bi_4V_2O_{11-\delta}$ (BIVOX) and related compounds with metallic substitutions, $Bi_4(Me_xV_{1-x})_2O_{11-\delta}$ (BIMEVOX; ME= Cu^{2+}, Co^{2+}, Ni^{2+}, Ti^{4+} etc.) have high oxygen conductivity, much greater than that of YSZ, and even higher than delta (cubic) stabilized Bi_2O_3 at 500°C and below.[4,5] As proposed by Abrahams and Krok,[6] the high conductivity is a consequence of the polarizability of the bismuth ions, the variable coordination of vanadium ions and the large concentration of oxygen vacancies.

The structure of BIVOX, shown in Figure 1, is an alternating stack of $(Bi_2O_2)^{2+}$ and $[VO_{3.5}\ _{0.5})]^{2-}$ perovskite layers.[4,7,8] Structure parameters can vary due to superstructure modulations and variable defect concentrations, but the room temperature α-phase of BIVOX is orthorhombic with cell parameters *a*=5.53Å, *b*=5.61Å, and *c*=15.3Å.[7] The compound undergoes a series of transitions with increasing temperature (α–β–γ) with the γ-phase having tetragonal symmetry and high ionic conductivity above around 570°C. Refinement of the high-temperature x-ray data gives space group *I4/mmm*, with *a*=3.988 Å and *c*=15.42 Å.[7] Using neutron diffraction, Muller's group determined the concentration of oxygen vacancies, $[V_O]$, and proposed oxygen diffusion paths in BICOVOX.[9] While little to no V_O are present in the $(Bi_2O_2)^{2+}$ layers (O(1) sites), $[V_O]$ is 9% in the O(2) site and 26% in the O(3) site. Helical diffusion paths between O(2) and O(3) is inferred from Fourier maps of nuclear density.

The high concentration of disordered oxygen vacancies $[V_O]$ serves to conduct oxide ions. The low-temperature phases have higher order and lower conductivity, but the γ-phase can be partially stabilized by the aliovalent substitutions mentioned above. Generally, most of the published stabilizations result from a 10-15 at% substitution for V^{5+}. BICOVOX and BICUVOX have received most attention,[10,11] but equally high conductivity has been reported for $Bi_2V_{0.9}Cu_{0.05}Ti_{0.05}O_{11}$ (BICUTIVOX).[12] Figure 2 compares the temperature dependence of conductivity for four compositions, all prepared using identical conventional ceramic fabrication procedures, but at individually optimized sintering temperatures for optimal density.

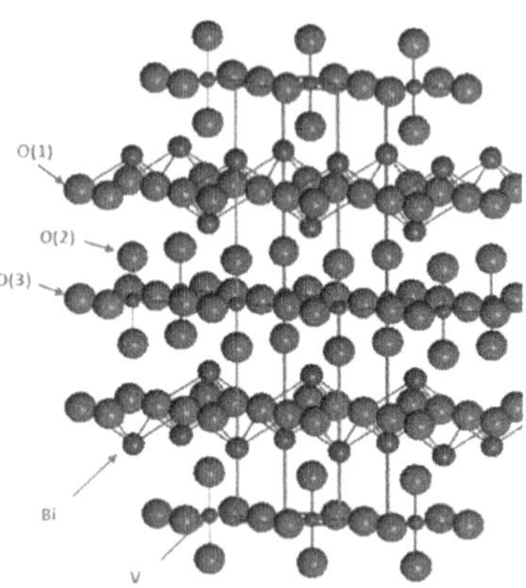

Figure 1. Tetragonal unit cell of the high temperature γ-phase of $Bi_4V_2O_{11}$ (BIVOX). (Structural data from F. Abraham et al. [4])

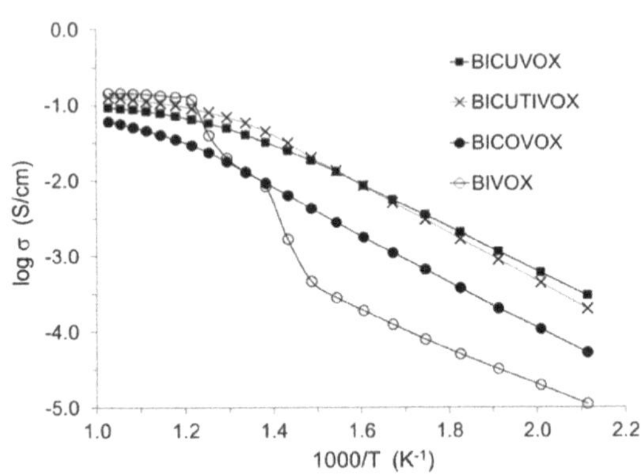

Figure 2. Total conductivity for BIVOX and BIMEVOX ceramics (after Fuierer, et al. [13])

As might be expected from the layered structure, anisotropy has been demonstrated in single crystals of BICOCOX, BICUVOX and BINIVOX, with conductivity parallel to the *a-b* plane from 10 to 10^2 higher than along the *c*-axis.[14,15] Microstructural texturing has been considered for maximizing the conductivity of polycrystalline ceramics, however limited work has been done in this regard. Muller et al.[16] were the first to obtain texture in BICOVOX via pressing and application of strong magnetic field. Shantha and Varma[17,18] produced textured BIVOX via splat-cooling and liquid phase-aided sintering with focus on dielectric and ferroelectric properties. More recently, molten salt synthesis (MSS) was used to exaggerate the anisotropic crystal habit of γ-BICOVOX powders.[19]

Conventional sintering has been the usual preparation method for BIMEVOX. Simner[20] focused on optimization of BICUVOX with resultant densities of 95%, d_g~7μm, and conductivity at 400°C of $2x10^{-2}$ $(\Omega cm)^{-1}$. Steil[21] concluded that small grain size stabilizes the γ-phase and reduces hysteresis in the temperature dependent conductivity of BICOVOX. Oxygen stoichiometry is a concern due to volatility of Bi, and possibility of V^{5+} reduction.[22] Coprecipitation,[23] sol-gel,[24] citrate,[25] and mechano-chemical-activation[26,27] have been reported to achieve small particle size in attempts to enhance sinterability at lower temperatures.

A few reports of thin film BIMEVOX can be found. PLD was used to deposit BIVOX films on $SrTiO_3$, MgO and YSZ substrates[28]. Smooth and *c*-axis oriented BICUVOX films on $SrTiO_3$ were obtained, although no electrical properties were reported. The same group reported sol-gel films on quartz and platinum.[29] Simner et al.[30] used solution deposition for good quality BICUVOX on MgO, but no transport or other properties were reported. Recently, BICUTIVOX thick (4μm) films were deposited via aerosol deposition, AD (Figure 3).[31] As- deposited films were essentially free of porosity. Annealing of films released residual strain and increased grain size and *c*-axis texture, as total conductivity approached that of bulk ceramics.

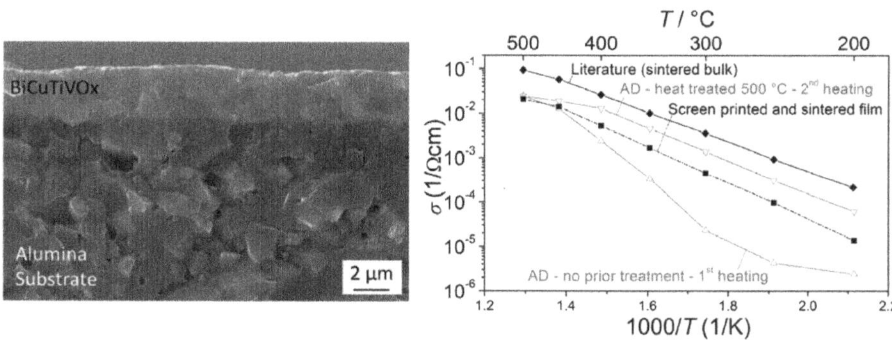

Figure 3. Aerosol deposition (AD) of BICUTIVOX: Cross-sectional view of a 4 μm thick film on an alumina substrate. Conductivity of an AD film before and after heat treatment, compared to a bulk ceramic and screen print film. (Reprinted from [31] with permission from Elsevier.)

Some have suggested that BIMEVOX will not work as a fuel cell electrolyte due to high chemical reactivity, increase in electron transference number, t_e, at pO_2 $<10^{-2}$ atm and high T >550°C, low strength, and high thermal expansion.[5,32] Researchers have tried to address these issues by chemical substitution, both on the Bi- and V-sites.[33-35] A trade-off between stability and conductivity seems to exist, but this has not been explored for all compositions, nor have all strategies been explored to maximize density or exploit the two-dimensional conductivity. Relative to other functional ceramics, little work has been done regarding processing-property relationships, especially for such a complex oxide. Furthermore, the benefits of an SOFC operating at such low temperatures have not been fully weighed against the possible material shortcomings.

This paper describes efforts in re-examining the processing of bulk BICU(TI)VOX polycrystalline ceramics, paying particular attention to crystallographic texture and bulk density. It assesses key properties for fuel cell electrolytes, namely thermal expansion and stability of the ionic conductivity, relative to microstructure.

EXPERIMENTAL PROCEDURE

An overview of the ceramic processing variants used in this work is shown in Fig. 4. Stoichiometric quantities of reagent grade oxides corresponding to $Bi_2(V_{0.9}Cu_{0.1})O_{5.5-\delta}$ (BICUVOX) and $Bi_2(V_{0.9}Cu_{0.05}Ti_{0.05})O_{5.5-\delta}$ (BICUTIVOX) were dry ball milled. For standard powders, calcination was carried out in covered alumina crucibles heated to 630°C for 12 hours. Phase content was confirmed by X-Ray diffraction (PANalytical X'Pert, $Cu_{K\alpha}$ source). Post calcination, a wet (DI water) three-stage milling was used for effective particle size reduction. Each stage used progressively smaller zirconia media and longer times: First stage used 13 mm media for 6 hrs., second stage used 10 mm for 16 hrs., and third stage used 7.5 mm for 36 hrs. For ordinary fired (OF) ceramics, 1-2 grams of powder were cold pressed using a co-vinyl polymer binder, and a ½ inch diameter steel die at a pressure of 105 MPa. Green compacts were sintered at 725°C and 750°C, for BICUVOX and BICUTIVOX respectively, for 10 hrs. For hot-forged (HF) ceramics reported here, green bodies of 4 to 8 grams were pressed and pre-sintered at 515°C for 5 hrs. The pre-sintered body was re-heated to 725°C or 750°C, at 10°C/min, loaded at a constant rate over the course of 30 minutes, sintered under load (125-150 kg) for 5 hrs, unloaded, and then cooled to room temperature at a maximum rate of 10°C/min.

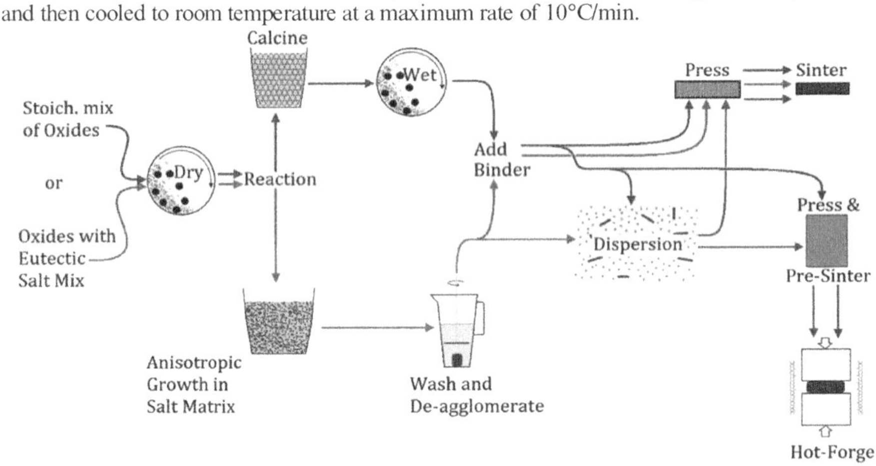

Figure 4. Schematic of processing schemes used for BICU(TI)VOX ceramics.

The lower pathway of Figure 4 depicts a unique approach using a combination of templated grain growth (tgg) and hot forging to increase the crystallographic texture of BICUVOX ceramics. High aspect ratio platelets were produced by molten salt synthesis (MSS), involving mixing the reagent oxides with an equal mass of the eutectic ratio of KCl and NaCl, and heating the mixture to temperatures between 610°C and 700°C for times between 2 and 8 hrs. to control platelet size. A comparison of MSS versus conventional powders is seen in Fig. 5. For tgg-hot-forging, platelets were mixed in with equiaxed matrix grains created by the conventional synthesis. Contents up to 10% platelets by weight were successfully forged (i.e. produced a monolithic, dense sample). For the results reported here, the content was 5%, and the sample was forged at 750°C for 3 hrs with a load of approx. 175 kg. For comparison, a compact was also prepared from 100% platelet powder, and pressure-less sintered at 775°C for 10 hrs.

Figure 5. BICUVOX powders produced via (A) conventional calcining and (3-stage) milling, and (B) molten salt synthesis (650°C/8 hrs.).

Density was determined by the Archimedes method (Sartorius R160P) or by helium pycnometry (Micromeritics AccuPyc 1330) and confirmed by geometric measurements made with a micrometer. For the hot forged samples, the poorly densified exterior (visually lighter in color) was ground off before the density measurement. Texture of sintered and forged bodies was estimated by Lotgering's method[36] using relative peak heights in diffractograms. For forged samples, the top 100 microns of the specimen were removed prior to XRD to eliminate any surface effects. Cross-sectional surfaces, revealed by cutting near the center of forged samples, were polished and thermally etched for examination with a scanning electron microscope, SEM (Hitachi S-3200N). Central sections of samples were cut and ground to regular prisms. Sample sets included pairs of samples cut from identical forge conditions to allow correlation between tests as well as perpendicular and parallel orientation pairs. The designations "perp" and "para" are used to distinguish a tensor property measured in the direction perpendicular and parallel to the forging axis respectively. A dilatometer (L75, Linseis) was used to measure the dimensional change of the length of bar samples both on heating and cooling at a rate of 5°C/min.

For 4-point impedance measurements, a porous firing, fritless platinum paste was applied to opposing ends of the bar on the longest dimension. Guide trenches were cut about the perimeter approximately 1/3 distance from each end of the bar for location of platinum wires to act as inner electrodes. Details are described in a previous publication.[37] The impedance spectra of each sample were measured over a bandwidth of 10 MHz to 1 Hz using a Nova Control Alpha-analyzer, and a custom built sample holder with shielded leads and spring loaded rod to make contact with the end electrodes. Total conductivity of the sample was calculated from sample dimensions, inner electrode spacing and the intersection of the Nyquist complex impedance plot with the real axis. Measurements were made with varying temperature, from 150°C to 650°C on both heating and cooling (10°C/min, 50°C step, 20 min equilibration time), and also with varying pO_2 (N_2/O_2 gas mixing), from 100% to 100 ppm O_2 (verified by a lambda probe).

RESULTS

Sample descriptions and their corresponding microstructural characteristics are listed in Table I. Scanning electron micrographs are shown in Figures 6 and 7. Ordinary fired (OF) ceramics have little to no texturing, with small grain size. The OF MSS sample has high texture due to settling and orientation of the large platelets during pressing, and very low sintered density as expected due to large starting particles. In both compositions, hot forging of conventional powder yields higher densities, small to medium grain size, and a small amount of texture. Inclusion of platelet grains increased both density and texture of hot forged BICUVOX. BICUTIVOX sinters and forges to a higher density than BICUVOX.

Table I: Sample descriptors and microstructural characteristics of bulk density, Lotgering orientation factor, and relative grain size. (OF=ordinary fired; HF = hot forged; MSS- molten salt synthesized powder; tgg= MSS powder dispersed to induce templated grain growth)

Sample description	Bulk density (%)	Orientation, f (%)	grain size (relative)
BICUVOX OF	83	-	small
BICUVOX MSS-OF	78	55	large
BICUVOX HF	91	6	med
BICUVOX tgg-HF	94	16	med
BICUTIVOX OF	94	-	small
BICUTIVOX HF	98	11	small/med

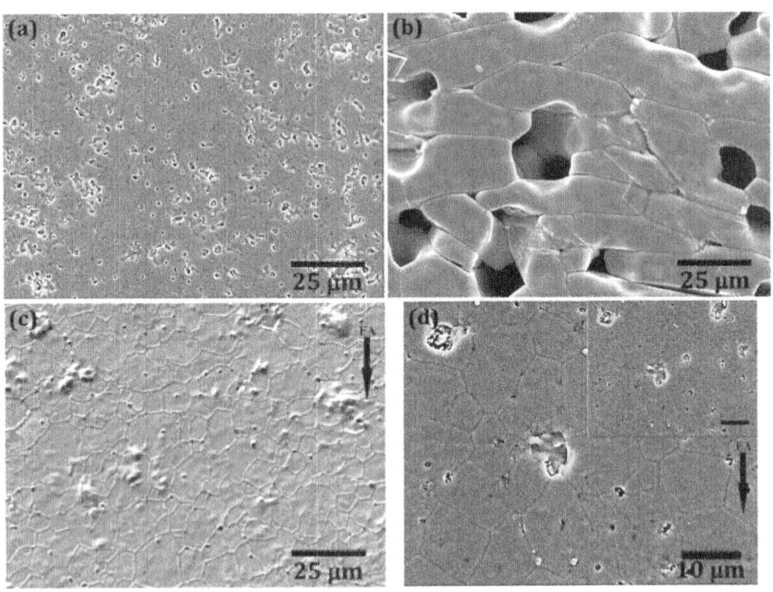

Figure 6. BICUVOX microstructures, a) OF b) MSS-OF c) HF d) HF-tgg. (inset matches the magnification of (a)-(c)) *Appearance of grain boundaries in (d) were enhanced for visibility using non-linear adjustments to contrast and brightness.

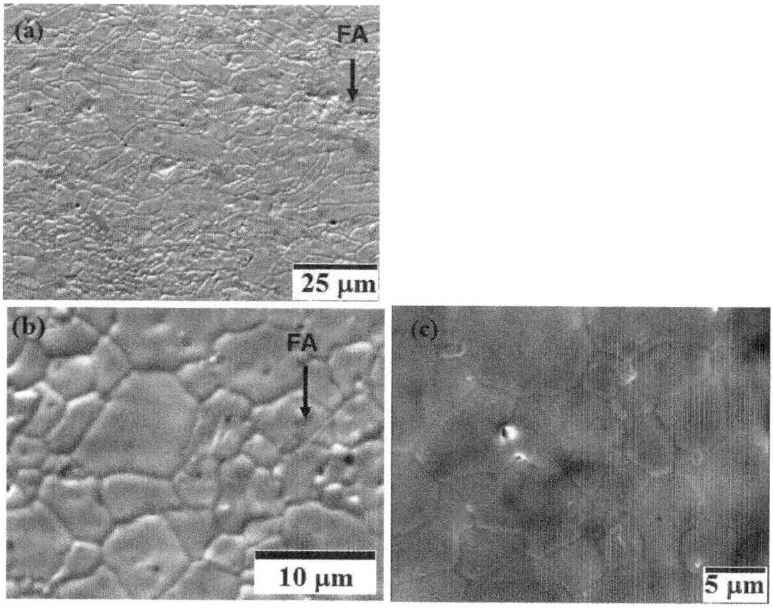

Figure 7. BICUTIVOX microstructures, a) HF (low mag) b) HF (high mag) c) OF (high mag). (Reprinted from [37] with permission from Elsevier.)

The temperature dependence of the thermal expansion coefficient is shown in Figure 8. A strong anomaly occurs on heating, corresponding to the γ'(order)-γ(disorder) structural transition. The peak occurs at an average of 475°C for BICUVOX and 450°C for BICUTIVOX samples. The TCEs are 16-17 ppm/°C below, and 17-19 ppm/°C above the transition. Small anisotropy is observed in BICUTIVOX, with greater TCE for the perp configuration than the para. But little difference is observed between OF and HF perp samples for either composition. The BICUVOX MSS-OF sample shows a minimum at 300°C, and somewhat reduced TCE and amplitude at the transition, presumably due to the high porosity.

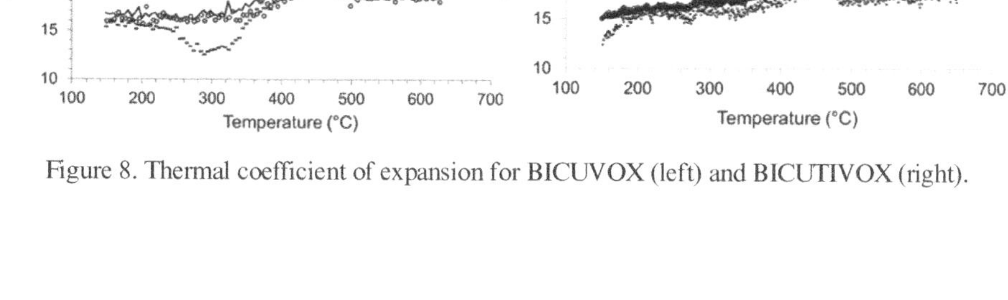

Figure 8. Thermal coefficient of expansion for BICUVOX (left) and BICUTIVOX (right).

Figures 9 and 10 show the temperature dependence of conductivity. Both BICUVOX and BICUTIVOX display modest anisotropy (factors of 2-4) in hot forged samples with modest texture. Generally, higher conductivity correlates with higher texture and density in HF and HF-tgg BICUVOX. Heating–cooling hysteresis below the transition temperature is seen in all BICUVOX samples, which has been reported in the literature.[38-40] The hysteresis appears to be more pronounced in the textured samples, and moderated in the OF sample. This complements the previous claims that the low temperature ordering of the γ'-phase may be impeded in fine-grained microstructures.[40] BICUTIVOX did not display significant hysteresis outside the error of the measurement. Despite the lower density, OF samples are close or equal in conductivity to HF perp samples, within the error of measurement. Cooling data is not shown for clarity. (See reference [37] for details.)

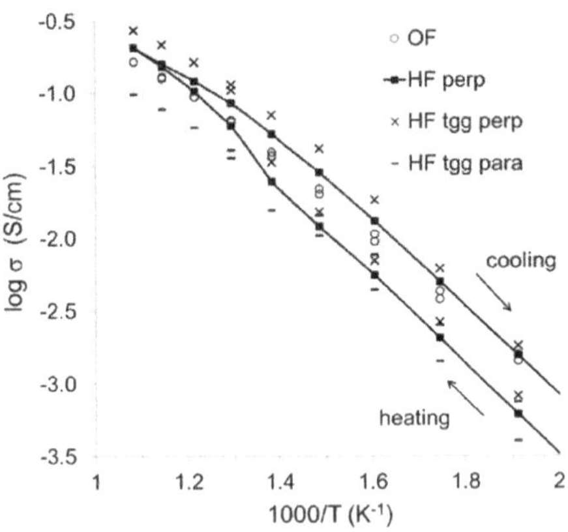

Figure 9. Hysteresis in conductivity for BICUVOX ceramics. The lower branch is heating, upper is cooling for each sample. Line aids the eye for the HF perp sample.

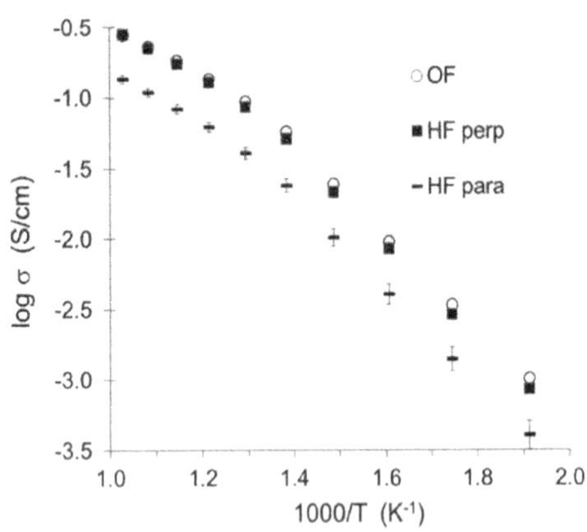

Figure 10. Conductivity of BICUTIVOX ceramics displays no measureable hysteresis, but small anisotropy between HF perp and HF para is observed.

Thermal aging experiments with BICUVOX (Figure 11) corroborate earlier work.[39] The γ' phase appears to be quasi-stable, and when held at a temperature just under the transition (i.e. 400°C) can quickly (hours) undergo significant additional oxygen vacancy ordering, reverting to an α-like phase. Like the hysteresis, the aging was more pronounced in the textured samples. On the other hand, very little change in conductivity occurs at 500°C, just above the transition in the γ-phase. Oxygen vacancy disorder is maintained for the extent of the 50 hour measurement. The pO_2 stability tests tell a similar story. At 500°C, no significant drop in conductivity is observed. At 400°C, a noticeable drop in conductivity occurs in the HF and particularly HF-tgg samples as pO_2 was lowered. Given the rate and direction of change, it is likely that this decrease in conductivity was a result of thermal aging rather than a response to the changing pO_2 (tests began in oxygen and then went to low pO_2). With BICUTIVOX (not shown), negligible drop in conductivity was detected over 50 hour hold at 425, 450, and 500°C. Likewise, no change in conductivity was detected from 100% to 100 ppm O_2.

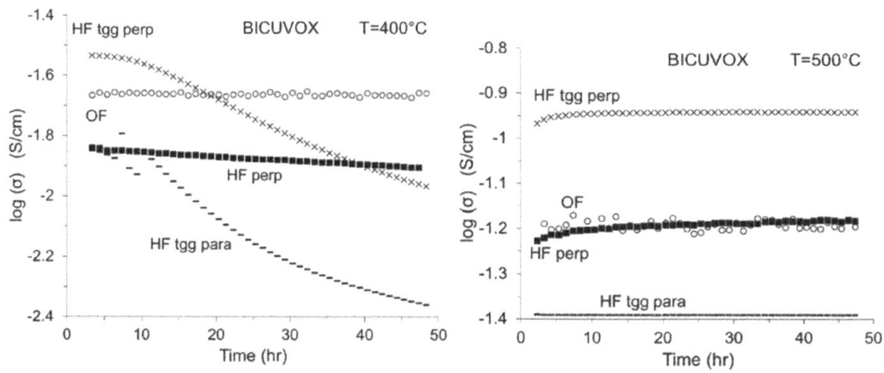

Figure 11. Thermal aging of BICUVOX held at 400°C (left) and held at 500°C (right)

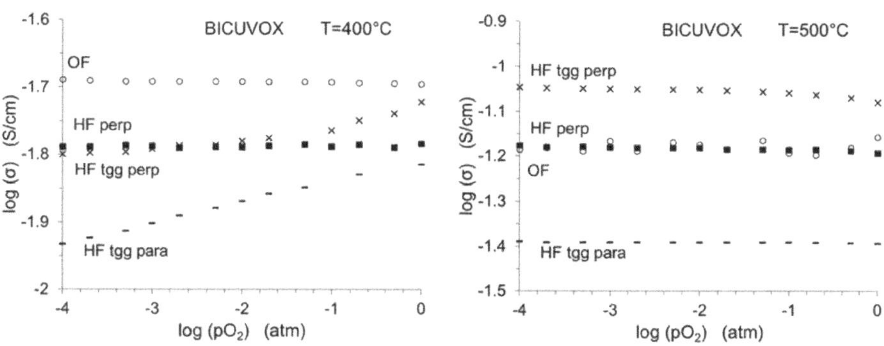

Figure 12. pO_2 stability of BICUVOX held at 400°C (left) and held at 500°C (right).

DISCUSSION AND CONCLUSIONS

Hot forging of BIMEVOX ceramics (ME = Cu/Ti) can yield high densities, but gives only modest preferred grain orientation. A combination processing method of templated grain growth – hot forging can give added texture. The modest preferred orientation leads to anisotropy in tensor properties, TCE and total conductivity.

Texturing does not appear to be advantageous for BICUVOX, since it is necessarily accompanied by larger (exaggerated) grains. This, in turn, increases thermal hysteresis and aging of the conductivity in the low temperature γ' phase. This finding is consistent with previous investigations for both BICUVOX and BICOVOX, which suggest that the oxygen vacancy ordering and accompanying degradation in conductivity is very pronounced in single crystals, but may be suppressed in finer grained polycrystalline samples.[20,21,38,39] The hot forged samples described herein can be regarded as intermediate between these two states. Both small grain size and titanium co-doping were found to reduce or eliminate the sluggish γ' to α-like ordering transition.

Both BICUVOX and BICUTIVOX exhibit time and pO_2 stability at 500°C (in the γ-phase). However, preliminary unpublished results measuring the partial electronic conductivity at this temperature suggest a significant electronic transference number (t_e). Previous work in the literature[34] indicates a large temperature dependence of the t_e, suggesting a rapidly widening electrolytic zone with lower temperature. Use of BICUVOX or the more stable BICUTIVOX in the γ' phase at 350-400°C (below the transition) where the ionic conductivity is still high, may therefore be feasible.

The TCE (16-17 ppm/°C) below the transition is very close to that of 304 stainless steel, which is a choice material for PEM fuel cells and BOP components due to its superior corrosion/oxidation resistance and formability into complex shapes. This TCE match would make 304 SS a possibility for the bipolar plate and electrodes in a BIMEVOX based SOFC stack. Thermal and electrochemical compatibility studies are of interest. Preliminary findings on aerosol deposited BICUTIVOX suggest that it yields small nano-grained films with very high density,[31] and may be a good way to integrate BIMEVOX with electrode materials. Further investigation of BIMEVOX ceramics for consideration as a solid electrolyte for low to intermediate temperature SOFCs is warranted.

ACKNOWLEDGMENT

Funding from the National Science Foundation (NSF) and the Deutsche Forschungsgemeinschaft (DFG), in the framework of the Materials World Network, is gratefully acknowledged (NSF grant # 1108466 and DFG grant # MO 1060/16-1).

REFERENCES

[1]Fergus, J. "Electrolytes for solid oxide fuel cells," J. Power Sources, 162, pp.30-40, 2006.

[2]Wachsman, E., Singhal, S. "Solid Oxide Fuel Cell Commercialization, Research, and Challenges" The Electrochemical Society Interface, pp 38-42, 2009.

[3]Wachsman, E. "Lowering the Temperature of Solid Oxide Fuel Cells," Science, 334 , pp. 935-939, 2011.

[4]Abraham, F. et al. "The BIVEVOX Series: A New Family of High Performance Oxide Ion Conductors", Solid State Ionics, 40/41, pp. 934-937, 1990.

[5]Kharton, V. et al. "Transport properties of solid oxide electrolyte ceramics: a brief review", Solid State Ionics 174, pp. 135-149, 2004.

[6]Abrahams and Krok, F. "A model for the mechanism of low temperature ionic conduction in divalent-substituted -BIMEVOXes," Solid State Ionics, 157, pp.139-145, 2003.

[7]Abraham, F., Debreuille-Gresse, M.F., Mairesse, G., Nowogrocki, G. "Phase Transitions and Ionic Conductivity in $Bi_4V_2O_{11}$ an Oxide with Layered Structure", Solid State Ionics 28-30, pp.529-532, 1988.

[8]Varma, K., et al. "Synthesis and Characterization of Layered Bismuth Vanadates," J. Mater. Res. 5 [11], pp.2718-22, 1990.

[9]Muller, C., et al. "Structural Studies of the Fast Oxygen Ion Conductor BICOVOX.15 by Single Crystal Neutron Diffraction at Room Temperature," J. Sol. State Chem., 141, pp.241-247, 1998.

[10]Krok, F., et al. "Studies on preparation and electrical conductivity of BICOVOX," Solid State Ionics, 70/71, pp. 211-214, 1994.

[11]Dygas, J., et.al. "Impedance study of BICUVOX ceramics," Solid State Ionics, 70/71, pp. 239-247, 1994.

[12]Paydar, M., et al. "Ionic Conductivity and Crystal Structure Relationships in Ti/Cu substituted $Bi_4V_2O_{11}$", J. Mat. Sci., 39, pp.1357-1361, 2004.
[13]Fuierer, P., et al. "Processing Issues Related to the Bi-dimensional Ionic Conductivity of BIMEVOX Ceramics," J. Mater. Sci. 46 [16], pp. 5447-53, 2011.
[14]Pernot, E., et. al "Structure and Conductivity of Cu and Ni-substituted $Bi_4V_2O_{11}$ compounds", Solid State Ionics, 70/71, pp. 259-263, 1994.
[15]Kim, S. and Miyayama, M. "Anisotropy in oxide ion conductivity of $Bi_4V_{2-x}CoxO_{11-\delta}$," Solid State Ionics, 104, pp.295-302, 1997.
[16]Muller, C., et al. "Pressureand magnetic field effects on the crystallographic texture and electrical conductivity of the $Bi_4(V_{0.85}Co_{0.15})_2O_{11-\delta}$ compound," J. Phys. D: Appl. Phys., 29, pp.3106-12, 1996.
[17]Shantha, K., and Varma, K. "Fabrication and characterization of grain-oriented bismuth vanadate ceramics," Mater. Res. Bull., 32 [11], pp.1581-91, 1997.
[18]Shantha, K. and Varma, K. "Dielectric, piezoelectric and pyroelectric anisotropy in KCl-modified grain-oriented bismuth vanadate ceramics," J. Materials Research, 14 [2],1999.
[19]Roy, B. and Fuierer, P. "Molten Salt Synthesis of $Bi_4(V_{0.85}Co_{0.15})_2O_{11-\delta}$ (BICOVOX) Ceramic Powders", J. Amer. Ceram. Soc., 92 [2], pp. 520-523, 2009.
[20]Simner, S., et al. "Synthesis, Densification, and Conductivity Characteristics of BICUVOX Oxygen-Ion-Conducting Ceramics", J. Am. Ceram. Soc., 80 [10], pp.2563-68, 1997.
[21]Steil, M., et al. "BICOVOX: Sintering and grain size dependence of the electrical properties," J. Eur. Ceram. Soc., 19, pp.815-18, 1999.
[22]Lee, C., et. al. "Stoichiometry and stability of bismuth vanadate, $Bi_4V_2O_{11}$, solid solutions," Solid State Ionics, 62, pp.193-198, 1993.
[23]Sabolsky E, et al. "Microstructural. electrical and mechanical characterization of $Bi_2Cu_{0.1}V_{0.9}O_{5.35}$ (BICUVOX) ceramics fabricated from co-precipitated precursor powders." J Mater Sci, 48, pp.733–43, 2013.
[24]Pell, J., et al."Sol-gel synthesis of α-$Bi_2VO_{5.5}$ using a soluble bismuth precursor," Mater. Lett., 25 [3-4], pp.157-60, 1995.
[25]Simner, S, et al. "Solution processing approaches for solid electrolytes and electrode materials", J. Mate. Res., 13 [4], pp.866-874, 1998.
[26]Shantha, K. and Varma, K. "Characterization of fine-grained bismuth vanadate ceramics obtained using nanosized powders", J. Am. Ceram. Soc., 83 [5], pp. 1122-1128, 2000.
[27]Castro, P., et al. "Room temperature stabilization of γ-$Bi_2VO_{5.5}$ and synthesis of the new fluorite phase f-Bi_2VO_5 by a mechanochemical activationmethod," J. Mater. Chem., 10, pp.767-71, 2000.
[28]Pell, J., et al. "Laser Deposition of oriented BiMeVOx films," Thin Solid Films, 300, pp.154-158, 1997.
[29]Pell, J., et al. "Sol-Gel Route to Oxygen Ion Conductors in Thin Film Form," Mate. Res. Soc. Symp. Proc., 403, pp.459-464, 1996.
[30]Simner, S., et al. "Solution processing approaches for solid electrolytes and electrode materials", J. Mate. Research, 13 [4], pp.866-874,1998.
[31]Exner, J., et al. "Aerosol deposition of (Cu,Ti) substituted bismuth vanadate films," Thin Solid Films, 573, pp.185-190, 2014.
[32]Iharada, T., et al. "Electrochemical characterization of BIMEVOX oxide-ion conductors," Solid State Ionics, 48, pp.257-265, 1991.
[33]Yaremchenko, A., et al. " Physiochemical and Transport Properties of BICUVOX-Based Ceramics," J. Electroceram. , 4, pp.235, 2000.
[34]Yaremchenko, A., et al. " Structure and electronic conductivity of $(BiLa)_2(VCu)O_{5.5}$", Mater. Chem. Phys., 77, pp.552-558, 2002.

[35]Steil, M., et al. "Thermal stability and preparation of dense membrane ceramics of BIMEVOX," Solid State Ionics, 176, pp.2305-2312, 2005.
[36]Lotgering, F. "Topotactical Reactins with Ferrimagnetic Oxides Having Hexagonal Crystal Structures," J. Inorg. Nucl. Chem., 9, pp.113-123, 1959.
[37]Fuierer, P., et al. "Anisotropy and thermal stability of hot-forged BICUTIVOX oxygen ion conducting ceramics", J. Eur. Ceram. Soc., 34, pp.943-951, 2014.
[38]Dygas, J., et al. "Structure-dependent impedance of BICUVOX," Electrochimica Acta, 40, (10), pp. 1545-1550, 1995.
[39]Watanabe, A, and Das K. "Time-dependent degradation due to the gradual phase change in BICUVOX and BICOVOX oxide-ion conductors at temperatures below about 500°C," J Solid State Chem, 163, pp.224–30, 2002.
[40]Pirovano C, et al. "Impedance study of the microstructure dependence of the electrical properties of BIMEVOXes," Solid State Ionics, 176, pp.2079–83, 2005.

SYMBOLIC ANALYSIS OF MULTI-STAGE ELECTROCHEMICAL OXIDATION FOR ENHANCEMENT OF ELECTRIC EFFICIENCY OF SOFCs

Y. Matsuzaki[a),b)], Y. Tachikawa[c)], T. Hatae[a)], H. Matsumoto[c)], S. Taniguchi[b),d)], K. Sasaki[b),c),d),e)]

[a)]Fundamental Technology Dept., Tokyo Gas Co., Ltd., 1-7-7 Suehiro-cho, Tsurumi-ku, Yokohama City, Kanagawa 230-0045, Japan.
[b)]Next-Generation Fuel Cell Research Center, Kyushu University, 744 Motooka, Nishi-ku, Fukuoka City, Fukuoka 819-0395, Japan.
[c)]International Institute for Carbon-Neutral Energy Research (WPI-I^{2}CNER), Kyushu University, 744 Motooka, Nishi-ku, Fukuoka City, Fukuoka 819-0395, Japan.
[d)]International Research Center for Hydrogen Energy, Kyushu University, 744 Motooka, Nishi-ku, Fukuoka City, Fukuoka 819-0395, Japan.
[e)]Faculty of Engineering, Kyushu University, 744 Motooka, Nishi-ku, Fukuoka City, Fukuoka 819-0395, Japan.

ABSTRACT

SOFCs have the solid-state ceramic construction and operate at high-temperatures, with flexibility in fuel choice, high efficiency, stability, and reliability. The most attractive characteristics of SOFCs should be high fuel-to-electricity conversion efficiencies of as high as 50 to 60 percent LHV. For further improving the electrical efficiencies, preceding studies on multi-stage electrochemical oxidation with SOFCs have been reported. However, there are many parameters for the multi-stage oxidation, and effects of the parameters on the efficiency remains to be identified. We have investigated the multi-stage oxidation by using a symbolic analysis method. In the case of n-stage electrochemical oxidation, the fuel utilization ratio in the individual stage was found to decrease with increasing the n value at a fixed fuel utilization ratio of an entire system, resulting in the enhancement of robustness against the operation at a high fuel utilization ratio of the entire system as well as against a gas-leakage.

INTRODUCTION

In order to increase the electrical efficiency of SOFCs, fuel utilization ratio (*Uf*) of the entire system should be increased up to a maximum possible value without increasing the risk for degradation. Nickel particles in the anode layer of SOFCs have a risk for oxidation at high *Uf*, which leads to large and irreversible degradation. Therefore, the *Uf* should to have an upper limit. The capability of SOFCs for high electrical efficiencies[1] is thus limited by the upper limit of *Uf*. Preceding studies have been reported on multi-stage electrochemical oxidation with SOFCs[2-5], which is one of the promising approaches to achieve higher fuel utilization ratios. However there are many parameters for the multi-stage electrochemical oxidation, and influence of the parameters remains to be identified. In this study, a symbolic analysis of the system with multi-stage electrochemical oxidation has been conducted to investigate the effects of variable parameters on the characteristics of the system.

METHOD OF SYMBOLIC ANALYSIS

Figure 1 shows the schematic illustrations of the examples of the multi-stage electrochemical oxidation design: (a) segmented-in-series type cells, and (b) planar type stacks with a series connection of fuel supply. There are multiple parameters that dictate the characteristics of the multi-stage electrochemical oxidation system in which the fuel supplied to an earlier stack placed upstream of the fuel flow is partially consumed through electrochemical oxidation in the stack and successively supplied to a latter stack placed downstream.

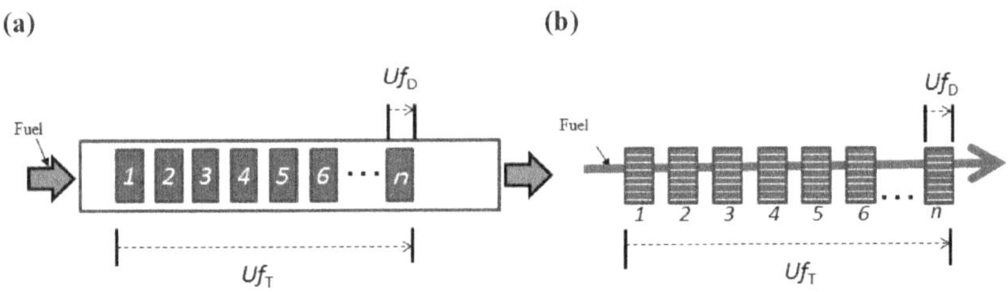

Figure 1. Schematic illustrations of the examples of the multi-stage electrochemical oxidation design: (a) segmented-in-series type stack, (b) planar type stacks with a series connection of fuel supply.

Symbolic analyses under the assumption of a uniform temperature are applied to the investigation of the multi-stage electrochemical oxidation. In order to simplify the discussion, most of the multiple parameters are given fixed values to clarify the influence of the other parameters. Fuel utilization ratio (*Uf*) of individual stack in the multi-stage design, $Uf(1), Uf(2), Uf(3), \cdots Uf(n)$, were expressed as algebraic equations of the parameters.

RESULTS AND DISCUSSION

Influence of *n* value in *n*-stage configuration

As shown in Fig. 1 (b), *n*-stage configuration with an equal fuel consumption at all stacks is equivalent to the segmented-in-series type design as shown in Fig.1 (a) in terms of multi-stage electrochemical oxidation. Table 1 shows *Uf* of the individual stacks for the *n*-stage configuration (*n* is from 1 to 5) with Uf_T of 80%. In this case, *Uf* at the individual stacks has the maximum values at the most downstream stack, which is depicted as hatching in Table 1. The *Uf* at the most downstream stack, Uf_D, should be expressed as a function of *n* and Uf_T.

When fuel is supplied to the inlet of the most upstream stack at a rate of Q[mol sec^{-1}], the fuel is consumed by the upstream stacks before reaching the inlet of the most downstream stack with a rate obtained by a following formula.

$$Q \cdot \frac{Uf_{\mathrm{T}}}{100} \cdot \frac{n-1}{n}$$

Therefore, the rate of the unutilized fuel is expressed as follows.

$$Q - Q \cdot \frac{Uf_{\mathrm{T}}}{100} \cdot \frac{n-1}{n}$$

Under the assumption of the equal fuel consumption at all the stacks, the rate of fuel consumption in the most downstream stack is as follows.

$$Q \cdot \frac{Uf_{\mathrm{T}}}{100} \cdot \frac{1}{n}$$

Consequently, Uf_{D}, which is defined as the fuel consumption ratio in the most downstream stack to the unutilized fuel at the inlet of the stack, should be expressed as Eq. [1].

$$Uf_{\mathrm{D}} = \frac{Uf_{\mathrm{T}}}{n \cdot (100 - Uf_{\mathrm{T}} \cdot \frac{(n-1)}{n})} \cdot 100 \ [\%] \qquad [1]$$

As shown in Table 1, Uf_{D} increases with increasing n, and has the values less than 60% at n values of higher than 3 even at $Uf_{\mathrm{T}} = 80\%$.

Table 1. *Uf* of the individual stacks in the n-stage configuration at Uf_T = 80%, under the assumption of an equal fuel consumption at all the stacks (n is from 1 to 5). *Uf* values at the individual stacks have the maximum at the most downstream stack, which are depicted as hatching.

	Cell Number (count from upstream to downstream)				
n	1	2	3	4	5
1	80%				
2	40%	67%			
3	27%	36%	57%		
4	20%	25%	33%	50%	
5	16%	19%	24%	31%	44%

With a variety of *n* values of from 2 to 20, the Uf_D is shown in Figure 2 as a function of Uf_T. This figure indicates that Uf_D can be controlled by the n value and Uf_T. In this manner, Uf_D, which is the highest *Uf* among the individual stacks, can be controlled to be adequate values even at high Uf_T. When n = 10 and Uf_T = 85%, Uf_D is calculated to be 36%, which is much smaller than Uf_T. Therefore if the most downstream stack has sufficient diffusivity of fuel in an anode layer and high robustness against lean fuel gas, Uf_T could be set to be higher values without serious degradation by using this type of multi-stage configuration. "UltraFuelCell" proposed by U.S. Department of Energy (DOE), which combines the multi-stage SOFC with a gas turbine[5], would be an example of the effective application of this mechanism.

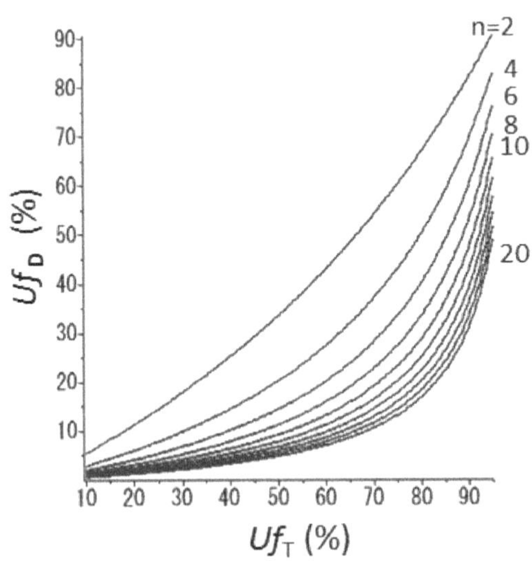

Figure 2. Uf_D as a function of Uf_T (n is from 2 to 20).

Robustness against gas-leakage

If there is gas-leakage in the stacks, the fuel available for the most downstream stack is reduced and a real *Uf* in the most downstream is given by the following equation.

$$Uf_{DL} = \frac{Uf_T}{n \cdot (100 - Uf_T \cdot \frac{(n-1)}{n} - L)} \cdot 100 \ [\%] \qquad [2]$$

where L [%] is the ratio of the gas-leakage rate to Q, and Uf_{DL} is a real Uf in the most downstream with the leakage. Figure 3 shows the Uf_{DL} as a function of n at a $Uf_T = 85\%$ with the gas-leakage. The Uf_{DL} increased with an increase of L and with a decrease of n. Under the condition of $L = 10\%$ and $Uf_T = 85\%$, Uf_{DL} was found to be 45.9% at n = 20, which is much lower than a Uf of 95% for the case of single-stack under the same condition. This suggests the robustness against the leakage would be enhanced by means of the multi-stage electrochemical oxidation configuration.

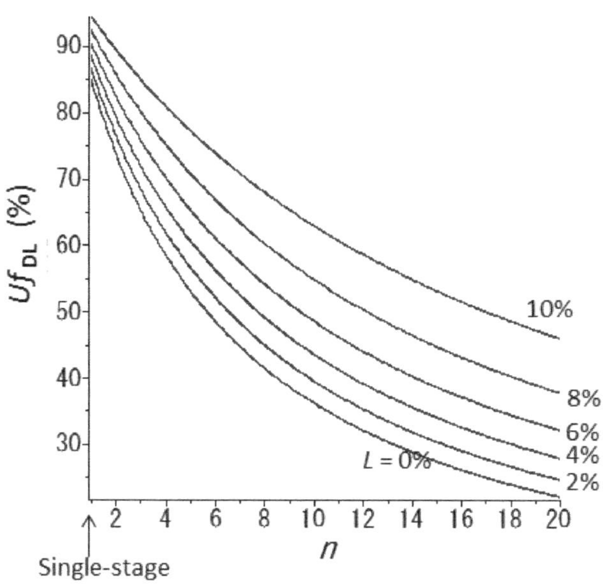

Figure 3. Uf_{DL} as a function of n at Uf_T =85% with the gas leakage or oxide-ion leakage.

CONCLUSIONS

SOFCs have a capability of fuel-to-electricity conversion efficiencies of as high as 60 percent LHV and more. For further increasing the electrical efficiency of SOFCs, fuel utilization ratio of the entire system should be increased to higher values. In this study, the symbolic analysis has been conducted to investigate *Uf* characteristics of the multi-stage electrochemical oxidation. The Uf_D, which is the highest *Uf* among the individual stacks in the multi-stage configuration, was found to have adequate values even at the high Uf_T. The Uf_{DL}, which is the Uf_D with the gas-leakage, also decreased with an increase of *n*, and could have a much lower values than that in the case of the single-stack under the same condition. These results suggest that by using the multi-stage configuration the Uf_T could have relatively high values without an increase of the risk for running out of fuel in some cell caused by the non-uniform fuel distribution in the individual stacks and/ or the gas-leakage.

ACKNOWLEDGEMENTS

This research is supported by The Japan Science and Technology Agency (JST) through its "Center of Innovation Science and Technology-based Radical Innovation and Entrepreneurship Program (COI Program)".

REFERENCES

[1]R. Payne, J. Love, M. Kah, *ECS Trans.* 35(1), 81-85 (2011).

[2]Y. Ando, H. M., M. Minemoto, *Kagakukougaku*, 37(3), 447-453 (2011).

[3]S. Elangovan, J. Hartvigsen, A. C. Khandkar, *J. Electrochem. Soc.*, 144(10), 3337-3342 (1997).

[4]L. J. Frost, R. M. Privette, A. C. Khandkar, *J. Power Sources*, 61(1-2), 135-139 (1996).

[5]J.H. Hirschenhofer, D. B. S., R.R. Engleman, and M.G. Klett. Fuel Cell Handbook Forth Edition, 7.29-7.33 (1998) Available at: http: //physics.oregonstate.edu/~hetheriw/energy/topics/doc/electrochemistry/fc/basic/fuel_cell_handbook_fourth_ed_fed_energy_tech_center_99.pdf (Accessed: 27th March 2015).

LOW TEMPERATURE AC ELECTRIC FIELD-ASSISTED SINTERING OF UNITARY ANODE-SUPPORTED SOLID OXIDE FUEL CELL

R. Muccillo[a)], E. N. S. Muccillo[a)], F. C. Fonseca[a)], D. Z. de Florio[b)]

[a)]Energy and Nuclear Research Institute, S. Paulo, SP, Brazil
[b)]Federal University of ABC, Santo André, SP, Brazil

Abstract

Unitary planar solid oxide fuel cells (cathode $La_{0.84}Sr_{0.16}MnO_3$ thin film/solid electrolyte ZrO_2:8 mol% Y_2O_3 thin film/anode ZrO_2:8 mol% Y_2O_3/NiO 55/45 vol.% 5.0 x 5.0 x 1.5 mm^3 thick pellet) were prepared by electric field-assisted sintering. The application of multiple 300 $V.cm^{-1}$ pulses at 1 kHz with 1-5 A limiting current at 1100 °C for 10 min produced single SOFC devices with dense electrolyte and porous anode and cathode. The electrical conductivity of the cells, measured at 590 °C by the impedance spectroscopy technique in the 5 Hz-13 MHz range in a HP4192A impedance analyzer, showed the increase in the electrical conductivity in comparison with the conventionally sintered cell. The microstructure of cross sections of the cells was observed in a scanning electron microscope. The images of the electric field-assisted sintered cells showed dense electrolyte and porous anode and cathode. Joule heating, provoked by the AC electric current pulses resulting from the application of the AC voltage, promoted NiO reduction at the anode, detected by EDX analysis in its intergranular region. $\Delta L/L_0$ dilatometry shrinkage curves, J x t electric current pulses, $Z'(\omega)$ x $Z''(\omega)$ impedance plots and scanning electron microscopy micrographs show how anode-electrolyte-cathode ceramic cells can be sintered at temperatures lower than the usually required.

INTRODUCTION

The development of planar unitary solid oxide fuel cells is still under way looking for electrically efficient and mechanically reliable single devices, not taking into account thermochemical and thermomechanical compatibility when dealing with a stack composite[1].

The manufacture of planar unitary anode supported solid oxide fuel cells requires several sintering steps depending on the way their conformation is done: a) one is sintering the anode to get a porous structure, coating with the electrolyte to be fired to get non-porous structure and coating with the cathode on top of the electrolyte to be fired again to get porous structure. However, find the suitable (temperature - heating rate - dwelling time - cooling rate - atmosphere) profile is not an easy task; b) another is single step co-firing the three components (electrodes, electrolyte, and cathode layers) conformed by tape casting, considered a low-cost SOFC manufacturing technique[2-7]. Electrolyte sintering is one important step for the co-firing process because the anode and cathode sintering temperatures are lower and consequently after sintering at these low temperatures the electrolyte would not reach the density required for avoiding fuel permeation at the operation temperature of the whole cell. The use of sintering aid to the electrolyte is one method to lower its sintering temperature[8-11]. After solving the problem of the choice of adequate temperature for co-firing anode and electrolyte, another problem remains to solved: to find an adequate temperature to co-fire the cathode to keep it porous and single phase[12,13].

The mismatch of the thermal expansion coefficients of the cell components causes warping and delamination of the cell and may increase porosity of the electrolyte in electrophoretic deposited anode supported cells[14]. During the manufacturing of a cell, thermal stresses are unavoidable due to temperature cycling from room temperature to the firing temperature. In proton conducting half-SOFCs it has been shown that the driving force for deformation is associated with a porosity gradient, produced during co-firing with a porous load to avoid warping[15].

Forcing flatness promotes microcracks and porosity in the electrolyte, which could be corrected by additional low viscosity electrolyte coatings followed by additional firings, a tedious and costly solution to the warping problem. A study of the prediction of warping conditions, based on fracture mechanics, has been reported[16].

Besides the warping problem, another difficulty is that during cell operation at temperatures higher than 700°C, the presence of O_2 might oxidize Ni of the Ni-electrolyte anode of the cell with concentration gradients appearing in the anode layer, leading to electrolyte cracking and warping[17].

Electric field-assisted sintering without application of a mechanical stress on polycrystalline ceramic materials, mainly in electroceramics, has been in the last few years largely reported. Sintering a green ceramic piece can be achieved in a short time (usually few seconds) with a setup consisting on a power supply and a laboratory furnace. This sintering method has been applied to Y_2O_3-stabilized ZrO_2[18-25], MgO-doped Al_2O_3[26], Co_2MnO_4[27], Gd_2O_3-doped $BaCeO_3$[28], SiC[29], Gd_2O_3-doped CeO_2[30], $SrTiO_3$[31], SnO_2[32] and Y_2O_3[33].

Anode/electrolyte half SOFCs have already been co-sintered by another electric field-assisted sintering under pressure (SPS, Spark Plasma Sintering). This technique differs considerably with the flash sintering technique for the following main reasons: a relatively low DC voltage (1 - 10 V) is applied under vacuum to a graphite die to produce extremely large currents (kA range) resulting in a very large heating rate in a powder specimen under pressure inside the die. This technique requires expensive equipment while flash sintering is carried out with a furnace and a power supply. A SPS experiment on a half SOFC allowed for reduction of the NiO with pore former in the anode, crack-free and well-bonded anode/electrolyte interface and, very important, no warping[34]. The flash sintering technique has also been proposed for manufacturing SOFC components[18,35] and anode-electrolyte multilayers[36].

Attempts for the application of the flash sintering technique on single planar anode-supported SOFCs to produce thick porous NiO-YSZ anode/dense YSZ thin film/porous LSM thin film are here reported.

EXPERIMENTAL

The assembling of the planar SOFC consisted on: anode - TZ-8Y/NiO (55/45 vol%) pressed pellets; electrolyte - slurry of commercial powder (cubic ZrO_2:8 mol% Y_2O_3 TZ-8Y, Tosoh, Japan) composed of 75 µm spherical granules diluted in isopropanol; cathode - lanthanum strontium manganite paste (LSM - $(La_{0.80}Sr_{0.20})_{0.95}MnO_{3-x}$).

A PC controlled vertical high temperature dilatometer (Anter model 1161) with PT grids on both parallel sides of the sample was used. Platinum wires connected these current collectors either to a custom-made power supply or to an Agilent 4294A (40 Hz-110 MHz frequency range, input signal 200 mV) or a Hewlett Packard 4192A (5 Hz-13 MHz, 200 mV) impedance analyzer. All experiments were performed under air. Further details may be found elsewhere[24].

The isothermal flash sintering experiment consists in heating the SOFC planar cell and applying an AC voltage (AC for only charge transport, DC could transport undesirable mass) when the specimen reaches a pre-established temperature. The shrinkage level is monitored at the dilatometer gauge while the applied voltage and the current through the cell data are collected in digital voltmeters. Figure 1 shows the possible experimental sequences for flash sintering: dynamic and isothermal.

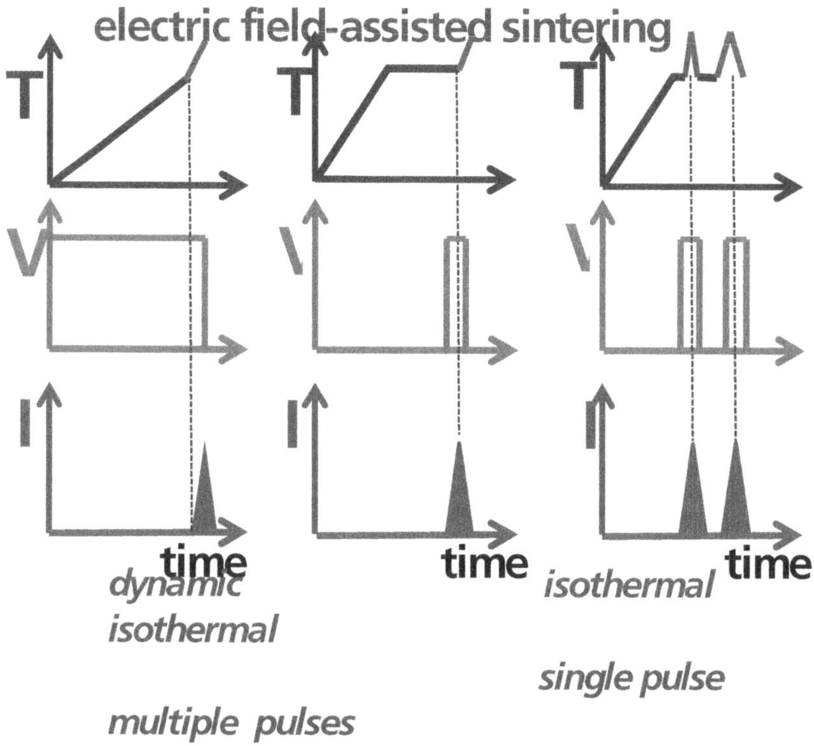

Figure 1: Sketch of temperature, voltage and current profiles during experiments on flash sintering.

Figure 2 shows the experimental arrangement for the flash sintering experiments: a simple spring-loaded tubular alumina sample cell with Pt connections and a type K thermocouple with its tip close to the sample position, a 0-60 V/0-6 A/60-1100 Hz power supply, a resistive furnace and DVMs.

Anode and cathode sides and fracture non-metalized surfaces of the electric field-assisted and conventionally sintered cells were observed in an Inspect F50 FEI FEG scanning electron microscope with EDS (EDAX) analysis.

The [-Z"(ω) x Z'(ω)] impedance spectroscopy data were obtained with a 4192A Hewlett Packard impedance analyzer with a 362 HP controller in the 5 Hz - 13 MHz frequency range [$f = (\omega/2\pi)$] under a voltage amplitude of 200 mV and at temperatures in the oxide ion conductivity region.

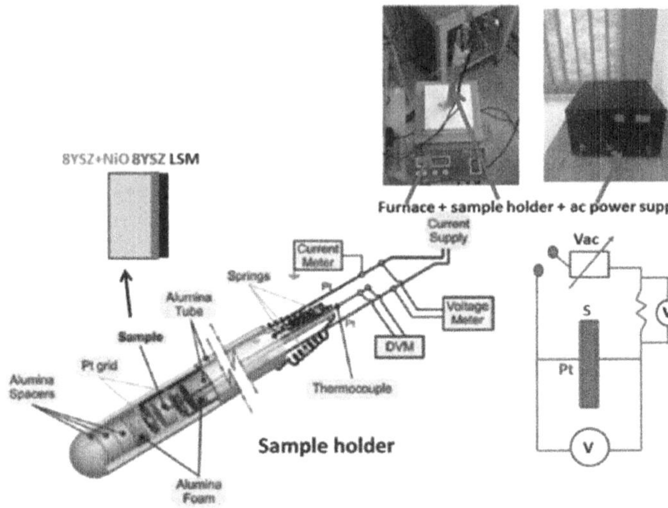

Figure 2: Simple setup for flash sintering experiments: sample holder, furnace and power supply. Electric circuit: V_{AC} - power supply; V - digital voltmeter; Pt - platinum electrode.

RESULTS AND DISCUSSION

Figure 3 shows the linear shrinkage behavior of a single SOFC upon heating to 1200°C and under the application of 310 $V.cm^{-1}$, 1000 Hz, 5 A limiting current, for 5 min at that temperature. The longitudinal (from the anode to the cathode) reduction of the cell thickness without the application of the electric field is 15% while under the applied electric field it reaches 26%. We must point out that this is the threshold level for 8YSZ reaching full density at 1450°C.

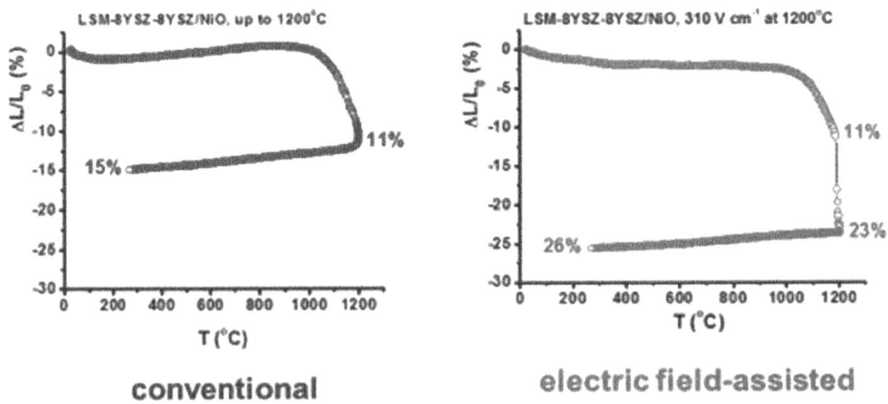

Figure 3: Dilatometric curves of anode-supported planar SOFCs from RT to 1200°C without (left) and with (right) application of an electric field (310 $V.cm^{-1}$) for 5 min.

Figure 4 shows a comparison of the shrinkage levels of the planar cell attained after flash sintering at 1100°C and at 1200°C with the same applied electric field (310 V.cm^{-1}), the same limiting current (5 A) for the same time (5 min). The lower is the temperature, the lower is the final shrinkage level (13%, 26%) as expected due to the lower Joule heating delivered to the cell at lower temperatures (low T, low density of charge carriers, low Joule current).

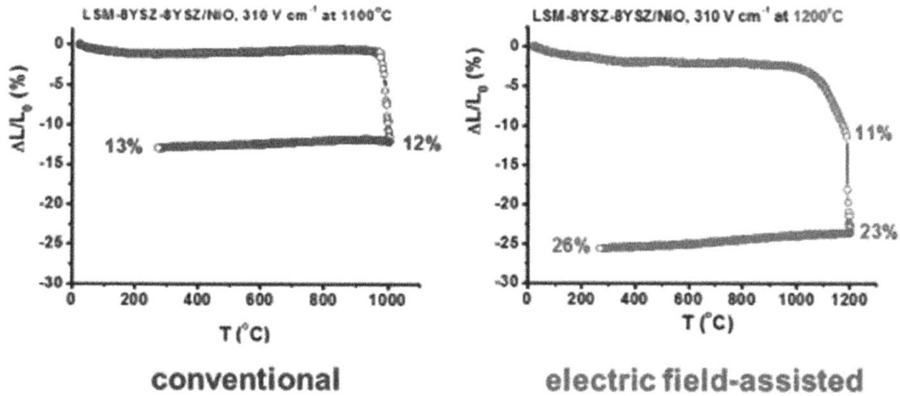

Figure 4: Dilatometric curves of anode-supported planar SOFC from RT to 1100°C (left) and to 1200°C with application of an electric field (310 V.cm^{-1}) for 5 min.

Figure 5 shows the applied voltage, the electric current pulse and the corresponding delivered power to the specimen during the flash sintering experiment. The power supply gives the appropriate voltage amplitude such as to keep the limiting current value, 1 A in this case.

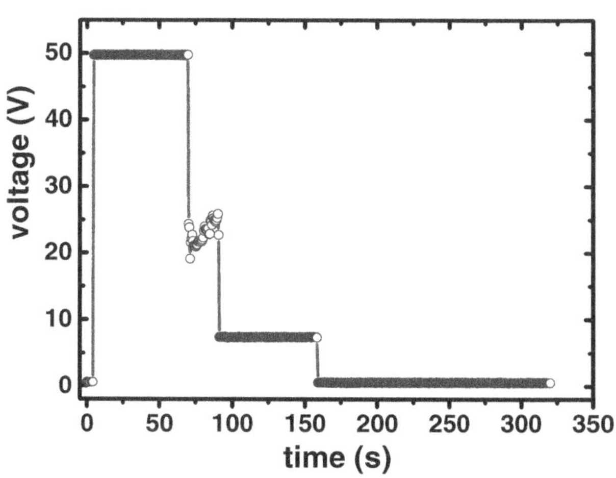

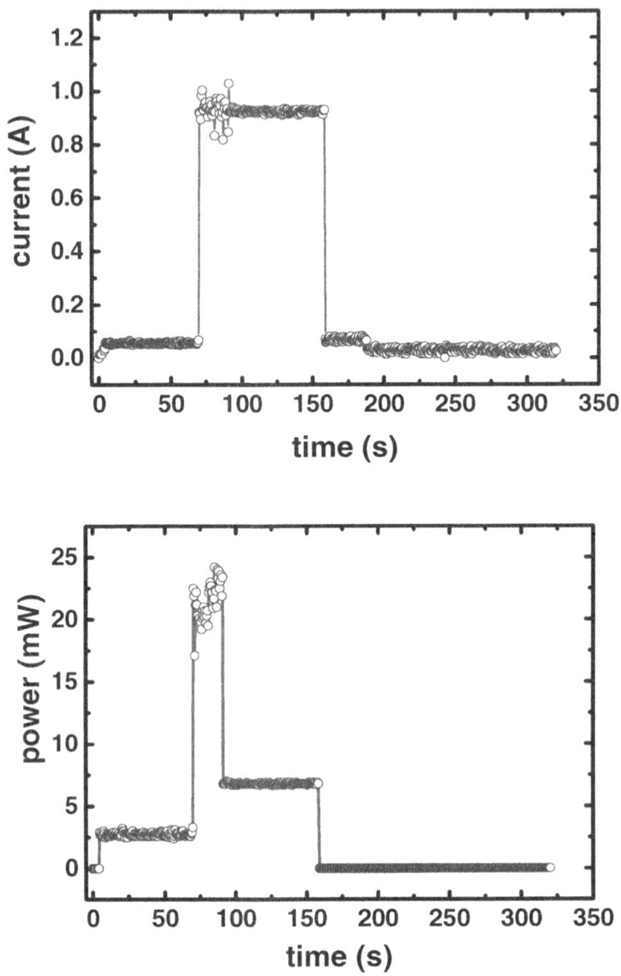

Figure 5: Typical voltage, current and power during the flash sintering experiment.

The cross section of the planar SOFC flash sintered at 1200°C (Figure 3, right) was observed in the FEG scanning electron microscope. The micrograph of one region is shown in Figure 6. The cathode is highly porous and all cell components are well welded, approximately 15 µm being the average thickness of the dense 8YSZ electrolyte. The EDS analysis on the anode and cathode shows Ni in the former and Mn, La and Sr in the latter. Zr is detected due to the limitation in the spot size of the scanning electron microscope.

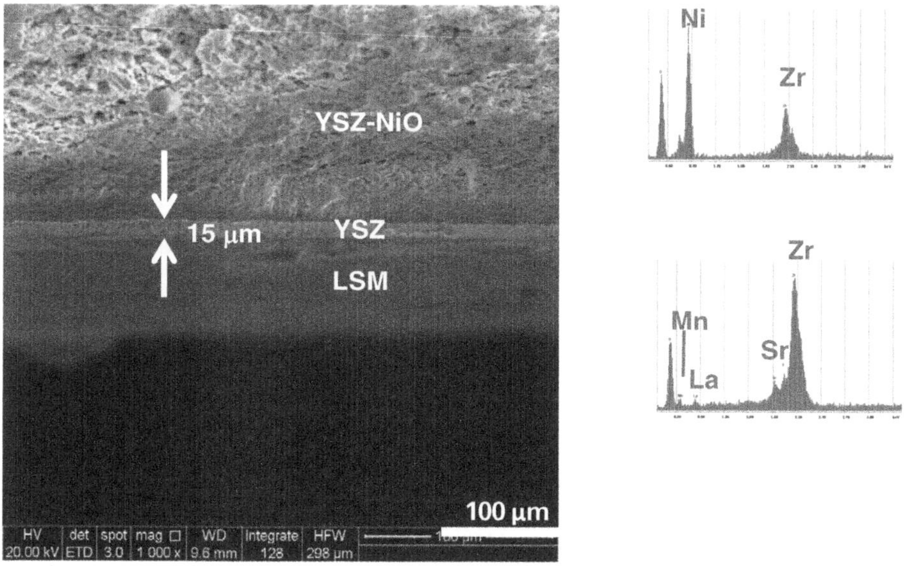

Figure 6: Left - scanning electron microscopy image of a cross section of a unitary planar SOFC after flash sintering at 1100°C. LSM: Lanthanum Strontium Manganite cathode; YSZ: 8 mol% yttria-stabilized zirconia solid electrolyte; YSZ+NiO: anode; right - EDS results on the anode and cathode sides.

Figures 7a and 7b (relatively higher magnification) show scanning electron microscopy micrographs of anode, electrolyte and cathode along with the corresponding EDS analysis. Anode and cathode are porous as required for percolation of the fuel and the oxidant. The electrolyte is dense, the porous region concentrated on its outer surface. The elemental analysis shows the occurrence of La and Mn from the cathode. The higher magnification view in Figure 7b shows that flash sintering at 1100°C did sinter the three components of the planar SOFC, with grain-to-grain neck formation.

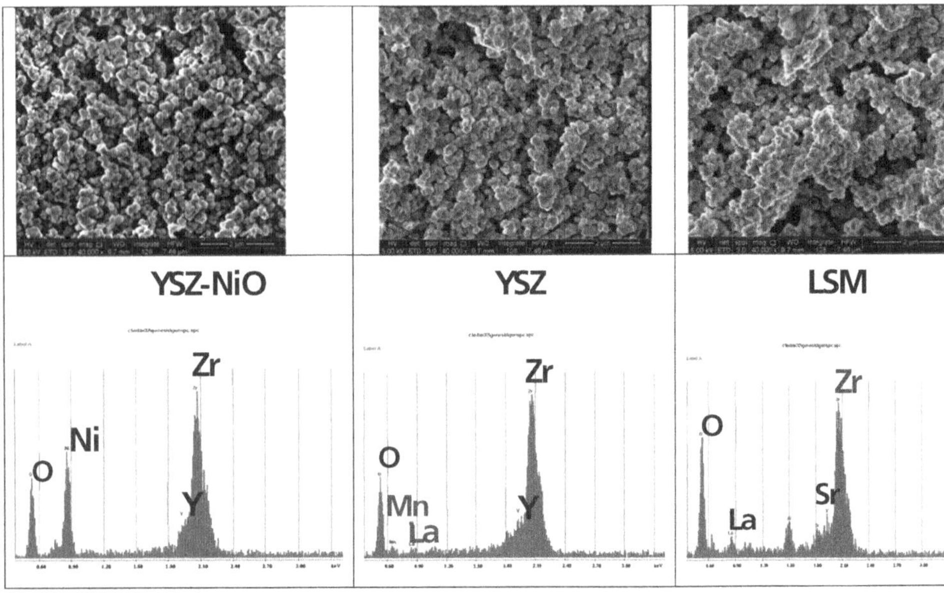

a

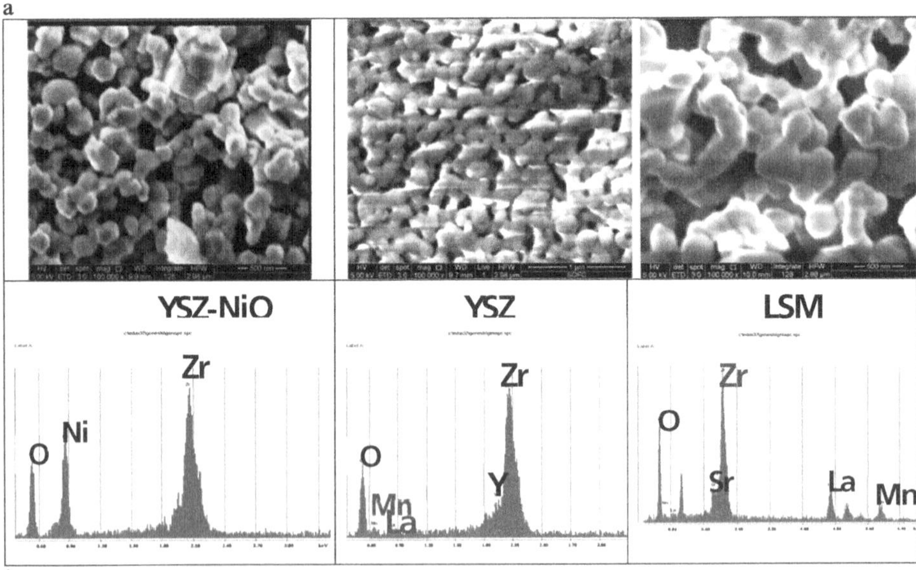

b

Figure 7 (a and b): Top - Scanning electron microscopy images of the components of a unitary planar SOFC after flash sintering at 1100°C. YSZ-NiO: anode; YSZ: 8 mol% yttria-stabilized zirconia solid electrolyte; LSM: Lanthanum Strontium Manganite cathode. Bottom - Results of EDS analysis.

Impedance spectroscopy data of planar anode-supported SOFCs flash sintered at 1200°C during 5 min with the application of 310 $V.cm^{-1}$ with a frequency of 1000 Hz and with limiting the current through the cell to 5A were collected at 590°C from 5 Hz to 13 MHz, input signal 200 mV. The results are shown in Figure 8. The conventionally sintered cell has a total resistivity of 33 kOhm (bulk conductivity 15 kOhm and grain boundary conductivity 18 kOhm). The flash sintered cell, on the other hand, after 5 min of 5 A electric current pulses at 1 kHz, had its total electric resistivity reduced to 12 kOhm (5 kOhm for the bulk and 7 kOhm for the grain boundaries). The deconvolution of the impedance diagrams was performed with a special software[37]. The major contribution to the electrical resistivity of the cell might be due to the anode and cathode. The reduction of the electrical resistivity in the flash sintered cell should be ascribed to, at least partially, the reduction of NiO to Ni. Figure 9 shows a SEM image of the anode region of the cell after flash sintering with 5 A electric current pulses. The Joule heat delivered to the cell is such that besides densifying cathode, electrolyte and anode, it reduces NiO to Ni which, upon melting, wet the YSZ grains.

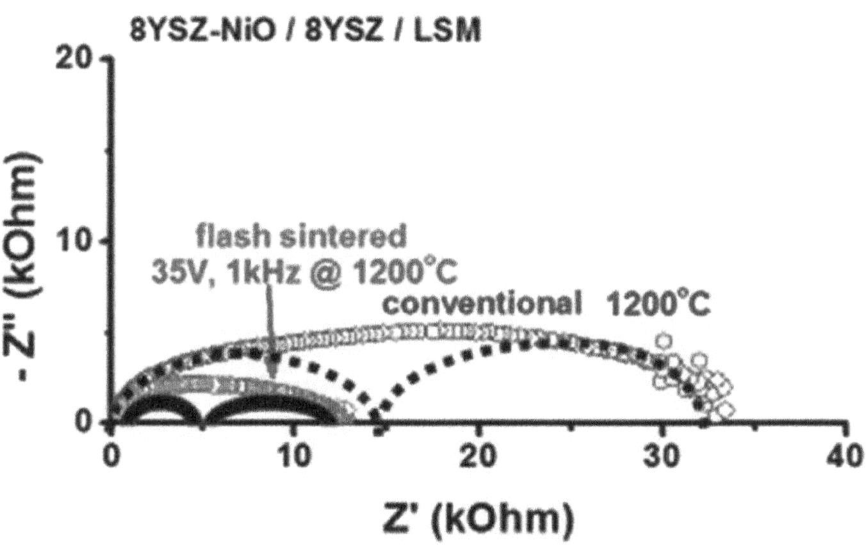

Figure 8: Impedance diagrams of of a unitary planar SOFCs after flash sintering at 1100°C and after co-sintering at 1200°C (conventional). Input signal 200 mV; frequency range 5 Hz-13 MHz; temperature 590°C. Green and red - experimental data; black - deconvoluted (bulk and grain boundary) data.

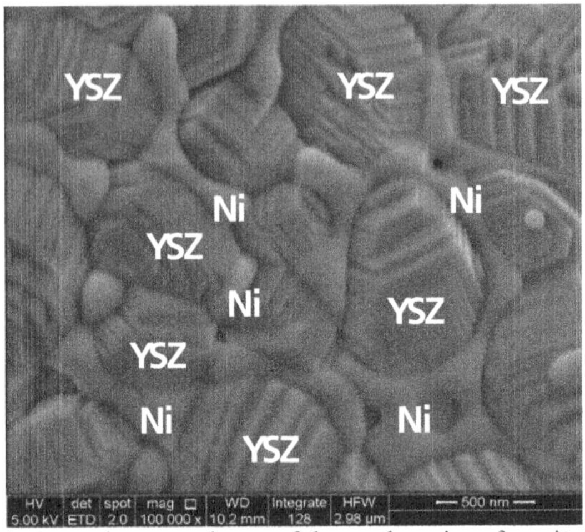

Figure 9: Scanning electron microscopy image of the anode region of a unitary planar SOFC after flash sintering at 1100°C with 310 V.cm^{-1}, 5 A. YSZ: fully dense 8 mol% yttria-stabilized zirconia; Ni: metallic nickel after reduction of nickel oxide.

Some experiments were also performed in commercial green 8YSZ-NiO/YSZ/LSM tapes. Figure 10 shows scanning electron microscopy cross sectional views of the tapes before and after flash sintering with 50 V, 1 kHz, 1A limiting electric current during 60 s at 800°C. The thickness of the tape decreased from ~ 770 µm to ~ 440 µm, while the YSZ electrolyte film thickness was reduced from ~ 90 µm to ~ 30 µm.

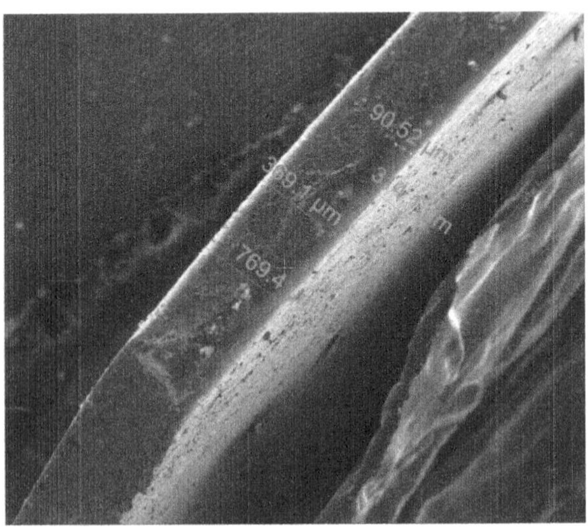

a

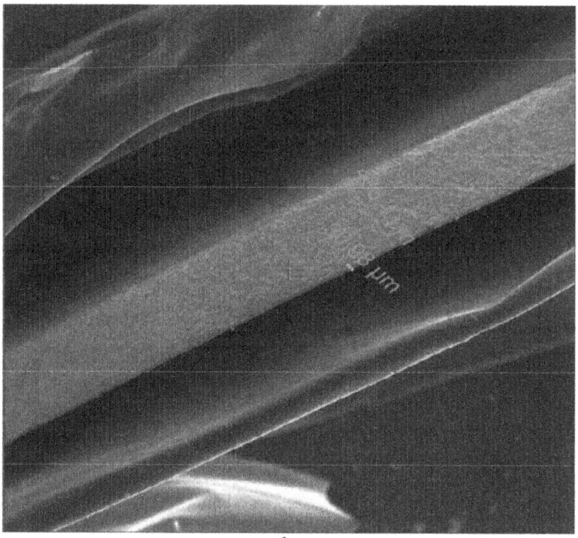

b

Figure10: SEM images of the cross section of green (a) and flash sintered (b) SOFC tape. See text for details.

CONCLUSIONS

Single anode-supported solid oxide fuel cells prepared by coating an anode composed of 8 mol% yttria-stabilized zirconia and nickel oxide (55/45 vol.%) with a 8 mol% yttria-stabilized zirconia film and on top of that a film of lanthanum strontium manganite were co-sintered at 1200°C and also at that same temperature under an applied AC voltage of 310 $V.cm^{-1}$ during 5 min. The electric current pulses through the cell promoted an enhancement of the shrinkage level from 15% to 26%. Moreover, the scanning electron microscopy images showed that the microstructure of the flash sintered cells present suitable densification with porous anode and cathode and dense electrolyte. The densification promoted by flash sintering lead also to a 63% decrease in the total electric resistivity of the cell. Joule heating is suggested as the main mechanism responsible the densification of the planar SOFC. The results show that adjusting properly the parameters for flash sintering, namely, electric field amplitude and frequency, electric current pulse amplitude and duration, and co-sintering temperature, make the flash sintering technique a candidate for co-sintering solid oxide fuel cells. Moreover, there is the possibility of reduction of the YSZ-NiO composite anode to YSZ-Ni cermet anode during flash sintering.

ACKNOWLEDGEMENTS

To FAPESP (Proc. 2013/07296-2) and CNPq (Procs. 405799/2013-6 and 470952/2013-3) for financial support. To Sabrina Gonçalves de Macedo Carvalho, Francisco Nobuti and Yone Vidotto França for technical support.

REFERENCES

1. O. Yamamoto, Solid oxide fuel cells: fundamental aspects and prospects, Electrochim. Acta 45 (2000) 2423-2435.

2. Jae-ha Myung, Tae ho Shin, Sun-Dong Kim, Hae-Gu Park, Jooho Moon, Sang-Hoon Hyun, Optimization of Ni-zirconia based anode support for robust and high-performance 5 x 5 cm^2 sized SOFC via tape-casting/co-firing technique and nano-structured anode, Int. J. Hydrogen Energy 40 (2015) 2792-2799.
3. H. Moon, S.D. Kim, S.H. Hyun, H.S. Kim, Development of ITSOFC unit cells with anode-supported thin electrolytes via tape casting and co-firing, Int. J. Hydrogen Energy 33 (2008) 1758-1768.
4. H. Moon, S.D. Kim, E.W. Park, S.H. Hyun, H.S. Kim, Characteristics of SOFC single cells with anode active layer via tape casting and co-firing, Int. J. Hydrogen Energy 33 (2008) 2826-2833.
5. J.H. Myung, H.J. Ko, H.G. Park, M. Hwan, S.H. Hyun, Fabrication and characterization of planar-type SOFC unit cells using the tape-casting/lamination/co-firing method, Int. J. Hydrogen Energy 37 (2012) 498-504.
6. G. Ye, F. Ju, C. Lin, S. Gopalan, U. Pal, D. Seccombe, Single-step co-firing technique for SOFC fabrication. In: N.P. Bansal, D. Zhu, W.M. Kriven, editors, Advances in Solid Oxide Fuel Cells, 4th ed. (2005) 25-32.
7. G. Ye, F. Ju, C. Lin, S. Gopalan, U.B. Pal, D. Seccombe, Low-cost single-step co-firing technique for SOFC manufacturing. In: Proc. IX Int. SOFC Symp., Quebec City, Canada, S.C. Singhal, J. Mizusaki, editors, ECS Trans. (2005) 451-9.
8. D.P. Fagg, J.C.C. Abrantes, D. Pérez-Coll, P. Nuñez, V.V. Kharton, J.R. Frade, The effect of cobalt oxide sintering aid on electronic transport in $Ce_{0.80}Gd_{0.20}O_{2-d}$ electrolyte, Electrochim. Acta 48 (2003) 1023-1029.
9. T.S. Zhang, J. Ma, Y.J. Leng, S.H. Chan, P. Hing, J.A. Kilner, Effect of transition metal oxides on densification and electrical properties of Si-containing $Ce_{0.8}Gd_{0.2}O_{2-d}$ ceramics, Solid State Ionics 168 (2004) 187-195.
10. C. Kleinlogel, L.J. Gauckler, Sintering of nanocrystalline CoO_2 ceramics, Adv. Mater. 13 (2001) 1081-1085.
11. A.L. Horovistiz, E.N.S. Muccillo, Microstructural and electrical characterization of chemically prepared $Ce_{0.8}Gd_{2-x}(Ag,Sr)_xO_{1.9}$ ($0 \le x \le 0.02$), Solid State Ionics 225 (2012) 428-431.
12. H.S. Song, W.K. Kim, S.H. Hyun, J. Moon, J. Kim, H.-W. Lee, Effect of starting particulate materials on microstructure and cathodic performance of nanoporous LSMeYSZ composite cathodes, J. Power Sources 167 (2007) 258-264.
13. H. Moon, S.D. Kim, S.H. Hyun, H.S. Kim, Development of ITSOFC unit cells with anode-supported thin electrolytes via tape casting and co-firing, Int. J. Hydrogen Energy 33 (2008) 1758-1768.
14. M.K. Lankin, K. Karan, Effect of Processing Conditions on Curvature of Anode/Electrolyte SOFC Half-Cells Fabricated by Electrophoretic Deposition, J. Fuel Cell Sci. Technol. 6 (2009) 021001.
15. R. Costa, J. Hafsaoui, A.P.A. Oliveira, A. Grosjean, M. Caruel, A. Chesnaud, A. Thorel, Tape casting of proton conducting ceramic material, J. Appl. Electrochem. 39 (2009) 485-495.
16. W.N. Liu, X. Sun, M.A. Khaleel, J.M. Qu, Global failure criteria for positive/electrolyte/negative structure of planar solid oxide fuel cell, J. Power Sources 192 (2009) 486-493.
17. J.L. Young, V.I. Birss, Crack severity in relation to non-homogeneous Ni oxidation in anode-supported solid oxide fuel cells, J. Power Sources 196 (2011) 7126-7135.
18. M. Cologna, A.L.G. Prette, R. Raj, Flash-sintering of cubic yttria-stabilized zirconia at 750°C for possible use in SOFC manufacturing, J. Am. Ceram. Soc. 94 (2011) 316-319.
19. R. Muccillo, M. Kleitz, E.N.S. Muccillo, Flash grain welding in yttria stabilized zirconia, J. Eur. Ceram. Soc. 31 (2011) 1517-1521.

20. A. Cordier, M. Kleitz, M.C. Steil, Welding of yttrium-doped zirconia granules by electric current activated sintering (ECAS): Protusion formation as a possible intermediate step in the consolidation mechanism, J. Eur. Ceram. Soc. 32 (2012) 1473-1479.
21. J.A. Downs, V.M. Sglavo, Electric field assisted sintering of cubic zirconia at 390°C, J. Am. Ceram. Soc. 96 (2013) 1342-1344.
22. R. Baraki, S. Schwarz, O. Guillon, Effect of electrical field/current on sintering of fully stabilized zirconia, J. Am. Ceram. Soc. 95 (2012) 75-78.
23. M.C. Steil, D. Marinha, Y. Aman, J.R.C. Gomes, M. Kleitz, From conventional AC flash-sintering of YSZ to hyper-flash and double-flash, J. Eur. Ceram. Soc. 33 (2013) 2093-2101.
24. R. Muccillo, E.N.S. Muccillo, An experimental setup for shrinkage evaluation during electric field-assisted flash sintering: Application to yttria-stabilized zirconia, J. Eur. Ceram. Soc. 33 (2013) 515-520.
25. J.-C. M'Peko, J.S.C. Francis, R. Raj, Impedance spectroscopy and dielectric properties of flash versus conventionally sintered yttria-doped zirconia electroceramics viewed at the microstructural level, J. Am. Ceram. Soc. 96 (2013) 3760-3767.
26. M. Cologna, J.S.C. Francis, R. Raj, Field assisted and flash sintering of alumina and its relationship to conductivity and MgO-doping, J. Eur. Ceram. Soc. 31 (2011) 2827-2837.
27. A.L.G. Prette, M. Cologna, V. Sglavo, R. Raj, Flash sintering of Co_2MnO_4 spinel for solid oxide fuel cell applications, J. Power Sources 196 (2011) 2061-2065.
28. R. Muccillo, E.N.S. Muccillo, M. Kleitz, Densification and enhancement of grain boundary conductivity of gadolinium-doped barium cerate by ultra fast flash grain welding, J. Eur. Ceram. Soc. 33 (2012) 2311-2316.
29. E. Zapata-Solvas, S. Bonilla, P.R. Wilshaw, R.I. Todd, Preliminary investigation of flash sintering of SiC, J. Eur. Ceram. Soc. 33 (2013) 2811-2816.
30. X. Hao, Y. Liu, Z. Wang, J. Qiao, K. Sun, A novel sintering method to obtain fully dense gadolinia doped ceria by applying a direct current, J. Power Sources 210 (2012) 86-91.
31. A. Karakuscu, M. Cologna, D. Yarotski, J. Won, J.S.C. Francis, R. Raj, B.P. Uberuaga, Defect structure of flash-sintered strontium titanate, J. Am. Ceram. Soc. 95 (2012) 2531-2536.
32. R. Muccillo, E.N.S. Muccillo, Electric field-assisted flash sintering of tin dioxide, J. Eur. Ceram. Soc. 34 (2014) 915-923.
33. H. Yoshida, Y. Sakka, T. Yamamoto, J.-M. Lebrun, R. Raj, Densification behavior and microstructural development in undoped yttria prepared by flash sintering, J. Eur. Ceram. Soc. 34 (2014) 991-1000.
34. O. Bezdorozhev, H. Borodianska, Y. Sakka O. Vasylkiv, Spark Plasma Sintered Ni-YSZ/YSZ Bi-Layers for Solid Oxide Fuel Cell, J. Nanosci. Nanotechnol. 13 (2013) 4150-4157.
35. Y. Liu, X. Hao, Z. Wang, J. Wang, J. Qiao, K. Sun, A newly-developed effective direct current assisted sintering technique for electrolyte film densification of anode-supported solid oxide fuel cells, J. Power Sources 215 (2012) 296-300.
36. J.S.C. Francis, M. Cologna, D. Montinaro, R. Raj, Flash sintering of anode-electrolyte multilayers for SOFC applications, J. Am. Ceram. Soc. 96 (2013) 1352-1354.
37. M. Kleitz, J.H. Kennedy, in: P. Vashishta, J.N. Mundy, G.K. Shenoy (Eds.), Fast Ion Transport in Solids, Elsevier/North-Holland, The Netherlands (1979) p. 185.

SOFC SYSTEM DEVELOPMENT AND FIELD TRIALS FOR COMMERCIAL APPLICATIONS

T. Pfeifer, S. Reuber, M. Hartmann, M. Barthel and J. Baade
Fraunhofer Institute for Ceramic Technologies and Systems IKTS
Dresden, Germany

ABSTRACT

At Fraunhofer IKTS, appropriate processes and methods for the development of customized, application-specific SOFC concepts and prototype systems have been established during various projects in the past, funded by public bodies and industrial customers. As an outcome of the cell and stack developments during the past years, two proprietary SOFC technology platforms for system integration are readily available at IKTS. The eneramic® technology can be utilized for SOFC devices in the power range between 50 and 300 W. CFY stacks are available in standard sizes between 10 and 40 cells with a power output between 300 W and 1.2 kW. Higher power levels can be achieved by integration of multiple stacks in joint HotBox modules. For the development of SOFC systems, a detailed plant specification and process design is prepared initially. Here, all application-specific requirements are considered and a system concept with special consideration of the proposed operating modes and fuel processing technology is derived. The proposed concept is validated afterwards by means of laboratory tests on component, HotBox and system level. Further design iterations and model-based analyses are used for the creation of optimized solutions to be implemented in prototype and demonstration systems. Two exemplary SOFC development projects are described in order to illustrate the development status and system engineering approach.

INTRODUCTION

In addition to their use in residential buildings and micro CHP applications, SOFC systems are especially suited for various industrial tasks and off-grid power supplies. In this context, the usability of different conventional fuels in SOFC systems, such as Natural Gas, LPG, biogas, ethanol, diesel and kerosene, provides crucial technological advantages for the development of application-specific and marketable power generators [1-4].

At Fraunhofer IKTS, appropriate processes and methods for the development of customized, application-specific SOFC concepts and prototype systems have been established during various projects in the past, funded by public bodies and industrial customers. During the implementation of these projects, professionals from different working groups of IKTS are working together in interdisciplinary project teams.

For the development of SOFC systems, a detailed plant specification and process design is prepared initially. Here, all application-specific requirements are considered and a system concept with special consideration of the proposed operating modes and fuel processing technology is derived. The proposed concept is validated afterwards by means of laboratory tests on component, HotBox and system level. Further design iterations and model-based analyses are used for the creation of optimized solutions to be implemented in prototype and demonstration systems. During the complete system engineering process the project partners, customers or users are directly involved. The transfer of knowledge and technology for a subsequent product development and commercialization might be included in the project, if desired. Figure 1 below shows the applied system engineering project structure in phases.

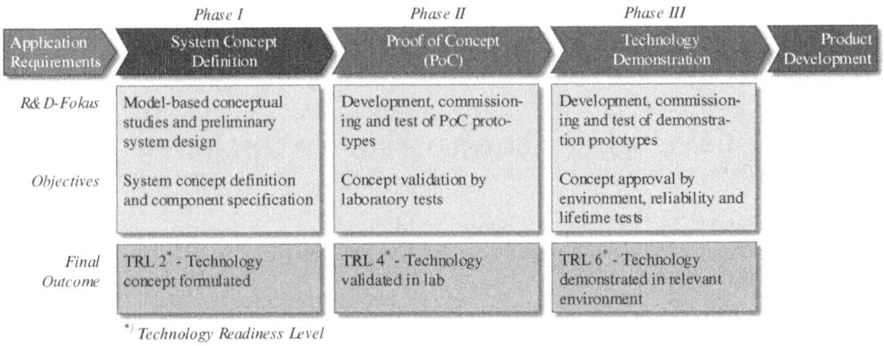

Figure 1. Project phases of the SOFC system development for commercial applications.

SOFC STACK TECHNOLOGY

As an outcome of the cell and stack developments during the past years, two SOFC technology platforms for system integration are readily available at Fraunhofer IKTS. The eneramic® technology can be utilized for SOFC devices in the power range between 50 and 300 W_{el}. CFY-stacks are available in standard sizes between 10 and 40 cells with a power output between 300 W_{el} and 1.2 kW_{el}. The basic specification of available SOFC stacks is summarized in Table 1 below. Higher power levels can be achieved by integration of multiple stacks in joint HotBox-modules. If required, proprietary stack technology of third parties or SOFC stacks of external suppliers can be used for system development projects, as well.

Table 1. Basic specification of SOFC stacks developed by Fraunhofer IKTS.

	eneramic® Stack (IKTS)	**CFY-Stack (IKTS / Plansee)**
Electrolyte	3YSZ (ESC), 90 µm	10ScSZ (ESC), 150 µm
Interconnect	CroFer 22 APU	CFY Alloy
Active Area per Cell	16 cm²	127 cm²
Rated Power per Cell	3,5 W_{el}	30 W_{el}
Standard Stack Size	20 / 30 / 40 Cells	10 / 20 / 30 / 40 Cells
Gross Power Range	70 .. 150 W_{el} per Stack	300 .. 1.200 W_{el} per Stack

The SOFC stacks manufactured by Fraunhofer IKTS are based on planar, electrolyte-supported cells (ESC), which are characterized by their long-term durability and cyclization capability. Figure 2 shows an exploded view of the SOFC stack configurations.

The CFY stack technology was jointly developed by Fraunhofer IKTS and Plansee SE. The robustness of CFY stacks has been demonstrated in various tests over more than 20.000 hours and 120 start/stop-cycles, showing a power degradation rate below 0,7 % per 1.000 hours and 0,8 % per ten cycles, respectively [5-7]. CFY stacks are available as a pre-commercial product, which has been adopted for system integration by various external research partners and companies, e.g. Convion Ltd. of Finland [8].

The eneramic® technology platform was developed by Fraunhofer IKTS with financial support by the Fraunhofer Future Foundation. The micro-stack is designed for a 100 W_{el} LPG-fueled portable power generator. Recent test results have proven the maturity of developed solutions in terms of long-term stability and cyclization capability. eneramic® stacks have been successfully operated for more than 16.000 hours and 110 start/stop-cycles with a power degradation rate below 0,6 % per 1.000 hours and 2 % per ten cycles, respectively [9, 10]. The eneramic® technology is currently commercialized by a corporate spin-off, aiming at SOFC-based products for off-grid power generation and stationary micro-CHP.

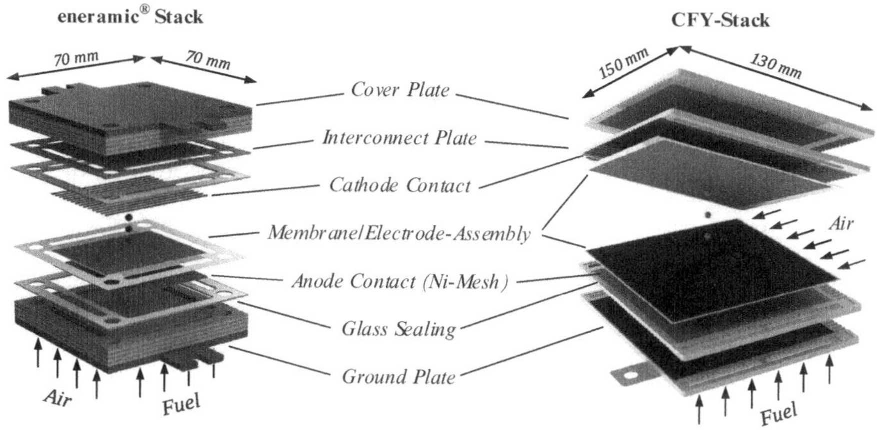

Figure 2. Exploded view of eneramic® SOFC stack (left) and CFY SOFC stack (right).

SYSTEM INTEGRATION OF SOFC STACKS

The development of application-specific SOFC systems requires the aligned engineering of system components and sub-modules according to the generic approach outlined in Figure 3. The basis for the conceptual design is an in-depth specification of desired system characteristics and application requirements.

The conceptual system design is usually initiated by considering the sizing and configuration of required SOFC stack(s) for the desired power output. Depending on the requirements, the integration of multiple stacks into a joint compartment or the integration of multiple stack modules into one system might be necessary. During the definition of the size and operating parameters of the SOFC stack, the desired system purpose – e.g. co-generation, tri-generation or electricity generation, only – has to be properly considered, amongst other requirements. As common SOFC stacks are operated at a temperature range between 700 and 850 °C, a sufficient thermal insulation is required.

For the conversion of provided hydrocarbons before entering the stack, a pre-reformer is usually integrated in SOFC systems, because a complete stack-internal reforming is not viable in most cases due to technical or cost-related implications. The reforming concept and reactor design may be based on partial oxidation (POX), steam reforming (SR), dry CO_2 reforming (DMR), any combination of before mentioned processes, or other known conversion technologies for special fuels. Particular solutions are to be implemented for proper mixing and pre-heating of reactants, residue- and pulsation free evaporation of liquid fuels, as well as for the removal of sulfur contaminants arising from natural impurities or odorants. In general, the reforming concept for SOFC systems has to be properly designed for each specific application, considering the used fuel, the system concept and operating requirements.

As the electrochemical conversion of provided reformate gases in the stack is limited by physical laws, the anode off-gas will in any case contain some remaining H_2 and CO. Those combustible residuals have to be fully depleted using an appropriate afterburner. Depending on the system concept and operating conditions, the concentration of reactants might be reduced to extremely low levels, which require specific afterburner solutions for a complete conversion. In some cases, a part of the anode off-gas is recirculated to support the reforming reaction, or the afterburner is integrated with the pre-reformer for proper heat exchange. Generally again, the afterburner concept for SOFC systems has to be properly designed for each specific application, considering the particular process conditions, system concept and operating requirements.

The pre-reformer, SOFC stack, afterburner and interlinked hot-gas devices are to be properly insulated due to their elevated operating temperature. Especially in small-scale systems, a very tight integration of above mentioned components into one thermally insulated compartment – the so-called HotBox – is beneficial. All system-internal heat sinks and sources have to be balanced by properly designed interconnections, based on convective, conductive or radiation heat exchange. Remaining excess heat can be extracted as usable heat, if a CHP system is desired – or has to be avoided to the possible extent, if a highly efficient electrical power generator has to be designed.

The variable-voltage DC power output of the SOFC stack needs to be fed into a power conditioning system, acting as the power distribution hub for internal and external electrical power flows during the start-up process and stationary operation. For achieving the minimum required operating temperature of reformer and stack, properly designed start-up devices have to be integrated in the system, which might be based on electrical heaters, burners or other devices. Those peripheral components are again to be designed in particular for each application-specific SOFC system concept, considering the given operational characteristics and user requirements.

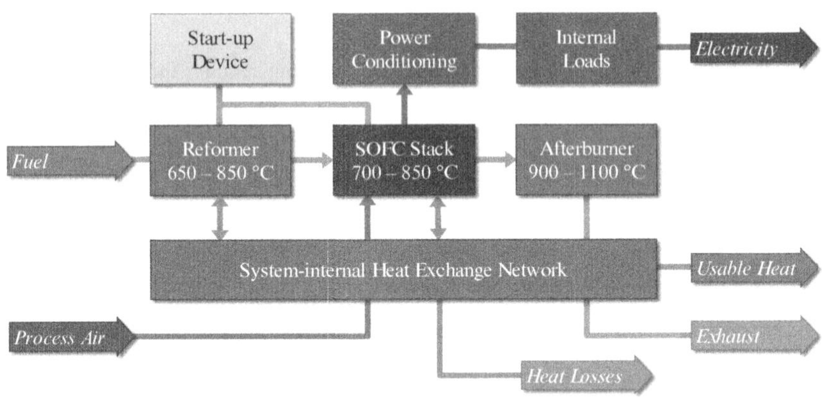

Figure 3. Generic SOFC system integration approach.

Based on manifold experience from prior development projects at IKTS, there is no preferential or standard solution for the technical implementation of stationary SOFC systems within the wide range of potential applications. However, the procedures and tools for the creation of application-specific SOFC systems can be standardized to a great extent, in order to provide safe and reliable, marketable solutions according to customer's intentions – using efficiently structured engineering processes.

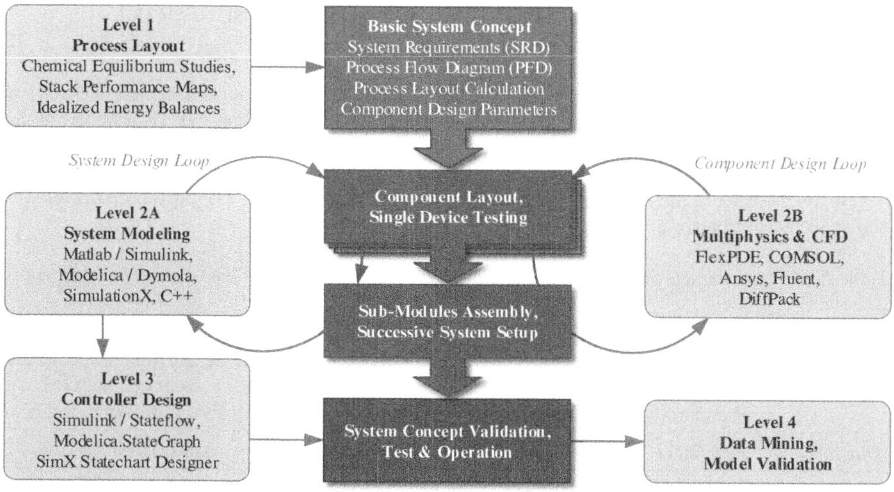

Figure 4. Schematic representation of the multi-level simulation supported SOFC system development concept.

Figure 4 above shows a schematic representation of the multi-level simulation supported SOFC system development concept, which was established at Fraunhofer IKTS based on a number of in-house developed software tools and commercial simulation environments. Proprietary function libraries (DLLs) have been implemented in terms of a freely configurable 0-D SOFC stack model, a computation library for thermo-physical properties of fluids and solid materials, as well as a solver for the approximate computation of chemical equilibrium states. Those function libraries are interlinked with various standard software packages, e.g. MS Excel, Matlab/Simulink, Modelica/Dymola, etc., in order to use a consistent database for the different modelling and simulation activities during SOFC system development.

As part of the system concept definition, a complete process layout calculation is drafted in MS Excel – or other applicable software – with interlinked DLLs. The process balance sheets are used for parameter studies, performance estimations, as well as for the specification of design parameters for all required system components (Level 1 in Figure 4). The actual system engineering and design of components and sub-modules is supported by modelling and simulation activities within two concurrent approaches. A dynamic system model, implemented in Matlab/Simulink or Modelica/Dymola, is created for the investigation of interdependencies between system components and balancing processes in the time domain (Level 2A). At the same time, detailed component and reactor models are implemented in commercial software environments for CFD-, FEA- or multi-physics simulations, in order to support the technical layout and design verification of selected system components (Level 2B). With the help of software state machines, the dynamic process model out of Level 2A may be further used for control logic development or the investigation and optimization of transient procedures, e.g. start-ups, load changes, etc. (Level 3). Finally, after the test operation of prototype systems has started, actual measurement data are used for model calibration – and the developed system models may be further used for in-depth process analyses and performance optimization (Level 4). As part of the contracted R&D services and technology transfer, the developed models are usually shared with customers and development partners during the execution of SOFC system development projects.

RECENT SOFC SYSTEM DEVELOPMENT PROJECTS AT FRAUNHOFER IKTS

Project Characteristics and Scope

The engineering approach and toolchain described above was successfully applied and further matured within two recent SOFC development projects at Fraunhofer IKTS. Funded by the Fraunhofer Future Foundation, the eneramic® technology platform was developed for a 100 W_{el}, LPG-fueled portable power generator. Secondly, the h2e® power generator, rated at 1 kW_{el}, is developed by order of the Indian company h2e Power Systems Pvt. Ltd. The project characteristics and scope is summarized in Table 2 below.

Table 2. Project characteristics and scope of the eneramic® and h2e® SOFC system developments.

	eneramic® by Fraunhofer	h2e
Funding	Internal program, funded by the Fraunhofer Future Foundation	Contract research, assigned by h2e Power Systems Pvt. Ltd., Pune, India
Duration	June 2007 – August 2015	January 2013 – March 2016
Market segment	Off-grid power generation for industrial applications (low power / long runtimes), e.g. traffic control systems, cathodic corrosion protection, surveillance systems, etc.	micro-CHP for continuous electrical power supply and hot water preparation in interference-prone grids, e.g. for small businesses, clinics, housing, rural applications, etc.
System description	LPG-fueled 100 W_{el} SOFC power generator (SOFC/battery-hybrid)	NG-fueled 1 kW_{el} SOFC µCHP system (SOFC/battery-hybrid with optional heat use)
Commercial exploitation	Commercialization through a corporate spin-off with third-party investment, NewCo to be operable in 2016	Contracted project includes IP- and technology transfer for local manufacturing and commercialization by the customer

Technical Implementation, HotBox Concepts

Governed by the application requirements, both system concepts are designed towards simplicity and robustness. At the same time, the electrical efficiency needs to be increased to the possible extent, as the primary product is the generated electricity in both cases. In order to reduce the number of start/stop cycles in practical operation, and to widen the range of potential applications, the power modulation range of the SOFC systems needs to be enlarged as much as possible. For all the given technical requirements, a number of proven, specific solutions might be utilized – in most cases having contrary effects on the system cost and complexity. Therefore, the actual system design is predominantly characterized by the necessary trade-off between system performance, equipment cost, as well as operation and maintenance efforts.

The pre-reforming of provided hydrocarbons prior to the stack is based on catalytic partial oxidation in both system concepts, being the simplest approach with respect to reactor design and process controls, and furthermore omitting any water supply and recovery. However, as partial oxidation considerably restricts the conversion efficiency on system level, appropriate measures for increased electrical efficiencies had to be included in the system concepts. In case of the h2e® power generator, the pre-heating of the air flow supplied to the reformer together with a stack fuel utilization in the range between 80 to 85 % leads to a sufficient electrical efficiency on system level in the range of 40 % (LHV, net), even with partial oxidation. In the eneramic® system, an external heat flow to the reformer for lowering the required air ratio – again leading to higher electrical efficiency – is achieved by proper design of the gas processing unit. In both systems, the reformer, SOFC stack, afterburner and supplemental devices are tightly integrated in a compact HotBox, surrounded by high-performance, multi-layered thermal insulation material. Figure 5 below shows the applied HotBox concepts by means of sectional drawings, derived from CAD-models.

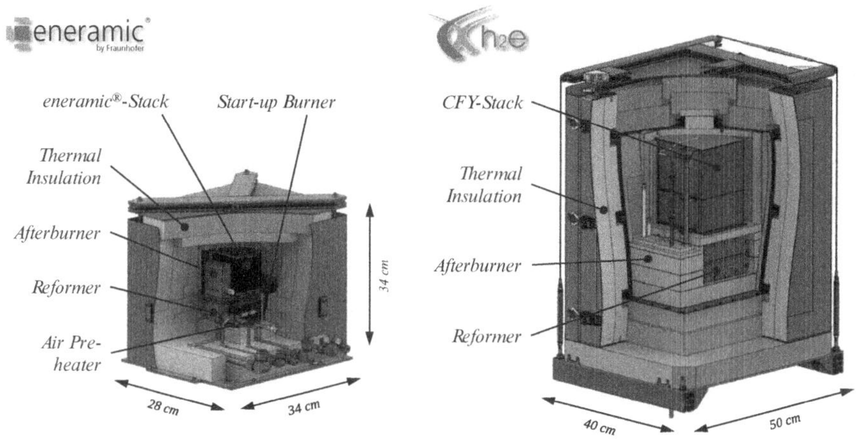

Figure 5. Sectional view of the eneramic® HotBox (left) and h2e® HotBox (right).

Achievements of the eneramic® SOFC System Development Project

After initiation of the eneramic® project in 2007, substantial research and development work was dedicated to the SOFC stack design and structural elements, functional layers, sealing glass, as well as joining technologies for hot-gas heat exchangers and reactors [9, 11]. System integration was started after the core components had proven reasonable maturity during pre-tests. Starting with an initial laboratory sample of the eneramic® system, an iterative improvement of prototype systems was achieved by design modifications in all contributing engineering fields, i.e. the process technology, BoP components, sensors and actuators, electronics, etc. The evolution of eneramic® prototype systems is summarized in Table 3. Apart from the evident reduction of weight and volume, the reliability and robustness of stacks, HotBoxes and prototype systems was successively improved.

The eneramic® system is designed for off-grid power generation in industrial applications. In typical usage scenarios, a total device lifetime of 10.000 hours including 100 to 200 start/stop cycles is sufficient. The durability of the eneramic® core module was demonstrated with a HotBox endurance operation test over more than 16.000 hours (cf. Figure 6). The test was interrupted by two shutdowns due to laboratory maintenance. After the second re-start, the output power of the stack recovered only very slowly with increasing stack temperature, but finally the initial gross power level of ~130 W_{el} was reached, again. From the slope of the stack voltage at the nominal operating temperature of 830 °C, prior to 12.000 operating hours, a power degradation rate of 0,5 % per 1.000 hours could be observed. Automated start/stop-cyclisation tests were carried out with various samples of each HotBox generation (cf. Figure 7). Stack performance degradation due to thermal cycling and repeated anode reduction and oxidation was compensated by increased current density and operating temperature. While earlier HotBox generations showed major faults after 40 to 70 start/stop-cycles, newer samples of the eneramic® HotBox have achieved 110 operating cycles (test with HB-4 is still ongoing). From the performance data with HB-3, an averaged power degradation rate of 0,2 % per each start/stop-cycle was determined.

Table 3. Sample views and technical data of developed eneramic® prototype systems.

Prototype Series	**E-1.0** Laboratory Sample	**E-1.1** Proof-of-Concept	**E-1.2 / E-1.3** Demonstrator Series	**E-2.0** Pre-Commercial Unit
Net Power Output	**95 W_{el}** 12 V_{DC}	**100 W_{el}** 12 V_{DC}	**100 W_{el}** 12 V_{DC}	**110 W_{el}** 12 or 24 V_{DC}
Fuel Input, Efficiency	35,3 g/h **21 %$_{LHV}$**	48,8 g/h **16 %$_{LHV}$**	35,5 g/h **22 %$_{LHV}$**	35,8 g/h **24 %$_{LHV}$**
Volume, Weight	175 l 71 kg	115 l 55 kg	94 l 36 kg	55 l 23 kg
No. of Units, Availability	1 05/2012	1 01/2013	9 09/2014	> 15 07/2015

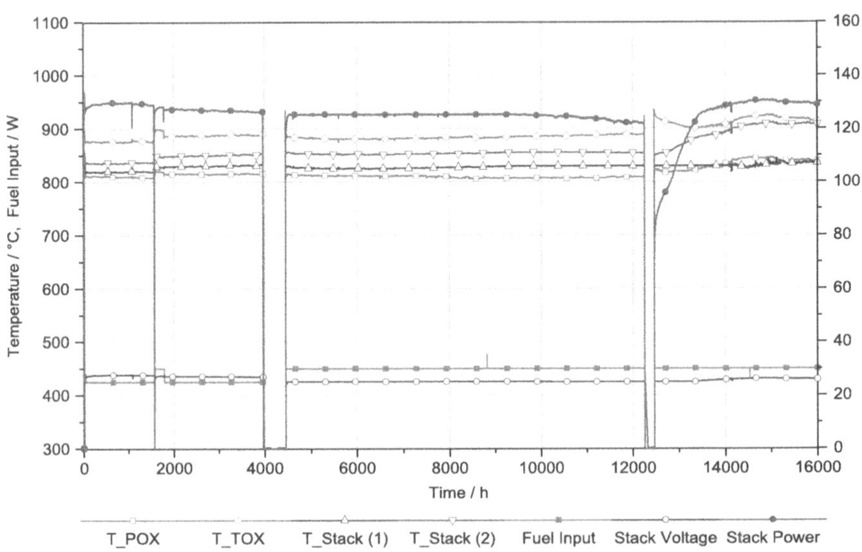

Figure 6. Endurance operation of an eneramic® HotBox, fueled by lab-grade propane.

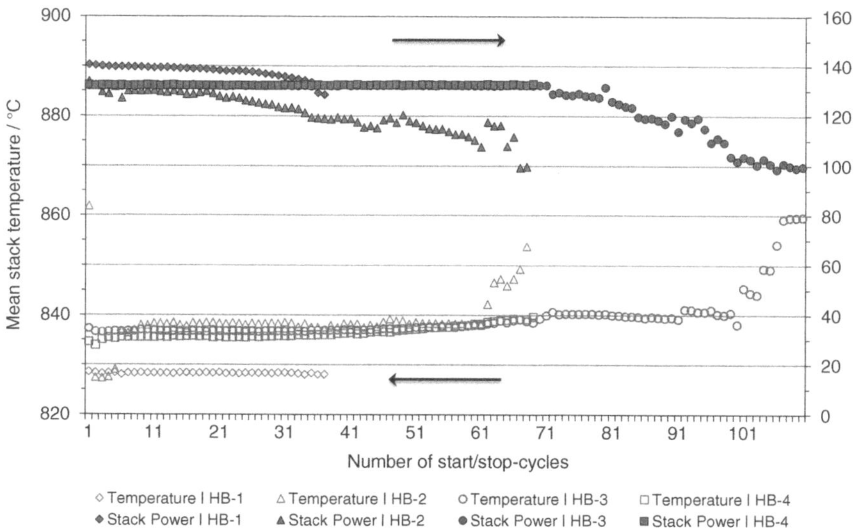

Figure 7. Power output and stack operating temperature during automated start/stop-cyclisation tests with sample HotBoxes of generation 1 to 4 (HB-4 test ongoing).

Following the extensive prototype testing and process optimization in laboratory environment, field trials with eneramic® systems were initiated in selected sample applications. One system was used for powering a LED traffic control panel; another one was installed for a cathodic corrosion protection device at an underground pipeline (cf. Figure 8). The SOFC power generators were installed together with a buffer battery and two 11 kg LPG gas cylinders including an automatic switching device inside a weather- and vandalism-proof housing. Figure 9 shows selected test data, obtained during the first half year of on-site operation with the LED traffic control panel. The eneramic® system was operated as a battery charger, where the actual operating cycles were triggered by the terminal voltage of the lead-acid battery. Due to varying power consumption of the LED panel and some interim performance decay of the system, the duty cycles varied to some extent. However, the continuous operation of the LED panel could be ensured with a service interval of four weeks on average for the exchange of emptied gas cylinders.

Figure 8. Field trials with eneramic® prototype systems, powering a LED traffic control panel (left) and a cathodic corrosion protection device at an underground pipeline (right).

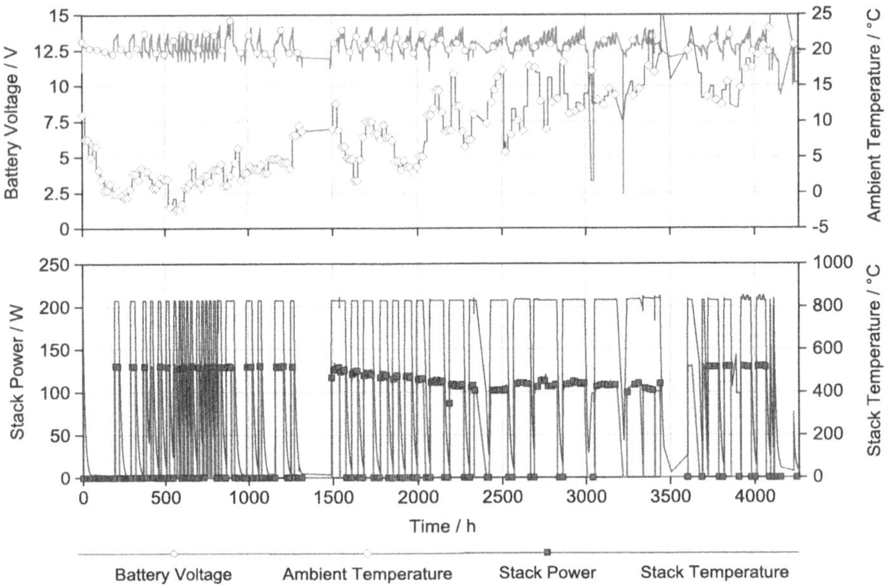

Figure 9. Sample data of the eneramic® field trial, powering a LED traffic control panel via a standard lead-acid battery (12 V_{DC}, 160 Ah).

Achievements of the h2e® SOFC System Development Project

The h2e® project was planned in two major phases. Phase I is aiming at the system concept definition and practical verification using two proof-of-concept prototype systems. During project phase II, the system design is to be optimized towards cost and manufacturability, and three improved prototype systems are to be delivered to the customer for on-site demonstrations and field trials. The project includes a complete technology transfer for subsequent product development, local manufacturing and commercialization in sole responsibility of the customer.

Project phase I was initiated with an in-depth system requirements definition, based on intended application characteristics. With respect to the given technical boundary conditions, an appropriate system concept was derived, backed by a detailed process layout calculation including parameter studies and a system performance estimation. With the created process design data, the detailed layout of required system components and sub-modules was initiated. The complete prototype system architecture was consolidated with a detailed 3-D CAD model, being the basis for actual component manufacturing and system assembly.

In order to achieve the desired net electrical efficiency of 35 to 40 $\%_{LHV}$ with a very simple process design based on partial oxidation, a special arrangement of HotBox-components was required and demanding process parameters for the reformer and SOFC stack have been specified. A stoichiometric air ratio of 0,27 was determined for the POX-reformer and the anode-side fuel utilization in the SOFC stack was intended to reach 85 % at rated power output. For the approval of the initial component design and HotBox arrangement with specified process parameters, it was decided to execute a verification test with a preliminary HotBox prototype on a test-rig at IKTS. The prototype HotBox included the POX-reformer, SOFC stack, afterburner, supplemental parts and thermal insulation as designed for later system integration (cf. Figure 10).

Figure 10. Sample pictures of the h2e® HotBox assembly; 1) ceramic base plates with integrated reformer and afterburner; 2) 50 cell CFY stack mounted on baseplates; 3) assembled HotBox with thermal insulation; 4) prototype HotBox mounted in laboratory test-rig for verification test.

With the HotBox prototype installed in a laboratory test-rig, the start-up procedure intended for later system operation was implemented and tested. After successful commissioning and approval of the process parameters for rated power output, the HotBox was operated under steady state conditions for 1.000 hours without any deviations. The endurance operation was followed by preliminary part-load tests, which indicated the possibility of a power modulation range between 400 W_{el} and 1.400 W_{el} (gross). Finally, an automated start/stop-cyclization test was executed. It covered 38 full operating cycles including start-up, seven hours of full load operation, shutdown and passive cooling down to 500 °C stack temperature. Figure 11 below shows an overview of the design verification test with the h2e® HotBox prototype, installed in a laboratory test-rig.

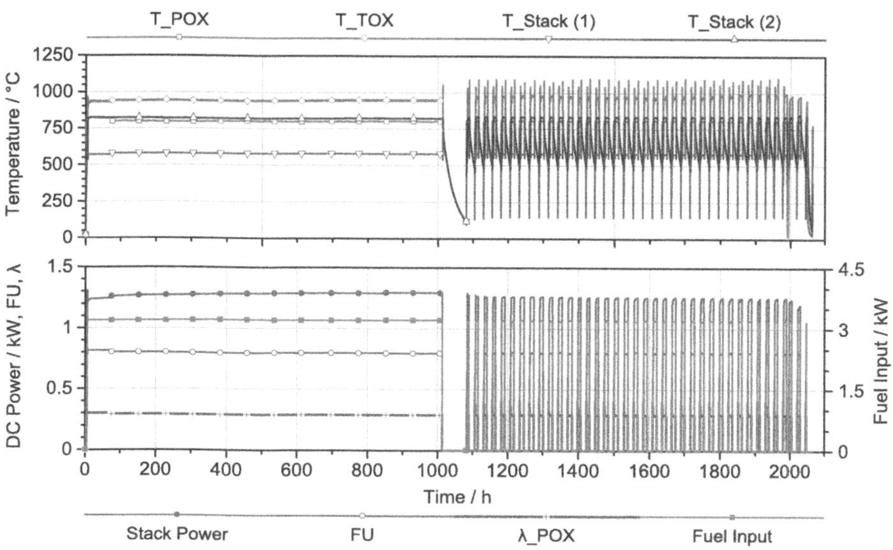

Figure 11. Design verification test with the h2e® HotBox prototype, installed in a laboratory test-rig (1.000 hours endurance operation, followed by automated start/stop-cyclisation test, fueled by lab-grade methane during all tests).

Following the HotBox design verification, the assembly and commissioning of two Proof-of-Concept prototype systems was initiated, based on a slightly modified HotBox design. Figure 12 shows sample views of the prototype manufacturing. The prototypes were commissioned in IKTS' SOFC laboratory (cf. Figure 13) and passed a characterization test for process analysis, including part load operation.

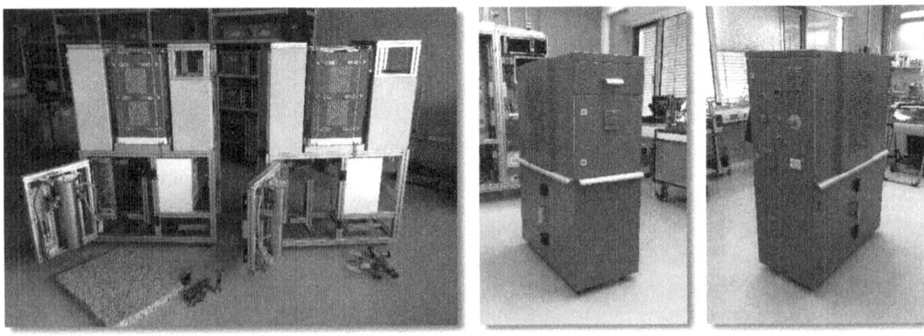

Figure 12. Sample views of h2e® Proof-of-Concept prototype systems during assembly.

Figure 13. h2e® Proof-of-Concept prototype systems installed in IKTS' SOFC laboratory (left) and at customer's site in Pune, India (right).

Figure 14 below shows an overview of the initial characterization test with one of the h2e® Proof-of-Concept prototype systems. The procedures included steady state operation at rated power output and several scheduled sequences of part load operation. The system was operated manually or by script-based, automated control sequences. Intermittent shutdowns were required for updating the control software and for other minor maintenance jobs.

The part load characteristics of the h2e® Proof-of-concept prototype systems is summarized in Figure 15. During the system characterization tests, a power modulation range between 1.420 W_{el} and 450 W_{el} (gross) was proven. The stack fuel utilization was kept constant at 80 % down to a power output of 820 W_{el}, and was reduced at lower power outputs in order to maintain the desired HotBox temperatures. The electrical efficiency showed, however, satisfying values, considering the fact that it is a water-less system based on dry partial oxidation. The balance of plant (BoP) power demand is dominated by the cathode air fan and the programmable logic controller (PLC). With a micro-controller based control unit and improved power conditioning system (PCS) in later system generations, an even higher net electrical efficiency can be expected.

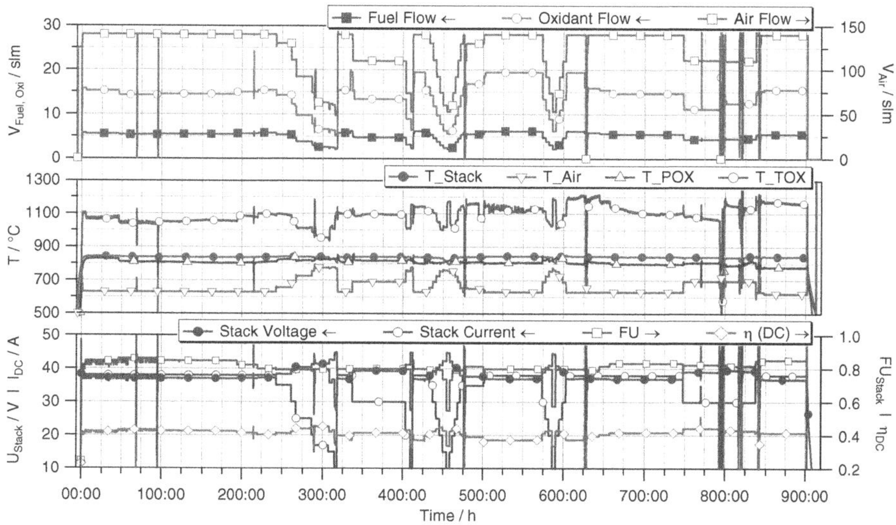

Figure 14. Overview of the initial characterization test with one of the h2e® Proof-of-Concept prototype systems. Test procedure includes steady state operation at rated power output and scheduled part load operation for process analysis.

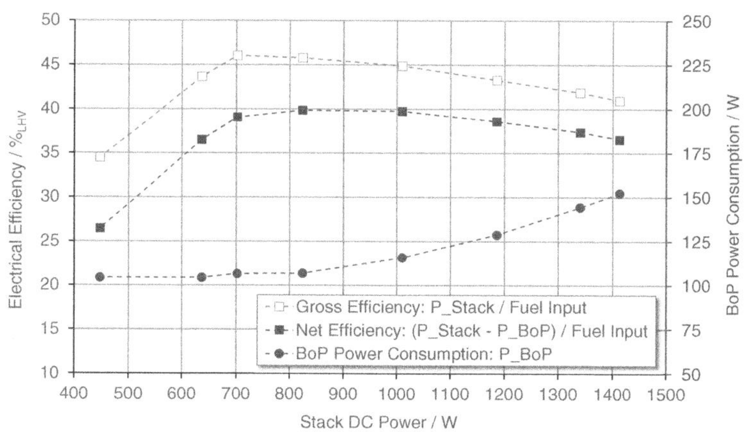

Figure 15. Part load characteristics of the h2e® Proof-of-Concept prototype system. Stack fuel utilization was kept constant at 80 % between 1.400 W_{el} and 800 W_{el}, and was reduced at lower power outputs to maintain desired HotBox temperatures.

Following the successful system commissioning and acceptance test, one of the h2e® Proof-of-Concept prototype systems was shipped to Pune, India, and installed in a laboratory owned by the customer for test, training and demonstration activities (cf. Figure 13). The second prototype system is remaining at IKTS for further technical development, i.e. process automation and parameter tuning, followed by endurance operation mimicking real-world conditions.

The commissioning of the Proof-of-Concept prototype systems marked the completion of project phase I according to the contracted work plan. Afterwards, project phase II was initiated, which includes a system design review towards cost and manufacturability. Three improved prototype systems are to be assembled and commissioned at IKTS until the end of year 2016. The systems are then to be shipped to India for test and demonstration activities. In parallel, the knowledge transfer, vendor sourcing and infrastructure development for local manufacturing and commercialization of h2e® systems in sole responsibility of the customer was initiated.

CONCLUSION AND OUTLOOK

Fraunhofer IKTS has developed SOFC stack technology platforms and system concepts for SOFC power generators – briefly, from materials development to near-product systems. A top-down requirements definition and bottom-up system engineering methodology was established – briefly, from customer ideas to prototype demonstrations. The proven engineering process, test infrastructure and tool-chain is available for application-specific SOFC development projects – from hand-held devices to stationary powerplants.

The general engineering approach and two recent SOFC development projects have been described in detail. Main achievements and upcoming activities related to prior discussed SOFC system developments are summarized in Table 4 below.

Table 4. Summary of the eneramic® and h2e® SOFC system development status.

	eneramic® by Fraunhofer	h2e
Project Status Summary	1. Proprietary stack technology, core components, system concept and control logic was developed 2. Prototype system development advanced to the pre-commercial stage	1. CFY-stack was successfully integrated into a compact and robust HotBox design 2. Two Proof-of-Concept (PoC) prototype systems were commissioned
Achieved Maturity Level	3. Readiness for marketing was demonstrated by field trials and business case validation 4. CE-Certification according to IEC 62282 and EMC directive is prepared	3. High-performance SOFC system based on POX was demonstrated 4. PoC prototype system was shipped and installed at customer's site for test and demonstration activities
Upcoming Activities	5. Pilot manufacturing of systems for demonstration activities and field trials starting in 2nd half of 2015 6. Corporate spin-off for commercial deployment to be operable in 2016	5. Project phase II (initiated): development, commissioning, shipment of improved demonstration prototypes 6. Technology transfer and start of local manufacturing / commercialization

ACKNOWLEDGEMENTS

The eneramic® project was partially funded by the »Fraunhofer Future Foundation«, Munich, Germany. The h2e® system development project is funded by the privately owned company »h2e Power Systems«, Pune, India, and New York, U.S.A.

REFERENCES

[1] Steele, B. C. H. (1999). Fuel-Cell Technology: Running on Natural Gas. *Nature 400*, 619-621, DOI: 10.1038/23144

[2] Ellamla, H., Staffell, I., Bujlo, P., et. al. (2015). Current Status of Fuel Cell based Combined Heat and Power Systems for Residential Sector. *Journal of Power Sources 293*, 312-328, DOI: 10.1016/j.jpowsour.2015.05.050

[3] Elmer, T., Worall, M., Wu, S., Riffat, S. B. (2015). Fuel Cell Technology for Domestic Built Environment Applications: State of-the-art Review. *Renewable and Sustainable Energy Reviews 42*, 913-931, DOI: 10.1016/j.rser.2014.10.080

[4] Hardman, S., Chandan, A., Steinberger-Wilckens, R. (2015). Fuel Cell Added Value for Early Market Applications. *Journal of Power Sources 287*, 297-306, DOI: 10.1016/j.jpowsour.2015.04.056

[5] Brandner, M., Bienert, C., Megel, S., et. al. (2013). Long Term Performance of Stacks with Chromium-Based Interconnects (CFY). *ECS Transactions 57*, 2235-2244, DOI: 10.1149/05701.2235ecst

[6] Megel, S., Dosch, C., Rothe, S., et. al. (2013). CFY-Stacks for Use in Stationary SOFC and SOEC Applications. *ECS Transactions 57*, 89-98, DOI: 10.1149/05701.0089ecst

[7] Bienert, C., Brandner, M., Skrabs, S., et. al. (2015). CFY-Stack Technology: The Next design. *ECS Transactions 68,* 2159-2168, DOI: 10.1149/06801.2159ecst

[8] Elsevier Ltd. (2015). Convion SOFC relies on Fraunhofer IKTS, Plansee Innovations. *Fuel Cells Bulletin 2015 (5)*, 5-6, DOI: 10.1016/S1464-2859

[9] Pönicke, A., Reuber, S., Dosch, C., et. al. (2013). Efficient Planar SOFC Technology for a Portable Power Generator. In: Advances in Solid Oxide Fuel Cells VIII, *Ceramic Engineering and Science Proceedings 33*, DOI: 10.1002/9781118217481.ch12

[10] Reuber, S., Megel, S., Jürgens, C., et. al. (2015). Application-oriented Design and Field Trial of the eneramic® Power Generator. *ECS Transactions 68,* 131-141, DOI: 10.1149/06801.0131ecst

[11] Schilm, J., Rost, A., Pönicke, A., et. al. (2012). Ceramic Integration Technologies for Solid Oxide Fuel Cells. *International Journal of Applied Ceramic Technology 9*, 688-699, DOI: 10.1111/j.1744-7402.2012.02789.x

TECHNOLOGY READINESS OF SOFC STACKS – A REVIEW

C. Wunderlich
Fraunhofer Institute for Ceramic Technologies and Systems IKTS
Dresden, Germany

ABSTRACT

The concept of technology readiness levels (TRL) had become recently popular both in Europe and in Canada to evaluate the maturity of a technology or product. The classification is mandatory in the European Union Horizon2020 Research Program and in the Industry "Build in Canada" Innovation Program (BCIP).

The charming simplicity of the TRL1 to TRL9 classification helps to label the maturity of a certain product concept or technology, especially in the early phases of research and development. On the other hand, it gives no or very limited information about the actual technology readiness. To overcome this limitation, IKTS has adopted and refined the technology readiness evaluation of Solid Oxide Fuel Cell (SOFC) Stacks.

The fuel cell stack is the key building block in any fuel cell system. Over the years, Fraunhofer IKTS has developed three different SOFC stack concepts, one is commercialized by the company Sunfire in Germany, a second concept is used in eneramic® systems, and a third concept, the CFY stack technology, is about to be commercialized with other partners. The technology readiness evaluation outlined in this paper is using three sets of parameters – technical performance and lifetime parameters, manufacturing and value chain parameters, as well as a market segment definition and commercialization aspects.

It is used to compare stacks on the market or under development, and to measure progress over time. Thus, an indication of SOFC technology readiness, segment specific time to market and R&D needs was developed.

BACKGROUND

A systematic approach to evaluate technology readiness levels (TRL) had been developed by NASA in the 1980s as presented by John C. Mankins[1], and is practically the base for any TRL definition given by European Union[2] or the "Build in Canada Innovation Program" (BCIP).

The approach had been originally developed for one of a kind solutions in space technology, and is thus not well suited to estimate technology maturity of critical components or a new technology (key enabling technology) as used by European Union. Limitations are obvious, as the self-evaluated technology readiness is not linked to any actual technical parameters required for a successful commercial application, and manufacturing and cost aspects are not included in any way. Furthermore, regulatory aspects, customer perception and component or supply chain readiness define to a considerable extent, if a new technology will be successful.

In 2010, Reinhart et al.[3] published an approach to evaluate technologies with a special focus on manufacturing aspects. This basic idea has been adopted in the present work and has been modified towards a consistent technology readiness estimation. In contrast to the study by Reinhart et al., three basic fields for technology readiness have been chosen – a) technical readiness, b) manufacturing and supply chain readiness, and c) aspects of the cost and business model.

In each of these three sub-fields, qualitative or quantitative parameters were standardized to a percent scale, assuming 100 % representing the ideal or target value (either low or high). The single worst value defines the overall readiness in the fields a) to c). The mean value of readiness parameters from fields a) to c) may be considered as the calculated technology readiness level. Suggested percent values for the classification of a technology's maturity status are correlated to common TRL definition as shown in Table 1 below.

Table 1. Suggested percent values for TRL calculation and correlation to common TRL definition.

Scale of Technical Maturity	Percent Value	TRL
Basic principles	1 %	1
Concept or application formulated	10 %	2
Analytical or experimental proof of concept	20 %	3
Key component breadboard validation in lab	30 %	4
Key component breadboard validation in field	40 %	5
System validation in lab	50 %	6
System validation in field	60 %	7
Field testing positively ongoing	80 %	8
Field-proven commercial product	100 %	9

The actual calculation of readiness levels in the sub-fields a) to c) as illustrated in detail further below is also different to Reinhardt et al. The herein presented approach is considered more logical, assuming that the single worst parameter is defining whether the product or technology will work or fail. In order to prioritize different assessment parameters, e.g. for various market segments or applications, an optional weight factor was included as well. The basic approach is explained in detail below and applied to SOFC stacks.

ADVANCED CONCEPT FOR TRL ESTIMATION – APPLIED TO SOFC STACKS

To evaluate the Technology Readiness of Fuel Cell systems – herein delimited to Solid Oxide Fuel Cell (SOFC) systems – the primary focus should be on the SOFC stack for a couple of reasons. First, the SOFC stack is the key element of any SOFC system, as sketched in the schematic boundary diagram in Figure 1 below, and accounts for 30 to 50 % of system cost.

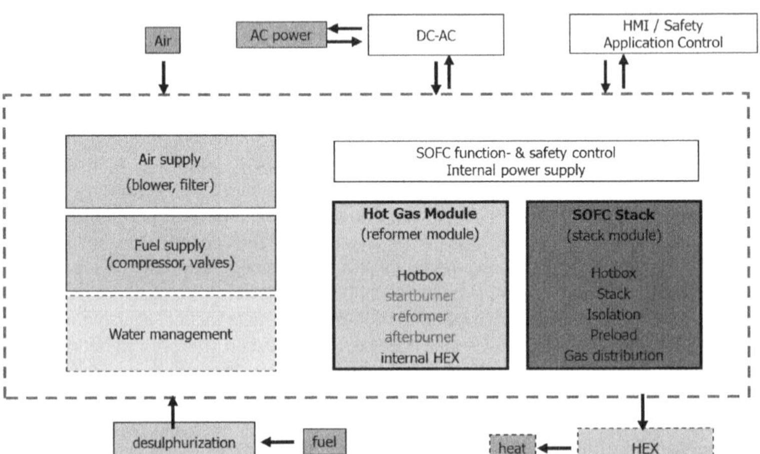

Figure 1. Generic boundary diagram of a SOFC system.

Secondly, the SOFC stack is a demanding combination of high temperature materials (ceramics, steels, glasses) and its development requires a unique combination of electrochemistry, material science, electrical, process and mechanical engineering skills, which can hardly be found in this combination in any other product. Therefore, there are no industrial standard equipment manufacturers available and no supply chain is established, so far. Today, the SOFC stack performance limits the efficiency and durability of most systems currently installed in the field.

Fraunhofer IKTS has developed over the last two decades three different SOFC stack concepts based on planar, electrolyte supported cells (ESC). One was transferred to a commercial supplier staxera GmbH (today Sunfire GmbH) and the two others are currently in the transition phase to commercialization. Thus, the institute is in a unique position to evaluate the different stack technologies with regard to their technical maturity. In this paper, two recently developed stack technologies are evaluated. Sample views of a CFY stack and an eneramic® stack are shown in Figure 2 below.

Figure 2. Sample views of CFY stack (left, mounted in handling device, footprint 15 x 13 cm), and eneramic® stack (right, with transport lock, footprint 7 x 7 cm).

In the CFY stack, the interconnect material (powder processed Chromium/Iron-alloy of Plansee SE) and cell type (Scandia doped ESC) differs significantly from the ferritic steel sheet metal interconnect in combination with partially yttrium stabilized zirconia (3YSZ) cells in the eneramic® stack. Detailed technical reports about the design, development and performance of the stacks have been published elsewhere[4, 5].

The technology readiness is evaluated for three different stationary fuel cell market segments, which are of high relevance for these stacks, i.e. micro-CHP in the range of 1 to 5 kW_{el}, industrial CHP between 20 and 500 kW_{el}, as well as off-grid power generation in the range of 100 W_{el} to 5 kW_{el}.

The following parameters were chosen for the technical readiness definition:

a) Volumetric power density in W/l; it scales proportionally to the amount of material used for the stack and the thermal insulation. It is also an indication for the long-term cost potential, but on the other hand, not a top priority for stationary products, as it would be the case e.g. for automotive applications.

b) Durability measured in a lab test or in the field by hours of continuous or combined operation demonstrated; the rationale behind this parameter is, that it defines the stack replacement cycles and thus the total cost of ownership (TCO) of the system. This parameter is related to stack design, manufacturing robustness and material effects.

c) Performance degradation rate, expressed by the growth rate of area specific resistance (ASR) in mΩ · cm² / 1.000 h; this parameter is the loss of power or efficiency of the stack and the system induced by material degradation. The parameter is more related to material than engineering issues, especially if long-term stable trends can be observed.

d) Combined cyclisation capability measured in system cycles from ambient to operation temperature; reasonable start/stop-cyclisation capability is required for most, especially smaller systems, and defines together with the degradation rate the lifetime of the stack. Combined cyclisation capability means here, that there is no protective gas on the anode side during start-up and shutdown, and therefore a certain RedOx impact will occur, affecting the Ni/NiO anode structure.

These parameters can be measured exactly in lab tests or during system operation. However, the test conditions (temperature, fuel composition, fuel utilization, details of cycling operation, etc.) and even the test-rig design will have significant impact on the results. Therefore, exact comparisons can be made only for those tests, which were executed in the same lab under carefully defined test conditions.

As mentioned before, the parameters described above were standardized to a percent scale p_i with 100 % representing the ideal or target value, which can be either low or high. For calculating the technical readiness tr, a weight factor or priority value q_i is applied, cf. Equation 1.

$$tr = 1 - \max[(1 - p_i) \cdot q_i] \quad (1)$$

Results of the TRL estimation for the CFY stack and the eneramic® stack are shown for different market segments in Table 2 and Table 3, respectively. It should be noticed that the technical readiness of stack technologies in each market segment is defined by individual target values for each market segment and a relative priority factor. In Table 3, the eneramic® stack is not evaluated for the market segment of industrial CHP, as the stack is too small to be reasonably integrated into such systems.

Table 2. Technical readiness of CFY stack technology in three different market segments.

Assessment Parameter			**Micro-CHP**			**Industrial CHP**			**Off-grid Power**		
	Unit, Description	*Status SOFC*	q_i	*Target*	*single TRL*	q_i	*Target*	*single TRL*	q_i	*Target*	*single TRL*
TRL demonstrated	cf. Table 1	-	1	-	60 %	1	-	50 %	0,5	-	80 %
Volumetric power density	W / l	400	0,75	250	100 %	0,5	250	100 %	0,7	350	100 %
Durability	kh	19	0,75	40	48 %	0,5	60	32 %	1	10	100 %
Degradation rate (ASR)	mΩ cm² kh⁻¹	17	1	10	59 %	1	10	59 %	1	20	100 %
Cyclisation capability	cycles	60	1	200	30 %	0,5	100	60 %	0,7	150	40 %
Total TRL					**30 %**			**50 %**			**58 %**

Table 3. Technical readiness of eneramic® stack technology in two different market segments.

Assessment Parameter			Micro-CHP			Industrial CHP			Off-grid Power		
	Unit, Description	*Status SOFC*	q_i	*Target*	*single TRL*	q_i	*Target*	*single TRL*	q_i	*Target*	*single TRL*
TRL demonstrated	cf. Table 1	-	1	-	50 %				0,5	-	80 %
Volumetric power density	W / l	364	0,75	250	100 %				0,7	350	100 %
Durability	kh	19	0,75	40	48 %				1	10	100 %
Degradation rate (ASR)	$m\Omega\ cm^2\ kh^{-1}$	17	1	10	59 %				1	20	100 %
Cyclisation capability	cycles	100	1	200	50 %				0,7	150	67 %
Total TRL					**50 %**			**n/a**			**77 %**

The production readiness of CFY stack and eneramic® stacks as shown in Table 4 and Table 5 was again calculated according to Equation 1. Here, the only quantitative parameter is the actual production volume versus the target production volume in each market segment. The target value of the annual production volume is the cumulated power generation capacity of produced stacks, where a viable business case could be established and estimated target cost are achieved. The target production volumes identified by separate evaluation is 2 MW for micro-CHP, 20 MW for industrial CHP and 200 kW for the off- grid case. The eneramic® stack production is still on a very low level with respect to volume, which results in a low TRL for manufacturing. The principle assessment logic still works sufficient with the remaining qualitative parameters; however, subjective judgement is more dominant, here.

Table 4. Manufacturing readiness of CFY stack technology in three different market segments.

Assessment Parameter		Micro-CHP		Industrial CHP		Off-grid Power	
	Target	q_i	*single TRL*	q_i	*single TRL*	q_i	*single TRL*
Required manufacturing capacity	see right		2 MW p.a.		20 MW p.a.		200 kW p.a.
Annual production volume	100 % p.a.	1	15 %	1	2 %	1	100 %
Technology for volume production	proven	0,75	50 %	0,75	25 %	0,75	100 %
In-line quality inspection	proven	0,5	50 %	0,5	50 %	0,5	50 %
Equipment / plants for target volume	available	0,5	50 %	0,5	25 %	0,5	100 %
Competitive supply chain	established	1	90 %	1	90 %	1	90 %
Total TRL			**15 %**		**2 %**		**75 %**

Table 5. Manufacturing readiness of eneramic® stack technology in two different market segments.

Assessment Parameter		Micro-CHP		Industrial CHP		Off-grid Power	
	Target	q_i	*single TRL*	q_i	*single TRL*	q_i	*single TRL*
Required manufacturing capacity	see right		2 MW p.a.				200 kW p.a.
Annual production volume	100 % p.a.	1	1 %			1	8 %
Technology for volume production	proven	0,75	50 %			0,75	50 %
In-line quality inspection	proven	0,5	80 %			0,5	80 %
Equipment / plants for target volume	available	0,5	20 %			0,5	50 %
Competitive supply chain	established	1	100 %			1	100 %
Total TRL			**1 %**		**n/a**		**8 %**

The off-grid segment case in Table 4 demonstrates the effect of the priority or weight factor in the TRL calculation. Total TRL for manufacturing in this segment is 75 %, not 50 %, as the parameter for in-line quality inspection is prioritized with 50 %, only.

However, quality control including appropriate in-line inspection technologies is an especially important factor. A SOFC stack is a serial assembly of 15 to 60 thin ceramic sheets with printed electrodes, called membrane electrode assembly (MEA) or SOFC single cell. A single cell failure will result in a failed stack, something that can be detected only after final stack joining at the end of stack manufacturing. Without appropriate quality inspection technologies and stable processes, an acceptable yield rate in stack production is not achievable. Fraunhofer IKTS is therefore developing sophisticated quality inspection technologies for series production of component parts and complete SOFC stacks.

A third field considered for successful market deployment of SOFC technology is the business environment for the commercialization. Here, the evaluation is made from a German or European perspective. The missing coverage of fuel cell micro-CHP products in eco-design and energy labeling directives of the European Union is one aspect considered here. For Asian or American countries, potential results in this segment might by different.

The chosen assessment parameters include required legislation and public perception as well as target cost for the target volumes in different segments, considering the design and technology available today. Results are summarized in Table 6 and Table 7 below.

Table 6. Business readiness of CFY stack technology in three different market segments.

Assessment Parameter	Micro-CHP		Industrial CHP		Off-grid Power	
	q_i	*single TRL*	q_i	*single TRL*	q_i	*single TRL*
Legislation as required	0,75	80 %	1	90 %	1	100 %
Positive public perception / acceptance	0,5	100 %	0,5	100 %	0,3	100 %
Viable business models at target cost	1	100 %	1	90 %	1	90 %
*Stack target cost at target vol. for business case**	1	100 %	1	43 %	0,9	57 %
Total TRL		**85 %**		**43 %**		**61 %**

**) in market segment micro-CHP: 4.000 €/kW$_{el}$ @ 2 MW$_{el}$ p.a.*

Table 7. Business readiness of eneramic® stack technology in two different market segments.

Assessment Parameter	Micro-CHP		Industrial CHP		Off-grid Power	
	q_i	*single TRL*	q_i	*single TRL*	q_i	*single TRL*
Legislation as required	1	80 %			1	100 %
Positive public perception / acceptance	0,75	100 %			0,3	100 %
Viable business models at target cost	1	100 %			1	90 %
*Stack target cost at target vol. for business case**	1	100 %			1	88 %
Total TRL		**80 %**		**n/a**		**88 %**

**) in market segment micro-CHP: 4.000 €/kW$_{el}$ @ 2 MW$_{el}$ p.a.*

By calculating the mean values of the three TRL results representing technical maturity, manufacturing technology and business case viability, the overall technology readiness of the specific SOFC stack in a specific stationary market segment were evaluated. Results are shown in Table 8 below.

Table 8. Overall technology readiness of CFY stack technology and eneramic® stack technology in three different market segments.

	Priority	Micro-CHP 1 .. 5 kW_{el}	Industrial CHP 5 .. 100 kW_{el}	Off-grid Power 0,1 .. 1 kW_{el}
CFY stack technology				
Technical maturity	100 %	30 %	50 %	58 %
Manufacturing technology	80 %	15 %	2 %	75 %
Buisness case viability	70 %	85 %	43 %	61 %
Total TRL		**51 %**	**44 %**	**70 %**
eneramic® stack technology				
Technical maturity	100 %	50 %		77 %
Manufacturing technology	80 %	1 %		8 %
Buisness case viability	70 %	80 %		88 %
Total TRL		**52 %**	**n/a**	**65 %**

Based on the results in Table 8 it can be concluded that the summary figure might be less important than the specific results in different sub-fields. The extremely low readiness of eneramic® stack in manufacturing aspects is attributed to low production volumes and should be drastically improved in rather short term if required investment in production equipment can be realized.

RESULTS AND ANALYSIS

Technical readiness data for the off-grid market segments are the highest values as lifetime and degradation target values are lower in this segment. Results can also be visualized in radar diagrams as shown in Figure 3 and Figure 4.

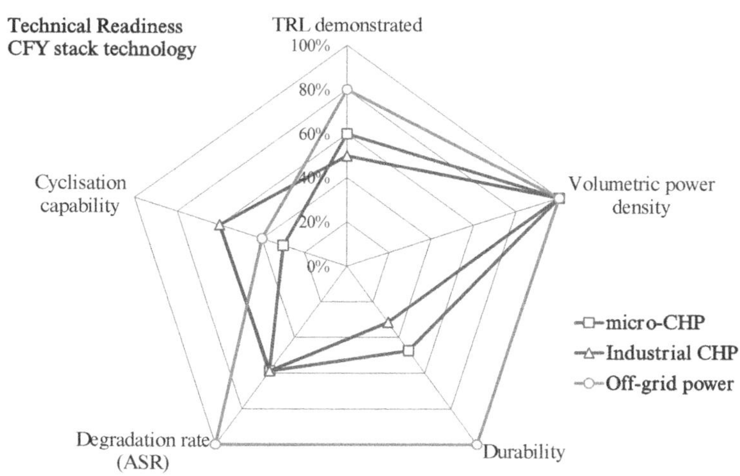

Figure 3. Visualization of technical readiness for CFY stacks in three different market segments.

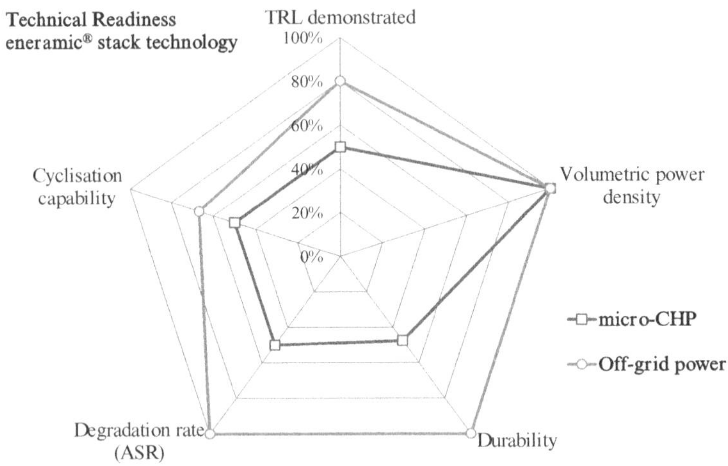

Figure 4. Visualization of technical readiness for eneramic® stacks in two market segments.

The comparison between the three considered market segments underlines that technical readiness of a particular stack technology might be significantly varying for different markets segments. For this reason and following the overall TRL results according to Table 8, Fraunhofer IKTS believes that a commercialization should start in the off-grid market segments. The recent initial market success of German company NewEnerday GmbH and successful field trials with the eneramic® system developed by IKTS underline this strategy.

The TRL analysis work can, however, reveal additional data of interest. Figure 5 shows how the technical readiness of the eneramic® stack technology platform has evolved over the last three years.

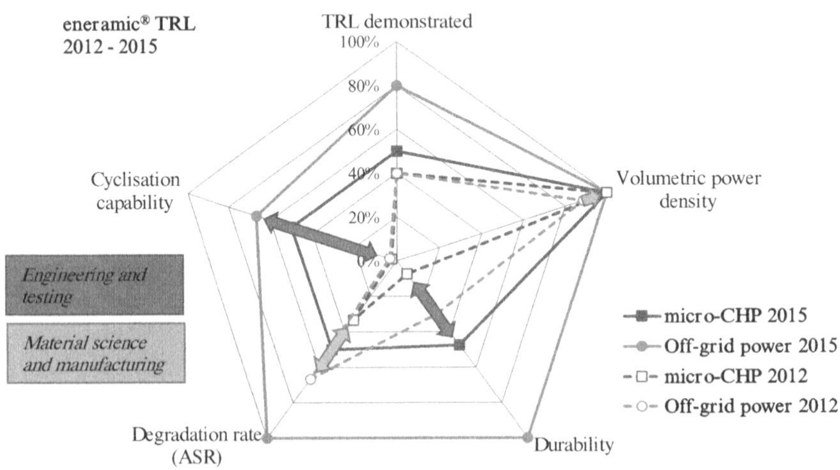

Figure 5. TRL evolution of the eneramic® stack technology platform over three years, 2012 – 2015.

Significant progress had been achieved in any of the considered parameters. This is a result of intensive testing and efficient engineering work. Relative progress however is less, where not engineering work but advances in material science are required to get substantial progress. This principle observation was to be seen even more clearly with other stack technologies not published here. R&D and business teams should focus in the early development phase on parameters like degradation rates or power densities for getting them right for their business case. Otherwise, there is a significant risk of failure.

CONCLUSIONS

Technology Readiness (self) evaluation is becoming an important element in Horizon2020 research program of the European Union and other similar programs. While this TRL definition helps to identify the relative maturity of a particular technology and to compare different technologies, it gives no information how far away from commercialization a technology really is.

In this paper, a more detailed analysis and comparison is proposed. The technology readiness calculated by the suggested method is always specific for a given solution and a targeted market segment or application.

The detailed technology readiness review has been applied to SOFC stacks, the heart of any SOFC system. The analysis published here is specific to the three stationary market segments micro-CHP, industrial CHP and off-grid power generation, and is made for the planar SOFC stacks developed at Fraunhofer IKTS.

The analysis makes clear, that if fundamental technical parameters, which are primarily driven by progress in material development, are not on the level required for commercialization yet, a near term commercialization is rather uncertain. This is true, even if a formal TRL level of 8 (tested successfully in the field) can be claimed with some evidence. A clear evaluation of SOFC technology readiness and a market segment specific gap analysis highlighting R&D needs is given above. The methodology can be easily applied to other stacks, if required test results and market data are available. Ideally, testing should be done in one lab.

Based on this analysis, the off-grid market segment offers best and short-term market opportunities for the SOFC technology. This market segment accepts highest cost, which can be reached with lower production volumes, and has at least in some applications lower lifetime and durability requirements, which are met by IKTS stacks today, already.

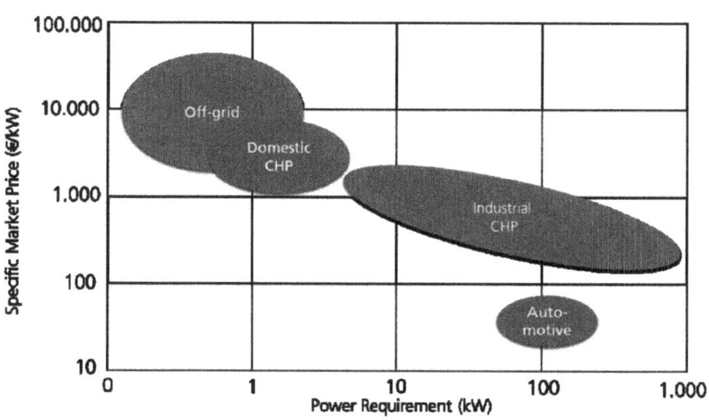

Figure 6. Specific market prices in different market segments for SOFC systems.

The analysis also reveals why the business case of a company, which is basically manufacturing stacks for different market segments, can be viable. A successful business in the off-grid market will help this company to gather learning experiences with lower volumes in a market that accepts significant higher prices while establishing volume and robust manufacturing for other even bigger market segments. The required specific cost level given in Figure 6 is one key indicator for the attractiveness of the off-grid market.

The herein presented analysis is made for SOFC stacks developed at Fraunhofer IKTS. However, our insight into other stack technologies and the general impression from many presentations in the numerous conferences lead to the impression that relative technology readiness of stacks at IKTS is at rather high level and that especially durability and performance degradation is in many other cases not in the target window for a successful commercialization.

There is one exception, however: Japanese SOFC stacks are outstanding with regard to reported lifetime and degradation rates. It should be noted that the early focus on the durability testing and analysis of failures at AIST by looking at different stacks from different manufacturers is one of the success factors of Japanese fuel cell industry. The project "Development of System and Elemental Technology on SOFCs" (2008-2012), sponsored by New Energy Development Organization (NEDO), included the important sub-task: "Fundamental Research for Improvement of Durability"[6]. Current market prices for Japanese SOFC system indicate however, that these systems and stacks are still too expensive as the sales price in 2015 is above 20.000 US$ for a system rated at 700 W_{el}, even while more than 10.000 systems have been sold (which is equivalent to a production volume of about 3 to 5 MW_{el} per year). It becomes obvious, that the target value of 4.000 €/kW_{el} at a production rate of 10 MW_{el} per year is not reached, yet.

The principle of technology readiness evaluation can be transferred easily to other technologies. In any case, a detailed understanding of the technology including internal testing facilities is required to do so.

ACKNOWLEDGEMENTS

Part of this research work has been funded by the Free State of Saxony and the European Union under contract 100116909. This funding is highly appreciated.

This work has been developed in long discussions with my colleagues at Sunfire GmbH and Fraunhofer IKTS. Most of the technical input is based on their test results and findings. Their careful work over the last ten years is the fundament of this summary paper.

REFERENCES

1 Mankins, J. C. (1995). Technology Readiness Levels – A White Paper. NASA Office of Space Access and Technology, Advanced Concepts Office,
URL: www.hq.nasa.gov/office/codeq/trl/trl.pdf

2 European Commission (2013). Technology Readiness Levels (TRL). Horizon 2020 – Work Programme 2014-2015, Annex G of General Annexes,
URL: http://ec.europa.eu/research/participants/data/ref/h2020/wp/2014_2015/annexes/h2020-wp1415-annex-ga_en.pdf

3 Reinhart, G., Schindler, S. (2010). A Strategic Evaluation Approach for Defining the Maturity of Manufacturing Technologies. World Academy of Science, Engineering and Technology, Vol. 4, 24.11.2010, URL: http://waset.org/publications/6394/evaluating-the-innovation-ability-of-manufacturing-resources

[4] Reuber, S., Megel, S., Jürgens, C., et al. (2015). Application-oriented Design and Field Trial of the eneramic® Power Generator. *ECS Transactions 68,* 131-141, DOI: 10.1149/06801.0131ecst

[5] Megel, S., Dosch, C., Rothe, S., et al. (2013). CFY-Stacks for Use in Stationary SOFC and SOEC Applications. *ECS Transactions 57,* 89-98, DOI: 10.1149/05701.0089ecst

[6] Yokokawa, H., Horita, T., Yamaji, K., et al. (2012). Degradation of SOFC Cell/Stack Performance in Relation to Materials Deterioration. Journal of the Korean Ceramic Society 49, 11-18, DOI: 10.4191/kcers.2012.49.1.011

HIGH-TEMPERATURE DIRECT FUEL CELL MATERIAL EXPERIENCE

Chao-Yi Yuh, A. Hilmi and R. Venkataraman
FuelCell Energy, Inc.
Danbury, CT, USA

ABSTRACT

FuelCell Energy (FCE) is meeting growing worldwide demand for clean energy with its Direct Fuel Cell products (DFC®). It is based on efficient high-temperature internal-reforming carbonate fuel cell technology. FCE has advanced the technology since mid-1970s by advancing fundamental technology understandings, practical and low-cost solutions, and multi-year field operational experience. The DFC stack material service life has been advanced to >5 years for the current commercial products and development continues to attain >7-10 year life. In this paper, an update to DFC materials experience is presented. Comparisons between DFC and SOFC materials are also discussed.

INTRODUCTION AND PRODUCT STATUS

FCE is an integrated high-temperature fuel cell company that conducts R&D, sales, manufacturing and project execution, as well as services, providing turn-key distributed power generation solutions (Figure 1). FCE began developing DFC since mid-1970s and anode-supported SOFC since 2003. These high-temperature fuel cells are ideally suited to distributed power generation applications due to their modest size, low emission profile, high-temperature operation and resulting simpler fuel-processing system and high efficiency[1]. DFC is based on efficient internal-reforming carbonate fuel cell technology using alkali carbonate mixtures as electrolyte and operates at approximately 550-650°C. Over the past four decades of research and development, the DFC cell technology has matured to a single DFC stack capable of producing 350kW net AC power and 5-year service life. The current scalable DFC products are rated in capacity at 350kW, 1.4MW, and 2.8MW. The DFC plants can achieve an electric efficiency of 45 to 49% (LHV) for simple-cycle system or 58-65% for hybrid systems. Using the high value heat byproduct for cogeneration, overall energy efficiency up to 90% can be achieved. FCE is currently commercializing the DFC power plants worldwide with global partners POSCO Energy, South Korea, and FCES (FuelCell Energy Solutions), a subsidiary company of FCE and IKTS (Fraunhofer Institute for Ceramic Technologies and Systems), Dresden, Germany. Presently the DFC power plants have attained more than 300 MW capacity installed/backlog. Over 100 DFC plants are generating power at about 50 locations worldwide. A 59MW plant has been operating in Hwaseong City, South Korea and a 14.9MW plant in Bridgeport, CT, USA (Figure 2). To date, DFC plants have generated over 3 billion kilowatt hours (kWh) of ultra-clean electricity. FCE is also developing technologies for distributed hydrogen generation and carbon capture, both based on the baseline DFC technology[2,3]. In terms of SOFC, FCE views it as complementary to the DFC product line, as DFC targets megawatt and multi-megawatt applications where SOFC future sub-megawatt power generation and storage applications. FCE can leverage current DFC infrastructure, expertise, and experience to accelerate the commercialization of the SOFC technology to expand market opportunities

DFC PRINCIPLES

Figure 3 illustrates the principle of internal-reforming DFC. It is capable of achieving a high electrical efficiency due to synergistic heat and mass transfer. The waste heat and reaction product water facilitates the endothermic steam reforming reaction, thus able to achieve near

100% methane conversion at load. The basic chemistry and electrochemistry of DFC involve the following reactions[4]:

<u>Electrochemical Reaction</u>:

Cathode Reaction	$CO_2 + \frac{1}{2} O_2 + 2e^- \rightarrow CO_3^=$
Anode Reaction	$H_2 + CO_3^= \rightarrow H_2O + CO_2 + 2e^-$
Overall	$H_2 + \frac{1}{2} O_2 \rightarrow H_2O$

<u>Chemical reactions</u>:

Dissociation	$CO_3^= \leftrightarrow CO_2 + O^=$
Hydrolysis	$H_2O + CO_3^= \leftrightarrow CO_2 + 2OH^-$

The dissociation reaction controls electrolyte basicity $[O^=]$, according to the Lux-Flood acid-base theory. Therefore, a higher melt basicity (higher $[O^=]$) is at a lower PCO_2. As opposed to SOFC where $O^=$ is transported from the cathode to the anode in the solid electrolyte, $CO_3^=$ is transported in DFC through the liquid alkali carbonate electrolyte (Figure 3). This is equivalent to transporting CO_2 as well as $O^=$. This unique electrochemical CO_2-transfer characteristic allows CO_2 capture/concentrating by utilizing DFC technology[3]. Since DFC is a very efficient internal reformer, essentially all the hydrocarbon feed can be converted to hydrogen. Therefore, a unique application is to utilize DFC to co-produce both power and hydrogen[2]. This distributed hydrogen generation technology can facilitate the deployment of hydrogen infrastructure.

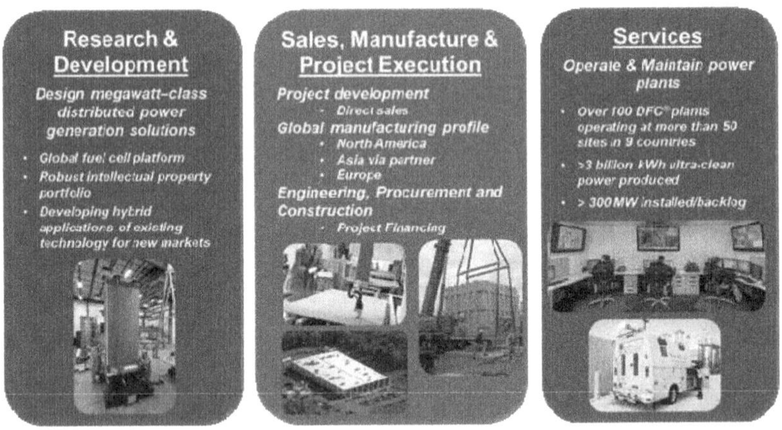

Figure 1. FCE is an integrated fuel cell company, providing turn-key distributed power generation solutions.

Figure 2. 14.9MW DFC fuel cell park in Bridgeport, CT: Five 2.8MW plants with Organic Rankine Cycle (ORC)

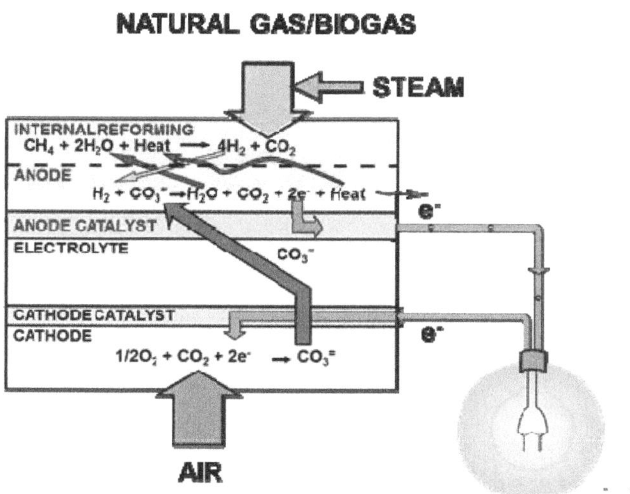

Figure 3. DFC schematics: High electric efficiency resulting from synergistic heat and mass transfer

Typical carbonate fuel cell environments are listed in Table I. SOFC has similar environments, but without involving CO_2 in the electrochemical reactions.

Table 1. Anode and cathode environments in DFC: The anode environment is reducing with high carbon activity, whereas the cathode environment oxidizing with low carbon activity.

Factor	Anode	Cathode
Atmosphere	Reducing	Oxidizing
Oxygen Activity	$\sim 10^{-22}$-10^{-24}	~0.1
Gas	H_2, H_2O, CO, CO_2, N_2	O_2, H_2O, CO_2, N_2
Carbon Activity	High (>0.1)	Low ($\sim 10^{-20}$)

Extensive carbonate fuel cell materials experience has been reported in the literature. Table 2 lists key current stack/BOP materials and improvement opportunities. Table 3 lists environments, processes, and similarities/differences between DFC and SOFC. The DFC cell construction is illustrated in Figure 4. The major characteristics of the DFC are: (1) large cell area (~1m^2), largest among various fuel cell types, (2) small seal area, more efficient use of active cell area, (3) ease of fabrication and (4) lower cost. Each DFC repeat cell package is composed of active components (anode, cathode, electrolyte matrix) and cell hardware (bipolar plate, current collectors, and seal area). The bipolar plate and the corrugated current collectors are made of appropriately specified austenitic stainless steels. The electrodes are made from porous nickel-based materials. The matrix is porous ceramic $LiAlO_2$, and traps an electrolyte mixture of alkali carbonate salts, which melt between 450 and 510°C. Direct internal-reforming catalyst is located inside the anode flow field. Very little costly precious metals or rare-earth materials (REM) are used in DFC.

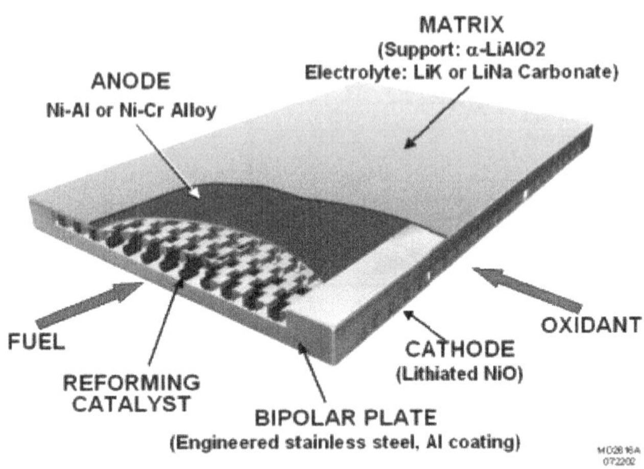

Figure 4. DFC cell construction employs appropriately specified stainless cell hardware and nickel-based electrodes: All are commercial commodity materials.

Table 2. Current carbonate fuel cell materials and improvement opportunities: Present development focuses are on cost reduction and life enhancement

DFC Component	Materials	Improvement opportunities
Anode	Ni-Al, Ni-Cr, Ni-Cr-Al	Alternate materials, low-cost processes, less materials
Cathode	NiO, Ni-Fe-MgO	Alternate materials or electrolytes
Matrix (Electrolyte Layer)	α or γ-$LiAlO_2$ with fiber or particulate reinforcement	Low-cost phase/ripening stable $LiAlO_2$, aqueous-base manufacturing system, advanced low-cost stable reinforcements
Internal Reforming Catalyst	Supported Ni catalyst	Low-cost carbonate-resistant catalyst
Bipolar Current Collector	Ni coated/non-coated austenitic stainless steels	Dual-atmosphere corrosion- and creep-resistant alloys
	Wet-seal aluminization	Low-cost coating processes
Electrolyte	Li/Na or Li/K carbonate with additives	Advanced additives
Stack/BOP	Austenitic stainless steels	Low-cost oxidation-resistant alloy/coating

SOFC is operated at average temperatures between 500 and 1,000°C, depending on the cell materials and design. Similar to DFC, Ni-base anode and stainless steel hardware materials are extensively utilized. SOFC utilizes similar fuel and oxidant contaminant cleanup processes to DFC.

Nevertheless, there are differences between DFC and SOFC. Due to the brittle and lower flaw-tolerance nature of the ceramics, SOFC cell area is typically much smaller than DFC. DFC stack adopted a simpler external manifold and wet-seal design, resulting in a simpler more manufacturable cells and a smaller seal area for more efficient use of the cell area for power generation. However, in DFC, the use of liquid alkali carbonate electrolyte requires electrolyte management to control electrolyte distribution/loss and electrolyte-accelerated corrosion (oxide dissolution/hot corrosion)[5]. The DFC electrolyte matrix is usually thicker compared to SOFC in order to store sufficient liquid electrolyte for required extended multi-year service. Fortunately, the much more conductive liquid carbonate electrolyte still results in an acceptable matrix ohmic resistance.

It is commonly known that all oxides dissolve to certain extents in molten salts. The dissolution processes are in many cases influenced by electrolyte basicity $[O^{=}]$. Oxides of Fe, Ni and Cr are extensively present in the DFC, as cell materials or corrosion scale. For example, NiO dissolves slowly into the liquid electrolyte in DFC as Ni^{+2} ions according to an acidic dissolution mechanism ($NiO \leftrightarrow Ni^{+2} + O^{=}$)[6,7]. However, at extremely low PCO_2 (a very basic condition, not typical in DFC), basic dissolution could become dominant. $LiFeO_2$ was found to have a relatively low solubility, independent of electrolyte basicity[8]. Under the oxidizing cathode environment, Cr_2O_3 (Cr^{+3}) is further oxidized into highly soluble $CrO_4^{=}$ (Cr^{+6})[8]. However, under the reducing anode environment, Cr_2O_3 (Cr^{+3}) is stable and has a drastically lower solubility.

Therefore, selection of DFC materials and operating conditions should take into account such oxide dissolution behavior.

For commercial viability, the fuel cell stack materials are desired to have a service life of >5 years and the BOP materials >20 years[5]. Material technologies play an important role in controlling performance, life and commercial success of fuel cell power plants. The selection of DFC materials is founded on intensive materials R&D carried out during the past four decades, validated by the extensive long-term multi-year field experience. FCE's current commercial DFC stacks had already demonstrated >40,000h field service. The current DFC development focuses are to further reduce cost and to extend stack service life to 7-10 years. In a previous paper DFC materials experience was reviewed[9] and a further update is presented here.

Table 3. DFC and SOFC: Environments, processes, and similarities/differences

	DFC	SOFC
Temperature	~550-650°C	500-1000°C
Atmosphere	Anode side: reducing & high carbon activity	
	Cathode side: oxidizing & low carbon activity (higher PCO_2 in DFC)	
Similarities	Ni-base anode and heat-resistant alloy hardware	
	Fuel processing and gas cleanup requirements	
	Oxidation, dual-atmosphere corrosion Redox Carburization Phase transformation Thermo-mechanical stress	
Differences	Liquid alkali carbonate electrolyte (LiK or $LiNaCO_3$), ionically more conductive than solid electrolyte	Oxide conducting solid electrolyte (YSZ, ScSZ, ceria, etc.)
	Austenitic cell hardware	Ferritic cell hardware or all ceramics
	Oxide dissolution (hot corrosion) Electrolyte management Thicker electrolyte layer (matrix)	CO_2 not involved in electrochemistry Cr poisoning
	Large cell area (~1m^2) External manifold Wet seal	Small cell area (0.01-0.1m^2) Usually internal manifold More complex cell sealing

ELECTRODE

Ni is still the preferred HOR (hydrogen-oxidation reaction) electrocatalyst for both DFC and SOFC. The DFC active anode material is porous metallic Ni (nickel) that is thermodynamically stable and insoluble in the liquid electrolyte. Oxide dispersion strengthening (ODS) has been universally adopted to manage mechanical creep. For SOFC, the Ni cermet anode is mechanically strengthened by the ceramic YSZ or other oxide conductors. DFC and SOFC have similar fuel processing and contaminant cleanup requirements. Since any carbon deposition (if using carbonaceous fuels) during operation could irreversible damage the Ni-base anode, process condition should be well controlled to avoid such deposition.

The baseline DFC cathode is *in-situ* oxidized and lithiated NiO with a dual-porosity microstructure[6]. During operation, NiO dissolves slowly into the liquid electrolyte as Ni^{+2} ions according to an acidic dissolution mechanism ($NiO \leftrightarrow Ni^{+2} + O^{=}$). Although the NiO solubility is only at ppm levels, the dissolved Ni^{+2} can be reduced to metallic nickel by the hydrogen diffused from the anode side, leading to continuous deposition of the metallic Ni in the electrolyte matrix and possible internal electric shorting. FCE's atmospheric-pressure DFC systems with proper operational control and electrolyte composition, cathode stability and nickel shorting are not a concern for operation beyond 5 years[5]. To attain 10-year DFC stack life, innovative cost-effective approaches are desirable.

For SOFC, poisoning of the perovskite cathodes by the Cr^{+6} [$CrO_2(OH)_2$] evaporated from Cr-containing hardware alloys in humid oxidant has been shown as a major contributor to performance decay. Therefore, numerous protective coatings are being developed to minimize the Cr evaporation. No Cr vapor poisoning has been observed for the DFC, likely because the Cr^{+6} vapor is absorbed into the liquid electrolyte as soluble $CrO_4^{=}$ and does not affect cathode structure or ORR (oxygen-reduction reaction).

ELECTROLYTE MATRIX

The electrolyte matrix, sandwiched between the electrodes, is a key cell component, isolating fuel from oxidant, storing liquid electrolyte and facilitating ionic transport. It has a microporous ceramic structure consisting of ultra-fine $LiAlO_2$ powders to immobilize the liquid electrolyte within the micropores via a >3 bar capillary force. The ~1m^2 matrix is generally manufactured by a slurry tape casting technique. The large-area flexible and durable green tapes (Figure 5) can be easily assembled into a DFC stack. The organic binder is eliminated during initial stack conditioning so that the resulting de-bindered fine-pore matrix structure can absorb electrolyte stored elsewhere. The slurry preparation requires adequate dispersion of the high surface area raw material powders in order to achieve a uniform dense packing and fine-pore structure for electrolyte retention. Long-term phase and pore-size stability are necessary for maintaining highly electrolyte fill for extended operation. Current effort at FCE is focused on optimizing powder phase purity, morphology, and particle-size distribution, reducing $LiAlO_2$ powder cost, and optimizing tape-casting slurry process to deliver a stable well-packed pore structure and consistent high production yield.

Figure 5. ~1m^2 flexible and durable DFC matrix produced by FCE: Continuous tape casting process is employed.

The $LiAlO_2$ support material has three allotropic phases (α, β, and γ)[10]. Ceramic fibers or coarse particulates are also incorporated for strengthening, crack attenuation and thermal cycleability. All these materials need to be sufficiently stable in the carbonate electrolyte to retain electrolyte for the intended life goal. The β phase is metastable and not used. The γ phase was extensively used prior to 2000; the α phase afterward. Like other oxide materials, $LiAlO_2$ dissolves slightly in molten carbonate, via basic dissolution $LiAlO_2 + O^{=} \rightarrow Li^{+} + AlO_3^{3-}$. Such dissolution-precipitation process has resulted in particle coarsening (Ostwald ripening). $LiAlO_2$ particles generally coarsen faster at higher temperature, in lower CO_2 gas atmosphere, and in strong basic melts, explainable by the solubility increase under such conditions. It was also found that the γ phase is stable at 700°C whereas the α phase at 600-650°C. Long-term cell and stack testing revealed significant $\gamma \rightarrow \alpha$ phase transformation with associated particle coarsening and reduced electrolyte retention. The solubility of the α phase is lower than the γ phase, probably explaining the better stability of the α phase. Therefore, α-$LiAlO_2$ has been the preferred matrix support material since 2000. It is important to maintain stack temperature uniformity to avoid undesirable higher temperatures near or above 700°C to prevent simultaneous $\gamma \leftrightarrow \alpha$ phase change and particle growth. FCE has demonstrated in long-term field operation virtually no phase change of the α-$LiAlO_2$ after ~5-year operation[11]. A uniform particle size is expected to slow down coarsening, according the Ostwald ripening theory. It has also been observed by FCE that the current α-$LiAlO_2$, although phase stable, showed faster coarsening under the reducing anode environment (Figure 6). It has been postulated in the literature that $LiAlO_2$ coarsening may be accelerated if more soluble intermediate phases form[12]. FCE is currently developing advanced phase- and pore-structure stable matrix under DOE sponsorship to enable DFC stack service life to 10 years[13,14]. Preliminary advanced matrix has been fabricated and enhanced pore-structure stability has successfully been demonstrated in accelerated single cells (Figure 7). Full-area stack validation will follow.

CELL HARDWARE

Heat-resistant alloys are extensively used in high-temperature fuel cells[15,16]. Sufficient high-temperature oxidation resistance in the presence of reducing fuel, oxidizing oxidant, or both (dual atmosphere) is needed. Due to the oxide scale dissolution process in DFC, hot corrosion resistance (liquid salt fluxing) is also important[15]. Active-passive corrosion behavior of stainless steels in molten carbonate has been extensively studied[17] and is important for material selection. The oxide scale conductivity, CTE match, mechanical properties (creep and ductility), cost and manufacturability should also be carefully considered for alloy selection. For DFC, electrolyte loss to the hardware surface should be minimized, whereas for SOFC, Cr-evaporation poisoning needs to be mitigated.

Heat-resistant alloys generally contain sufficient Cr (typically >16 wt%) and/or Al (up to 5 wt%) to promote protective chromia and/or alumina scales. For SOFC, standard or specialty ferritic alloys[16] such as AISI430, AISI441HP, Crofer 22APU, Hitachi ZMG232, Plansee ITM, Plansee PM/CFY and Sandvik Sanergy HT for better CTE match to other cell components are investigated. Conductive protective coatings such as MnCo spinel are being extensively studied to improve oxidation resistance as well as to reduce Cr evaporation. For DFC, commodity 300-series austenitic stainless steels are extensively used. It has been determined by FCE that the cell hardware metal loss rate is acceptable based on >5-year field service experience. To attain >7-10 year stack service life, the stability of the corrugated cathode current collector (CCC) needs to be improved. FCE has identified a CCC protective coating that demonstrated in full-area stack testing >80% reduction in metal attack (Figure 8). This coating approach is projected to extend the DFC CCC life to well beyond 10 years. This type of coating could also be used for module and BOP to reduce oxidation and Cr evaporation.

Figure 6. Faster coarsening of DFC matrix α-$LiAlO_2$ at reducing anode side observed in accelerated cells.

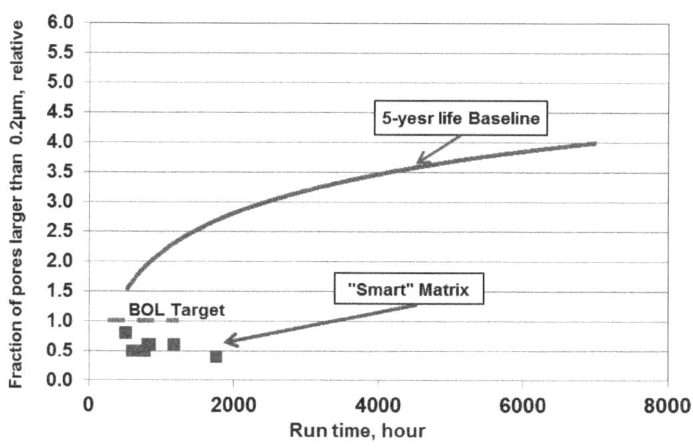

Figure 7. Advanced matrix demonstrated significantly enhanced stability of pore structure.

Bipolar separator plates and interconnects experience reducing anode gas on one side and oxidizing oxidant gas on the other side (dual-atmosphere condition). Anomalous accelerated corrosion has been observed on the oxidant side (in DFC as well as in SOFC)[18]. This accelerated attack can be mitigated using higher-Cr alloys, as demonstrated by FCE (Figures 9). FCE's current bipolar-plate alloy is projected sufficient for >7-year field service.

Precipitate phases such as carbide, σ,η, and χ tend to form in heat-resistant alloys, particularly in stainless steels[19], during extended high-temperature service. Such phase transformation could reduce creep strength, resulting in cell dimensional change and possible contact/seal loss under thermo-mechanical stress. The formation of all these brittle precipitates also embrittles the alloys, in certain high-stress situations causing fracture (stress corrosion cracking). FCE has demonstrated alloy design approaches with improved microstructure, alloy composition, and/or additives that reduced the formation of secondary phases to potentially increase creep strength for >7-year service.

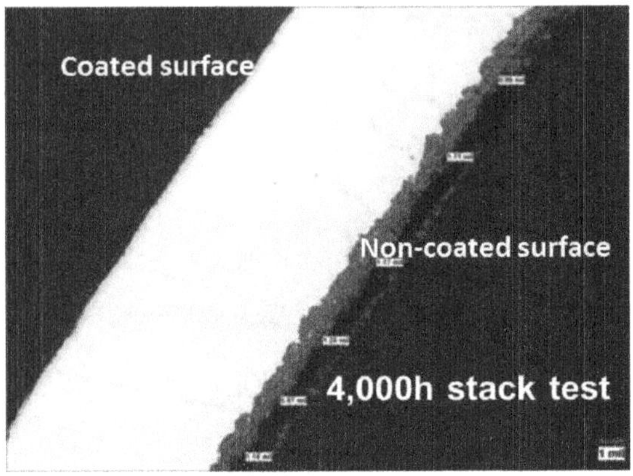

Figure 8. Protective coating effective in eliminating DFC cathode current collector corrosion

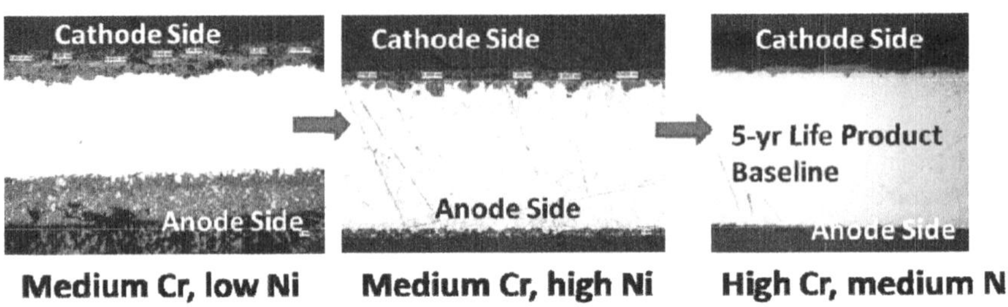

Figure 9. Excellent corrosion resistance of DFC high-Cr stainless steel bipolar plate under dual-atmosphere condition after 15,000h field operation (plate thickness ~250μm): A higher Cr content is more effective in mitigating dual-atmosphere corrosion.

CONCLUSION

Significant progress has been made in DFC technology development, manufacturing, product engineering, cost reduction and enhancement of field performance and life. Proper selection of high- temperature materials founded on many years of focused research, long-term operating experience and design improvements have extended the stack service life to > 5 years. The baseline DFC construction materials have been demonstrated in the field to meet the commercial endurance goals. All these vast materials expertise accumulated can be utilized for aiding SOFC product development.

ACKNOWLEDGEMENT

The authors would like to acknowledge FCE's technical, engineering and manufacturing staffs on the combined contribution to this paper. The financial support by US Department of Energy (DOE), Electric Power Research Institute (EPRI), Gas Research Institute (GRI), and others are greatly appreciated.

REFERENCES

[1] M. Farooque and H. Maru, Carbonate Fuel Cells: Milliwatts to Megawatts, *J. Power Sources,* **160**, 827-834 (2006).

[2] P. Patel, et al., Fuel Cell and Hydrogen for Smart Grid, *ECS Transactions,* The Electrochemical Society, Inc., **30**, 305-313 (2011).

[3] Hossein Ghezel-Ayagh, et al., A Novel System for Carbon Dioxide Capture Utilizing Electrochemical Membrane Technology, *ECS Transactions,* The Electrochemical Society, Inc., **51**, 265-272 (2013).

[4] J.R. Selman and H. Maru, Physical Chemistry and Electrochemistry of Alkali Carbonate Melts, *Advances in Molten Salt Chemistry,* Vol. 4, G. Mamantov and J. Braunstein, Editors, Plenum Press, 159-390 (1981).

[5] C. Yuh and M. Farooque, Materials and Life Considerations, *Fuel Cells-Molten Carbonate Fuel Cells, Encyclopedia of Electrochemical Power Sources,* Elsevier Science, 497-507 (2009).

[6] N. Minh, Technological Status of Nickel Oxide Cathodes in Molten Carbonate Fuel Cells - A Review, *J. Power Sources,* **24**, 1-19 (1988).

[7] K. Ota, et al., Solubilities of Nickel Oxide in Molten Carbonate, *J. Electrochem. Soc.,* **139**, 667-671 (1992).

[8] H. S. Hsu, et al., Solubilities of $LiFeO_2$ and $(Li,K)_2CrO_4$ in Molten Alkali Carbonates at 650°C, *J. Electrochem. Soc.,* **134**, 2146-2150 (1987).

[9] C. Yuh and M. Farooque, High-Temperature Direct Fuel Cell Material Experience, *Advances in Solid Oxide Fuel Cells X, Ceramic Engineering and Science Proceedings,* 11th International Symposium on Solid Oxide Fuel Cells (SOFC): Materials, Science and Technology in 38th International Conference and Exposition on Advanced Ceramics and Composites (38th ICACC 2014), American Ceramic Society, John Wiley & Sons, Hoboken, NJ, **35**, 9-21 (2015).

[10] C. Yuh, M. Farooque and H. Maru, Advances in Carbonate Fuel Cell Matrix and electrolyte, *in Proc. Symp. Carbonate Fuel Cell Technology V,* The Electrochemical Society, **PV99-20**, 189-201 (1999).

[11] A. Hilmi, C. Yuh, and M. Farooque, Carbonate Fuel Cell Matrix and Electrolyte Developments, *ECS Transactions,* The Electrochemical Society, Inc., **41**, 65-73 (2012).

[12] H.J. Choi, et al., Phase and Microstructural Stability of Electrolyte Matrix for Molten Carbonate Fuel Cell, *Fuel Cells,* **10**, 613-618 (2010).

[13] C. Yuh and A. Hilmi, Smart Matrix for Direct Carbonate Fuel cell, *Proc. 2015 U.S. Department of Energy (DOE) Hydrogen and Fuel Cells Program and Vehicle Technologies Office Annual Merit Review and Peer Evaluation Meeting,* Arlington, Virginia, June 12-16 (2015).

[14] A. Hilmi, A. Surendranath and C. Yuh, An Innovative Carbonate Fuel Cell Matrix, *Abstract #188, Joint General Session: Battery and Energy Storage -and- Fuel Cells, Electrolytes, and Energy Conversion,* 227th ECS Meeting, Chicago, IL, May 28 (2015).

[15] J. Hoffmann, C. Yuh and A. Godula Jepek, Electrolyte and Material Challenge, *Handbook of Fuel Cells - Fundamentals, technology and Applications,* Vol. 4, W. Vielstich, A. Lamm, and H.A. Gasteiger, Editors, John Wiley and Sons, 921-941 (2003).

[16] Z. Yang, Recent Advances in Metallic Interconnects for Solid Oxide Fuel Cells, *International Materials review*, **53**, 39-54 (2008).
[17] B. Zhu, G. Lindbergh, D. Simonsson, Comparison of electrochemical and surface characterisation methods for investigation of corrosion of bipolar plate materials in molten carbonate fuel cell: Part I. Electrochemical study, Part II. Surface analysis, *Corrosion Science*, **41**, 1497–1528 (1999).
[18] Z. Yang, G Xia, M. Walker, C. Wang, J. Stevenson, and P. Singh, High Temperature Oxidation/Corrosion Behavior of Metals and Alloys under a Hydrogen Gradient, *Int. J. of Hydrogen Energy*, **32,** 3770-3777 (2007).
[19] T. Sourmail, Precipitation in Creep Resistant Austenitic Stainless Steels, *Materials Science and Technology*, **17**, 1-14 (2001).

DEVELOPMENT OF HIGHLY-EFFICIENT ENERGY STORAGE SYSTEM USING SOLID OXIDE ELECTROLYSIS CELL

Masato Yoshino, Tsuneji Kameda, Hisao Watanabe and Masahiko Yamada
Power System R & D center, Toshiba Corporation Power Systems Company, 2-4 Suehiro-cho, Tsurumi-ku, Yokohama, 230-0045 JAPAN

ABSTRACT

Large-scale and Low-cost energy storage system is required to promote renewable energy utilization, especially for mega-solar or wind farm plants. Hydrogen will be the promising medium for large-scale storage. Toshiba has developed the hydrogen electric power storage system, which produces hydrogen using surplus electricity and generates electricity using hydrogen by fuel cell. This system will provide more cost-effectiveness than batteries and less-restricted plant location than pumped hydro. Especially, the SOEC/SOFC will contribute high efficiency for electrical power storage system. The most important part for this system will be the hydrogen generation process by high temperature steam electrolysis method using solid oxide electrolysis cell (SOEC), and the solid oxide cell and stack of various shapes have been developed and evaluated.

As one of results of cell performance evaluation, the current density at 1.3V as thermal neutral point achieved more than 1A/cm^2 in SOEC test condition using the flat-tube type single cell. And the SOEC/SOFC adopting electrical power storage system design and the system efficiency with charge-discharge have been considered on the basis of I-V properties of SOEC/SOFC. The charge-discharge efficiency, which is conducted on heat plant mass balance analysis of the system, was improved at higher temperature and lower current density, and it was found that charge-discharge efficiency obtained 80% or more in optimized condition.

INTRODUCTION

Utilization of renewable energy sources has been promoted internationally in terms of the environmental load reduction and the energy security enhancement. To integrate the large electricity which is supplied from the fluctuating renewable energy sources, such as solar and wind power, into the grid, energy storage technologies are required to maintain the power quality, leveling the load, and absorbing the surplus.

Hydrogen is the promising medium for large-scale storage, with more cost-effectiveness than batteries, and less-restricted plant location than pumped hydro. Figure 1 shows the conceptual scheme of the hydrogen power storage system. This system consists of electricity/hydrogen conversion device and hydrogen storage tank, mainly. Fluctuating electricity is converted to hydrogen by electrolysis and stored in form of hydrogen. And when required, hydrogen can be re-converted to electricity to minimize the difference of supply and demand of electricity. And also while supplying hydrogen itself; oxygen and heat supply are available. Such system is planned and demonstrated in many projects recently[1,2,3,4,5]. Many systems adapt alkaline water or proton exchange membrane cells for electrolysis, and proton exchange membrane cells for electrical power generation. These systems can be built easily, for such apparatuses are already on the market. However, their electrical efficiency is expected relatively low because these apparatuses work at low temperature of around 353K, and it is difficult to use their exothermic heat effectively. On the other hand, the system using solid oxide cells for both electrolysis and electricity generator would achieve high electrical efficiency by conveying the stored heat at the fuel cell exothermic reaction to the electrolysis endothermic reaction. Solid oxide EC (electrolysis cell) / FC (fuel cell) will contribute the power storage system with high efficiency

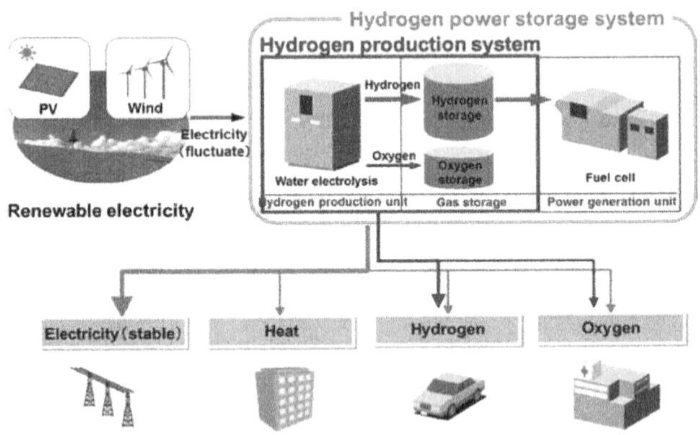

Figure 1. Conceptual scheme of hydrogen power storage system.

The most important elemental technologies for the hydrogen power storage system using SOEC/SOFC are the solid oxide cell for high temperature steam electrolysis and high temperature heat storage method. High temperature steam electrolysis with solid oxide cell was under development during early 1980s[6], and it has become increasingly investigated again in recent years. As for the technology of the high temperature heat storage, that development is pushed forward in fields such as the solar thermal power generation. For the hydrogen power storage system using SOEC/SOFC, heat storage technology at higher temperature around 800 degree C is required. Furthermore, based on the properties of the electrochemical cell and the heat storage method, it is necessary to examine the design of the power storage systems and charge-discharge efficiency.

The main purpose of our development is potential evaluation of SOEC/SOFC adapted power storage system. We carried on evaluation for three elemental technology items in our development. One is the evaluation of SOEC performance of proven SOFC cell using residential system. Another one is investigation of high temperature heat storage method around 800 degree C. And last one is system design based of the results of these investigations.

ENERGY STORAGE SYSTEM

Figure 2 shows the conceptual configuration diagram of the SOEC/SOFC electrical power storage system. The unique point of this system is to be able to use heat with electricity to make hydrogen. High charge-discharge efficiency is available by conveying the stored heat at the fuel cell exothermic reaction to the electrolysis endothermic reaction. The most important elemental technologies for this system are SOEC and high temperature heat storage method. We have evaluated the charge-discharge efficiency for our developing SOEC/SOFC system.

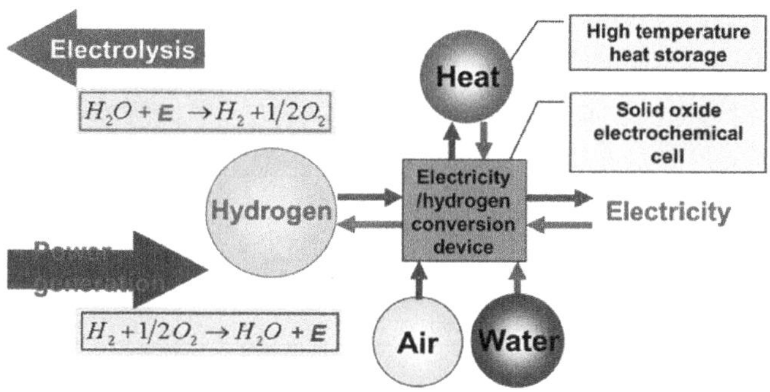

Figure 2. Conceptual configuration diagram of the SOEC/SOFC electrical power storage system.

ELEMENTAL TECHNOLOGIES FOR SOEC/SOFC ELECTRICAL POWER STORAGE SYSTEM

The evaluation for three elemental technology items in our development was carried on. One is the evaluation of SOEC performance of proven SOFC cell using residential system. Another one is investigation of high temperature heat storage method around 800 degree C. And last one is system design using results obtained these investigation.

Solid Oxide Electrolysis Cell

Solid oxide electrochemical cell is the key device for the power storage system using Hydrogen as a storage medium. And it has the function of hydrogen production with electrolysis of steam (power storage) and power generation with the fuel cell reaction (discharge). In this development program, the flat-tube type single cell was selected because this type cell was proven its durability in commercial residential SOFC system. The electrolysis properties of the prototype flat-tube type single cell in the mass production level were evaluated. Figure 3 shows the schematic diagram of a flat-tube type single cell. The flat-tube cell is supported by hydrogen electrode support which has inner channel. One end of a flat-tube type single cell was fixed to the stainless steel gas manifolds and the other end was opened. The mixed gas of steam and hydrogen feeds to inner channel formed support. Excess gas and product gas which passed an inner channel burns at the opened end. To measure electrolytic performance, the flat-tube single cell was set in an electric furnace. Two Pt wires for current supply and potential probe were contacted to Pt mesh attached to oxygen electrode and interconnector each other. Test condition for the electrolysis properties of the flat-tube type single cell was shown in Table I. The initial I-V properties, steam utilization, hydrogen generation performance were evaluated.

Table I: Test condition for the electrolysis properties of the flat-tube type single cell

Operation Temperature		700～800 degree C
Feed Gas to Hydrogen Electrode	composition	H_2O/H_2=50/50vol% or H_2=100%
	flow rate	0.1～1L/min
Feed Gas to Oxygen Electrode	composition	Air
	flow rate	3L/min
Current Density		0 ～ over 1A/cm^2

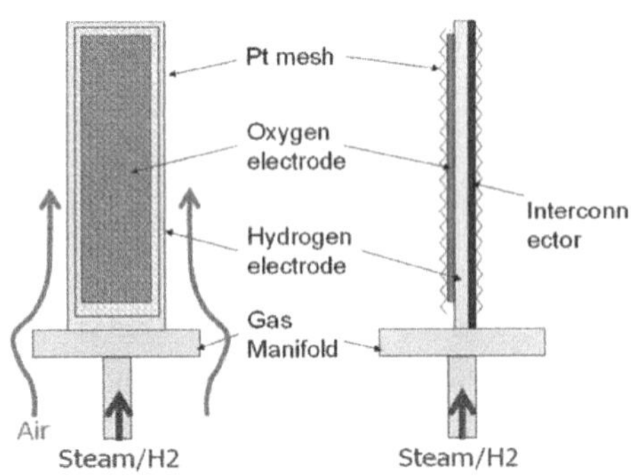

Figure 3. Schematic diagram of a flat-tube type single cell.

Figure 4 shows the initial I-V properties of the open-end flat-tube type cell at from 700 degree C to 800 degree C in both SOEC and SOFC mode test condition. This cell had good performance in SOEC mode. The current density at thermal neutral point of 1.3 V was achieved over 1.0 A/cm^2 at over 750 degree C. And it was confirmed that influence of the temperature for I-V properties was larger in the test condition of SOEC than compared with that of SOFC.

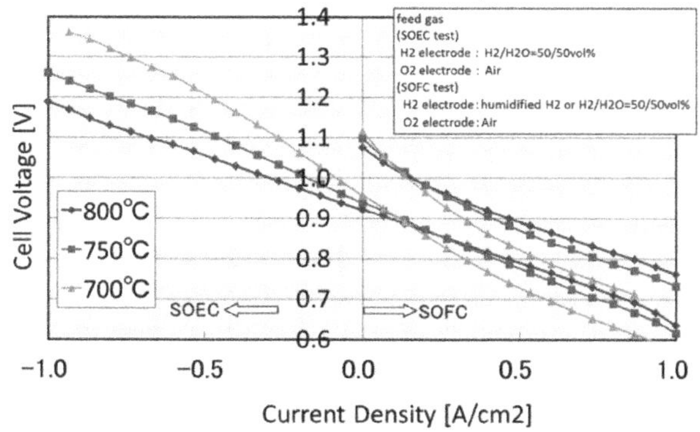

Figure 4. I-V properties in both SOFC and SOEC mode of flat-tube single cell.

Figure 5 shows the steam utilization characteristics at SOEC mode test condition. The supply flow rate was changed under the same current density conditions and the voltage rise was measured. At all current density condition from 0.2 to 0.8A/cm^2, the steam utilization was stable up to 80%. On the other hand, when the steam utilization was up to 90%, the increasing slope of the cell voltage became larger.

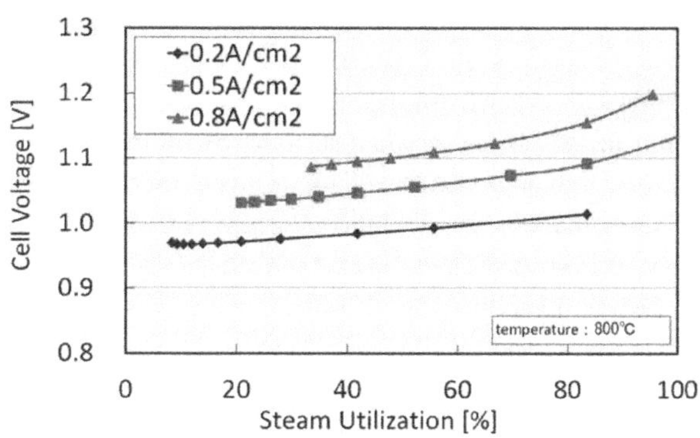

Figure 5. Steam utilization characteristics at the SOEC mode.

Figure 6 shows the hydrogen-producing properties. In this evaluation, the flat-tube single cell which capped its open end was prepared. The red line in this figure shows theoretical H_2 production rate. With the increase in the current value, and the amount has a tendency to increase linearly. Performance of H_2 production was almost equal to the theoretical value and the production rate with respect to the theoretical value showed 95-100%.

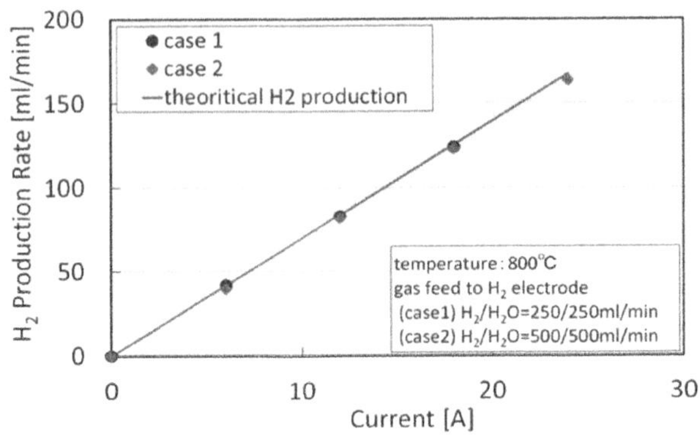

Figure 6. Hydrogen-producing properties.

High Temperature heat Storage method

The hydrogen energy storage system using the SOFC / SOEC is assumed the operating temperature of 800 degree C. For utilizing the heat generated in the SOFC exothermic reaction to the endothermic reaction in SOEC, it is necessary to develop a high temperature heat storage method. Phase change material (PCM) having high latent heat was selected as the heat storage material and a high-temperature heat storage device that the heat could be put in and out efficiently with nitrogen as a heat medium is studied. The heat storage unit capsule was fabricated and the repeated melting/solidification characteristics of the heat storage unit capsule were evaluated.

Figure 7 shows the heat storage unit capsule. From SOEC / SOFC operating temperature about 800 degree C, sodium chloride (NaCl) with the melting point of 800 degree C was selected for PCM. And NaCl enclosed in a capsule. Silicon carbide (SiC) was selected for the capsule material because of the thermal conductivity and the corrosion resistance to the molten salt NaCl. Mass of the PCM in the heat storage unit capsule was an amount of 90% of the maximum volume of capsule, and the heat storage unit capsule was encapsulated by using a ceramic adhesive. The melting and solidification repetition test was carried out on the prototype heat storage unit capsule to measure the temperature change of the PCM internal and external capsule.

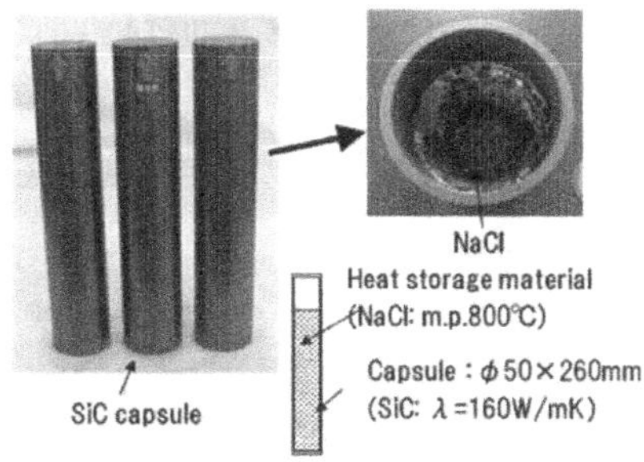

Figure 7. High temperature heat storage unit capsules.

Figure 8 shows the temperature changes of NaCl and outside wall of the capsule. In the heat cycle test, temperature plateau was confirmed at about 800 degree C in both temperature increasing and decreasing process. And it was confirmed that temperature difference between heat storage and dissipation process was under 2 degree C and temperature residence time of heat storage and dissipation process was the almost same. It might be revealed that phase change of NaCl from solidus to liquidus, or from liquidus to solidus was occurred. In addition, a significant reaction of SiC capsule and PCM (NaCl) was not observed.

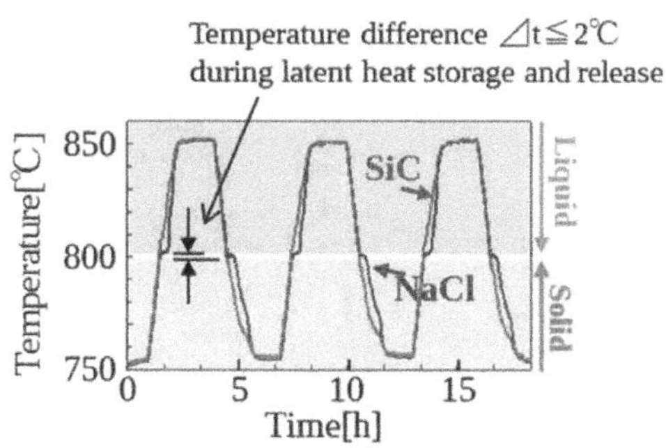

Figure 8. Temperature changes of NaCl and outside wall of the capsule in the heat cycle test.

System Design

Based on the properties of the developed electrochemical cell such as ASR and the heat storage method, running temperature, an effect of scale-up, the design and system efficiency of the hydrogen production and the power storage systems are studied with heat mass balance analysis, using VGMsim v7 simulator. Figure 9 shows the diagram of hydrogen power storage system and Figure 10 shows the image of the 5MWe-class hydrogen power storage system. That is mainly consisted from the electrochemical cell, the heat storage, the hydrogen tank, some heat exchangers and others. The plant size is recognized to be feasible comparing other power storage systems. The plant size is estimated 20m x 25m, feasible comparing other power storage systems. And power storage capacity can be easily increased with the number of H_2 gas tanks.

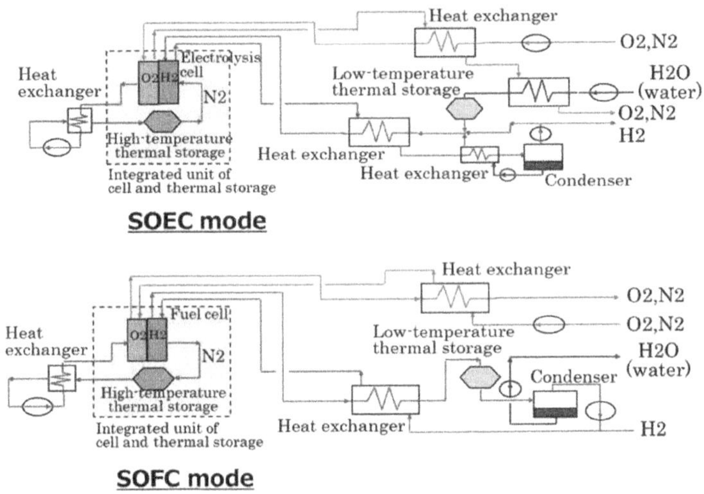

Figure 9. Diagram of hydrogen power storage system.

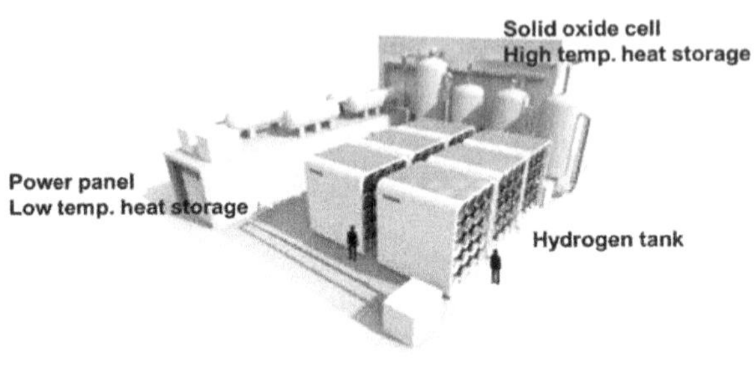

Figure 10. Image for 5MWe-class hydrogen power storage system.

The specific electricity consumption for the hydrogen production is shown in Figure 11, with the relation to the operating temperature and current density. Relatively low specific electricity consumption regions, such as under 3kWh/Nm^3, are recognized. Better property and cost are expected comparing conventional methods, alkaline and PEM electrolysis. And for the power storage system, the charge-discharge property is shown in Figure 12, with the relation to the operating temperature and current density. High efficiency of charge-discharge, such as over 70% is achieved by choosing the operating conditions.

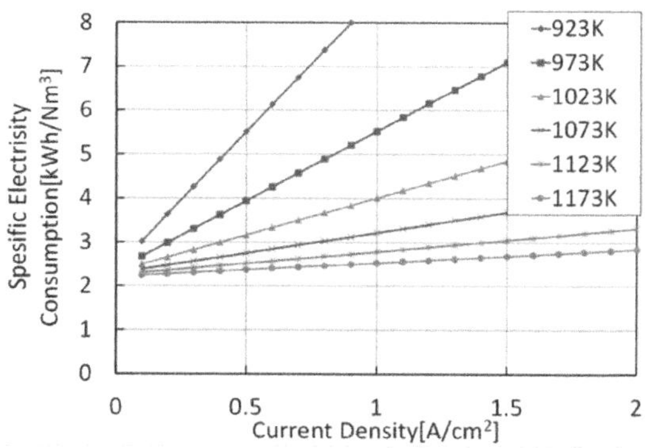

Figure 11. Specific electricity consumption for the hydrogen production.

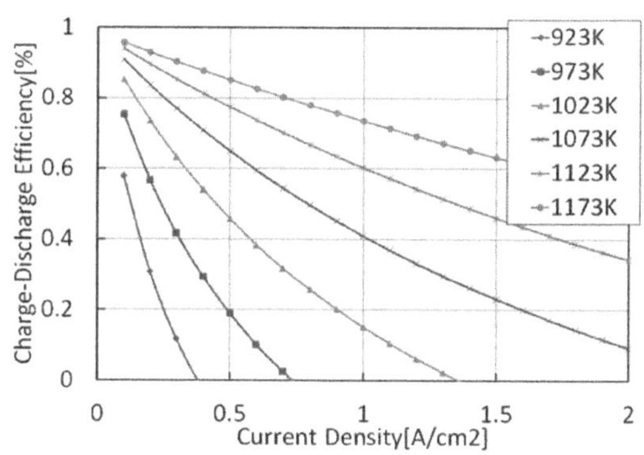

Figure 12. Charge-discharge properties for SOEC/SOFC power storage system.

Economical Evaluation

The power storage cost is evaluated as shown in Figure 12. For this comparison, some assumptions of the capital costs listed in Table 2 are used. The capital cost of NaS battery is expected to come down to the METI target cost in 2020. SOEC/SOFC power unit cost is also thought to become the expected cost around 2020 in the road map. Three lines of the hydrogen energy storage systems have small sloops in Figure 12, because these systems can increase the charging time, corresponding to the charging amount, by increasing the capacity of the relatively low cost hydrogen storage tank. For long charging time, the hydrogen storage systems are evaluated cost effective.

Table III shows the comparison of typical power storage systems, based on assumptions for capital costs of various power storage systems as shown in Table II. The SOEC/SOFC hydrogen power storage system has relatively high charge-discharge efficiency. And it will provide more cost-effectiveness than batteries and less-restricted plant location than pumped hydro.

Table II: Assumptions for capital costs of various power storage systems

system	System Cost (k¥/kWh)	reference
Pumped hydro	23	METI, target cost in 2020 for the battery technology development program, 2012.07
NaS battery	23	METI, target cost in 2020 for the battery technology development program, 2012.07
Li-ion battery	100	Future generation energy transmission and distribution network study group, 2010.03
Pb battery	50	Resources and Energy Agency, Current status and approach for battery technology, 2009.02
Hydrogen (Alkaline)	11.8	・Alkaline electrolyzer : 50k¥/kW (Hydrogen Production Costs-A Survey, DOE/GO/10170-T18(6)・・・$600/kW) ・Hydrogen tank : Cost based on current quotation
Hydrogen (SOEC/SOFC)	62.6	・SOEC/SOFC power unit : expected cost around 2020 ・Low temperature heat storage, heat exchanger, hydrogen tank : Cost based on current quotation
	45.2	・SOEC/SOFC power unit : expected cost around 2020 ・Low temperature heat storage, heat exchanger, hydrogen tank : Cost based on current quotation x 1/2

Table III: Comparison of power storage systems

System		Hydrogen		Nas battery	Li-ion battery	Pumped hydro	CAES
		Alkaline+PEFC	SOEC+SOFC				
Main device		Alkaline electrolyzer PEFC Hydrogen storage	SOEC SOFC Hydrogen storage Heat storage	Nas battery Heat retention	Li-ion battery	Water reservoir Power generation Pump Pipe arrangement	Air reservoir Power generation Air compressor Pipe arrangement
Output / Input		0.4	0.80	0.90	0.96	0.7	0.7
Development status		actual use for each device test for combined system	actual use for small FC under development for others	actual use	actual use	actual use	actual use
Economical view	leveling for short-period	△	△	○	○	× (○ for variable)	×
	leveling for day/night	○	○	○	○	○	○
	storage for long-period	○	○	△ ~ ×(heating)	△ ~ × (cost)	× (location)	× (location)
Weak point		Low O/I Durability for fluctuation	Scale up	Heat retention Hazardous materials	Cost	Restricted location Construction leading time	Restricted location Construction leading time
Good point		Low cost Large scale	Low cost High efficiency	High efficiency	High efficiency	Practial accomplishment	Improve GT efficiency

CONCLUSIONS

The flat-tube type cell applied in commercial residential SOFC system showed good performance in SOEC test condition. High temperature heat storage technologies around 800 degree C have been developed. The hydrogen production and power storage systems using SOEC were studied, and their potentials are confirmed to have high efficient performances. These systems would be cost competitive to other typical conventional technologies.

ACKNOWLEDGMENT

A part of this paper is based on results obtained from a project commissioned by the Ministry of Economy, Trade and Industry (METI) and the New Energy and Industrial Technology Development Organization (NEDO) in fiscal 2012 to 2014. We express our appreciation.

REFERENCES

1. Mustafa Serdar Genc, Muhammet Celik and Ilyas Karasu, Renewable and sustainable energy reviews 16 (2012) 6631-6646.
2. Christpher Graves, Sune D. Ebbesen, Mogens Mo-gensen and Klaus S. Lackner, Renewable and sustain-able energy reviews 15 (2011) 1-23.
3. J.M. Zolezzi, A. Garay and M. Reveco, Journal of Power Sources 195 (2010) 8236-8243.
4. K. Hedegaard and P. Meibom, Renewable Energy 37 (2012) 318-324.
5. Magnus Korpas and Christopher J. Greiner, Renew-able Energy 33 (2008) 1199-1208.
6. W. Donitz and E. Erdle, Int. J. Hydrogen Energy 10 (1985) 291-295.

Ceramic-Related Materials, Devices, and Processing for Heat-to-Electricity Direct Conversion

THERMOELECTRIC PROPERTIES HIGHER MANGANESE SILICIDE CONTAINING SMALL AMOUNT OF MNSI/SI NANO-PARTICLES

Swapnil Ghodke[1], A. Yamamoto[2], H. Ikuta[1], T. Takeuchi[2,3]

[1]Department of Crystalline Materials Science, Nagoya University, Furo-cho, Chikusa-ku, Japan
[2]Toyota Technological Institute, 2 Chome-12-1 Hisakata, Tenpaku Ward, Nagoya, Japan
[3]PRESTO, JST, Tokyo, Japan
E-mail : swapneelghodke@gmail.com

ABSTRACT

In this study, we prepared higher manganese silicides (HMS) containing small amount of MnSi or silicon nano-particles to study the energy dependent scattering effect, which had been reported to improve the power factor. The samples were prepared in induction furnace followed by liquid quenching technique, to obtain fine grain structure of HMS phase with precipitation of nano particles of secondary phases. The powder XRD data and SEM images indicate the presence of MnSi or Si nano particles in liquid quenched samples of fine grain size of 1~2μm. An effective increase of Seebeck coefficient and increase of electrical resistivity were observed for all the liquid quenched samples regardless of the nano particle precipitation. Though the electrical resistivity was increased, the effective increase in Seebeck coefficient led power factor to 120% from that of the samples prepared without liquid quenching. These facts suggested that the energy dependent scattering effect was introduced in HMS by the grain boundaries.

INTRODUCTION

The non-renewable fossil fuel in form of crude oil, coal, or natural gas is being used to fulfil the energy need of our modern society. The electrical energy comes with a cost for industrial as well as household usage. Considering the limited efficiency of energy generation and utilization, we need to utilize the energy that is given out to atmosphere in the form of heat. Here, thermoelectric generators (TEG's) can play key role to convert waste form of heat energy into useful electrical energy.

The performance of thermoelectric generator is measured by dimensionless figure of merit (*ZT*), which is increasing function of overall efficiency η [1].

$$ZT = \frac{S^2\sigma}{\kappa_e + \kappa_L} T$$

Here, S, σ, κ_e, κ_L and T are Seebeck coefficient or thermo-power, electrical conductivity, electron thermal conductivity, lattice thermal conductivity, and absolute temperature, respectively. To obtain materials possessing high *ZT*, the material should possess high power factor $S^2\sigma$ together with low electron thermal conductivity and lattice thermal conductivity. Theoretically all the physical quantities are strongly coupled to one another that make it more challenging to obtain most efficient thermoelectric material [1-3].

Many of practical thermoelectric materials possessing high *ZT* contain expensive and toxic elements like Lead, Selenium, Antimony, and Tellurium. Thus, it limits the access and usage of this technology for general applications. The most promising alternatives are higher manganese silicide (HMS, $MnSi_\gamma$) ($1.73 \le \gamma \le 1.75$), as the constituent elements are abundant, inexpensive, non-toxic, with low cost of production. The thermal and mechanical stability of HMS at high temperature gives capability to work in operational 600 – 800 K temperature range[1, 4]. The p-type HMS have shown large Seebeck coefficient ~ 220μVK^{-1}, low electrical resistivity ~ 2x10^{-3}Ωcm, and thermal conductivity ~ 2.5WmK^{-1} with maximum *ZT* of ~0.6 at 800 K[(5-9)].

The structure of HMS belongs to the group of chimney ladder crystal structure, and was reported to have four different forms, Mn_4Si_7[10], $Mn_{11}Si_{19}$[11], $Mn_{15}Si_{26}$[12], and $Mn_{27}Si_{47}$[13]. Mn sub-lattices and Si sub-lattices are arranged in such a way that they are stacked with Mn atoms forming chimney and Si atoms as helical ladder with long c-axis. The four different structures were interpreted as the products of different incommensurate-length-ratios of two sub-lattices[5].

Unfortunately, the ZT value is not competitive with the commercially available thermoelectric materials. The need to improve the performance of higher manganese silicide has drawn our attention to investigate new technique that can enhance the *ZT*. We focused, therefore, on the energy dependent scattering effect that has been reported by Zebarjadi et. al[15], Xie et. al [16], and Soni Ajay et al.[17] to enhance the power factor. It is generally accounted for on the principle of selective scattering low/high energy carriers and allowing high/low-energy carriers in n/p-type materials, respectively. This could be done by introducing grain boundaries or nano particles as additional scattering sites.

Thus, in this work we have tried to investigate the effect of carrier filtering due to secondary phases and grain boundaries. With reference of Mn-Si binary phase diagram[18] we synthesized different samples in silicon deficient and silicon rich regions, forming MnSi and Si precipitates, respectively. We employed liquid quenching technique because it helps to greatly reduce the grain size of secondary phases and also to obtain high density of grain boundary. The composition dependence of S, σ, and κ is discussed in this paper.

EXPERIMENTAL PROCEDURE:

Higher manganese silicide $MnSi_\gamma$ with different composition γ =1.67,1.73,1.79,1.81, and 2.11 were synthesized by melting manganese (99.99%) and silicon (99.99%) flakes in boron nitride crucible under pressurized argon atmosphere in induction furnace. The ingot was further used for melt spinning/ liquid quenching, where the sample is again melted in silica tube and injected on copper wheel which rotating at 4500 rpm in Argon atmosphere. The ribbon-shaped samples with 10~15 mm in length, 0.5~0.1 mm in width, and 10~15 μm in thickness were obtained. The ribbon sample grounded in mortar and pestle, the powder is sintered in graphite die (ϕ=10mm) by spark plasma sintering under 60 Mpa pressure at 950°C for 5min in vacuum.

We have prepared two types of samples for characterization; liquid quenched ribbons and bulk-sintered samples. The phase purity of compounds was identified by powder X-ray diffractometer of BRUKER D8 advance with Cu-Kα radiation. The microstructure was observed by Hitachi SU66000 scanning electron microscope (SEM). Seebeck coefficient was measured by steady state method using Seebeck measurement system manufactured by MMR technologies, electrical resistivity by conventional four-probe method. Thermal conductivity was calculated by formula $\kappa = D\rho Cp$, thermal diffusivity (D) was measured by laser flash technique (LFA457, NETZSCH), heat capacity (Cp) by using differential scanning calorimetry (DSC 8231, Rigaku), and the density (ρ) of sintered samples by Archimedes method.

Results

The XRD pattern plotted in a graph (Fig. 1) for $MnSi_{1.73}$ shows formation of phase pure compound, without any impurity. The small peak of MnSi can be observed in $MnSi_{1.67}$ sample.

Samples containing silicon above nominal composition showed small peaks of silicon, which was precipitated during the liquid quenching process. The intensity of peak for silicon increases with γ ($1.79 \leq \gamma \leq 2.11$), without forming any other impurity phases.

To investigate the particle size, grain size, and the distribution of secondary phases, we took SEM images, which are shown in Fig. 2. In Fig.2 (a) - (c), grains of 2 μm-3 μm are observed for liquid quenched samples with $\gamma = 1.67$, $\gamma = 1.73$, and $\gamma = 1.79$, respectively. Figure 2 (d) – (f) shows dense microstructures of bulk samples, the grain boundaries are highly diffused after spark plasma sintering process. By comparing (a) and (d) for the composition of $\gamma = 1.67$, we can clearly confirm the effect of processing on microstructure. The density of grain boundaries was definitely reduced to a large extent in SPS samples, though both the samples show the nano-size precipitates of MnSi. The similar effect can be seen throughout the whole composition range. The microstructure of LQ and SPS samples at $\gamma = 1.73$ showed no impurity phases. This fact is consistent with the XRD data. Figure 2 (c) and (f) for Si-rich composition ($\gamma = 1.81$) shows difference in grain boundary density, as for other composition samples. Nano precipitates in black color were observed in gray matrix of $\gamma = 1.81$ SPS samples shown in (f), which could be silicon nano particles, as small peak of Si was observed in XRD data. The similar microstructural behaviour for Si-rich compositions were also observed for $\gamma = 1.79$ and $\gamma = 2.11$ samples.

The Seebeck coefficient at 300 K is plotted as a function of composition in Fig. 3. The sign of Seebeck coefficient was positive regardless of the compositions, indicating that the present samples are p-type thermoelectric materials. It was clearly observed that the absolute value of Seebeck coefficient for LQ samples was definitely larger than that of SPS samples. The Seebeck coefficient is almost constant over compositions for liquid quenched or SPS samples.

Figure 4 shows the composition dependent electrical resistivity at 300 K. The resistivity of $\gamma = 1.67$ sample is lowest which might be due to presence of metallic MnSi as impurity phase in the matrix. The sample with Si-rich composition showed an increase in electrical resistivity with increasing silicon concentration because of the increasing amount of precipitated silicon nano-particles. Similar to Seebeck effect, here we observed small but finite process dependent difference in LQ sample and SPS sample in absolute electrical resistivity for a composition. The LQ samples always showed slightly higher electrical resistivity than that of SPS samples. This experimental fact is presumably brought about by the scattering effect of grain boundary because the former has much dense grain boundary than the latter.

The composition dependence of power factor ($S^2\sigma$) is shown in Fig. 5. The power factor showed ~20 % difference between LQ & SPS samples. The power factor is large for all the LQ samples, which is due to dominant contribution of Seebeck coefficient. The power factor is large for the nominal composition ($MnSi_{1.73}$) HMS, and it is reduced by presence of MnSi or Si as a secondary phase in the matrix.

Thermal conductivities of liquid quenched ribbon samples were not measured due to practical limitation. The composition dependent thermal conductivity for SPS samples is shown in Fig. 6. The results suggest that the addition of secondary phases MnSi/Si particles could increase the thermal conductivity. In the case of $\gamma = 1.67$ sample, both the contribution due to lattice and electron was definitely increased, whereas the thermal conductivity of Si-rich samples $\gamma = 1.79$ or $\gamma = 1.81$ was enhanced solely by lattice thermal conductivity and the electronic contribution is reduced as compared to $\gamma = 1.73$ of pure HMS.

The composition dependence of *ZT* at 300 K is plotted in Fig. 7. By assuming the same thermal conductivity values of SPS samples, we calculated the *ZT* even for the liquid quenched samples. The maximum *ZT* at room temperature was observed for $MnSi_{1.73}$ SPS sample that has no precipitation of secondary phase. The ZT of $MnSi_{1.67}$ was reduced due to the large thermal conductivity of precipitated MnSi. The large electrical resistivity of precipitated silicon nano-particles also reduced *ZT* of $MnSi_{1.79}$/$MnSi_{1.81}$ samples.

DISCUSSION

By considering (a) Seebeck coefficient is not dependent on the density of material, (b) the grain boundary density in LQ samples was definitely higher than that of SPS samples, and (c) the increase of Seebeck coefficient in LQ samples was observed regardless of compositions, we reached the conclusion that the finite difference observed in Seebeck measurement between LQ and SPS samples can be attributed to energy filtering at grain boundary. There was large difference in grain size between Lq and SPS samples, which means higher grain boundary density in LQ samples.

The enhancement of Seebeck coefficient in LQ samples would be closely related to the increase of electrical resistivity. The energy filtering effect should be brought about by the enhanced scattering at a limited energy range. This enhancement of scattering probability naturally leads to an effective increase of electrical resistivity.

The enhancement of Seebeck coefficient of p-type materials, such as the present HMS, should be realized with the strong scattering at the energy range above the chemical potential. Unfortunately we do not have any critical evidence that proves the enhanced scattering at this particular range. It should be investigated with new analytical techniques, and therefore we left this serious point for our future study.

The most important point we found in this study is that the high density of grain boundary is capable of increasing power factor. We should investigate this effect for other thermoelectric materials to know whether it is universal behaviour or not. We are now investigating this behaviour, and the results will be reported in near future elsewhere.

CONCLUSION

In this work we successfully synthesized higher manganese silicide compound with MnSi/Si as secondary phases in it. The secondary phases were precipitated in nanometer range due to rapid quenching. Simultaneously, this work was focused to investigate the effect of grain boundary density on the thermoelectric properties. We clearly observed the finite difference in Seebeck coefficient, which we attributed to energy filtering effect due to grain boundaries by comparing the result of liquid quenched and bulk sintered samples. The energy filtering/scattering effect could improve the Seebeck up to 120%. Further, the results obtained for $\gamma = 1.67$ SPS sample suggest that the presence of MnSi metallic phase could reduce the electrical resistivity, but simultaneous increase in electron thermal conductivity nullifies improvement in *ZT*. Similarly, the precipitation of silicon particles increases electrical resistivity and thermal conductivity, which largely reduces the *ZT* at room temperature. Hence, we want to conclude that increasing grain boundary density could enhance the power factor of thermoelectric material. The understanding of physical and electronic structure of grain boundaries is essential to improve the performance of thermoelectric materials.

REFERENCES

1. D.M. Rowe (Ed.), CRC Handbook of Thermoelectrics, CRC Press, Boca Raton, 1995.
2. 2. H.Julian Goldsmid-Introduction to Thermoelectricity-Springer (2009).
3. J.R. Sootsman, D.Y. Chung, M.G. Kanatzidis, Angew. Chem. Int. Ed. Engl. 48 (2009) 8616–8639.
4. G.S Nolas, Thermoelectrics: Basics principles and new materials developments, Springer, 2001.
5. Yuzuru Miyazaki, Dai Igarashi, Kei Hayashi, and Tsuyoshi Kajitani, Physical Review B **78**, 214104, 2008.
6. Itoh, Takashi, and Masataka Yamada. Journal of electronic materials 38, no. 7 (2009): 925-929.
7. Y. Sadia, L. Dinnerman, Y. Gelbstein, J. Electron. Mater. 42 (2013) 1926–1931.
8. D.-K. Shin, K.-W. Jang, S.-C. Ur, I.-H. Kim, J. Electron. Mater. 42 (2013) 1756–1761.
9. Z. Zamanipour, X. Shi, M. Mozafari, J.S. Krasinski, L. Tayebi, D. Vashaee , Ceram. Int. 39 (2013) 2353–2358.
10. P. Norouzzadeh, Z. Zamanipour, J.S. Krasinski, D. Vashaee , J. Appl. Phys. 112 (2012) 124308.
11. O.G. Karpinskii, B.A. Evseev, Izv. Akad. Nauk SSSR, Neorg. Mater. 5 (1969) 525.
12. O. Shwomma, A. Preisinger, H. Nowotny, A. Wittman, Monatsh. Chem. 95 (1964)1527.
13. H.W. Knott, M.H. Mueller, L. Heaton, Acta Crystallogr. 23 (1967) 549.
14. G. Zwilling, H. Nowotny, Monatsh. Chem. 102 (1971) 672.
15. Zebarjadi et al. Applied Physics Letters 94, no. 20 (2009): 202105.
16. Xie, W. J., et al., Acta Materialia 61.6 (2013): 2087-2094.
17. Soni Ajay et al., Nano letters12.8 (2012): 4305-4310.
18. H. Okamoto, A. B. Gokhale and G. J. Abbaschain: Binary Alloy Phase Diagrams, Second Edition, eds. T. B. Massalski, et al., (ASM, 1990).

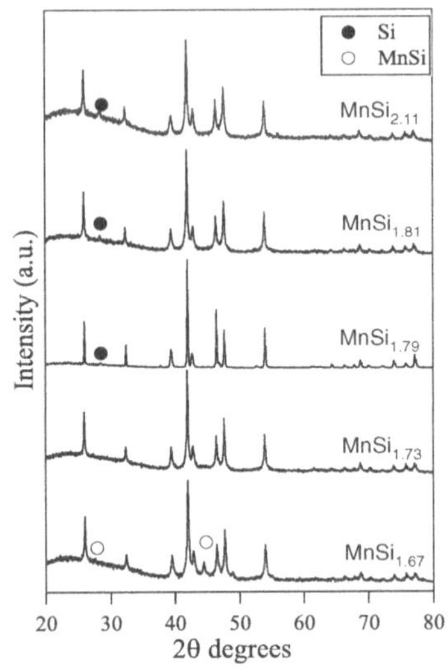

Fig. 1 XRD data plotted for $MnSi_\gamma$ with different composition γ =1.67, 1.73, 1.79, 1.81, and 2.11.

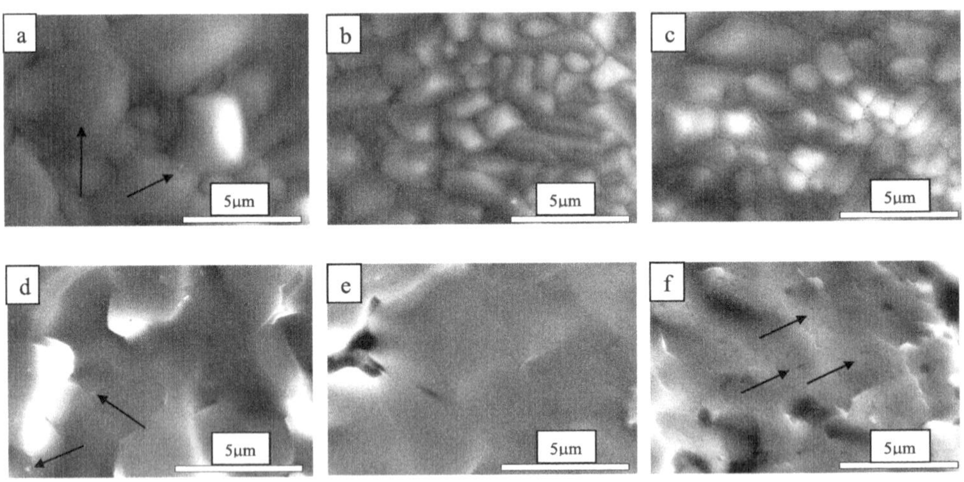

Fig. 2 SEM Images: (a), (b), and (c) shows microstructure of liquid quenched samples of γ=1.67, 1.73, and 1.81 respectively. (d), (e) and (f) shows microstructure of SPS samples γ=1.67, 1.73, and 1.81, respectively.

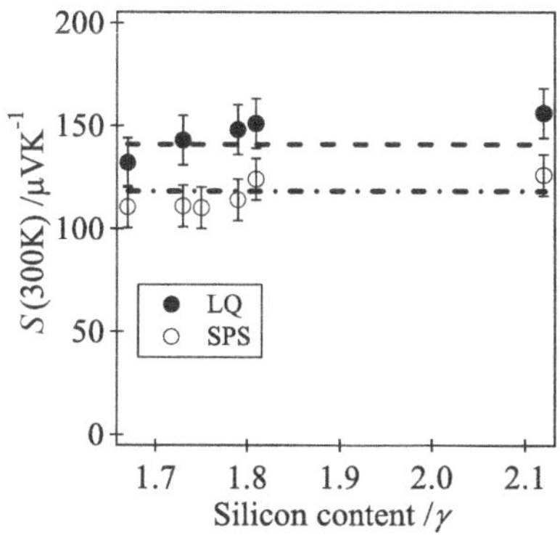

Fig. 3: The composition dependence of Seebeck coefficient for LQ and SPS ($MnSi_\gamma$) samples at room temperature.

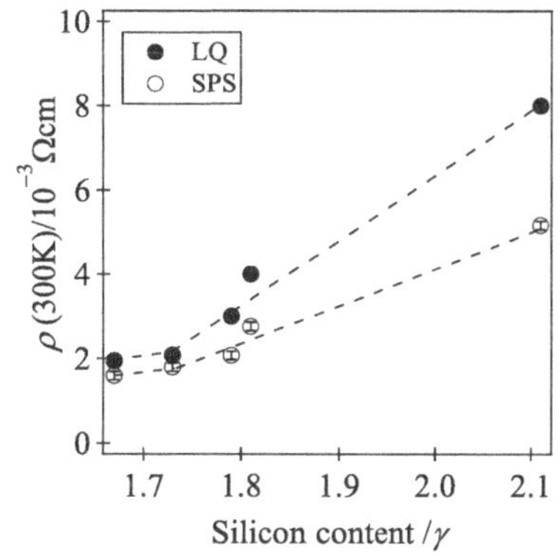

Fig. 4: The composition dependence of electrical resistivity for LQ and SPS ($MnSi_\gamma$) samples at room temperature.

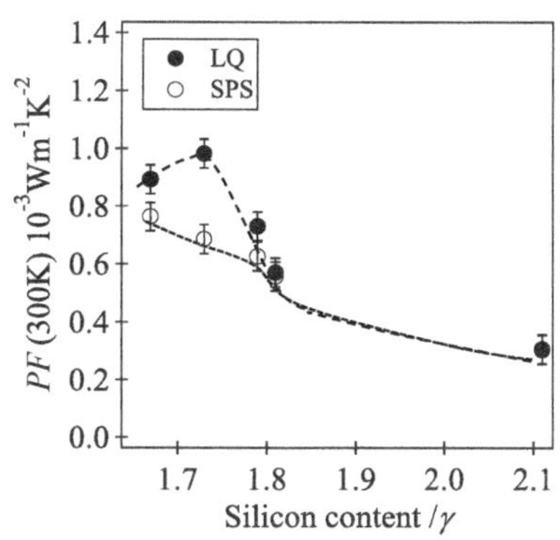

Fig. 5: The composition dependence of power factor for LQ and SPS ($MnSi_\gamma$) samples at room temperature.

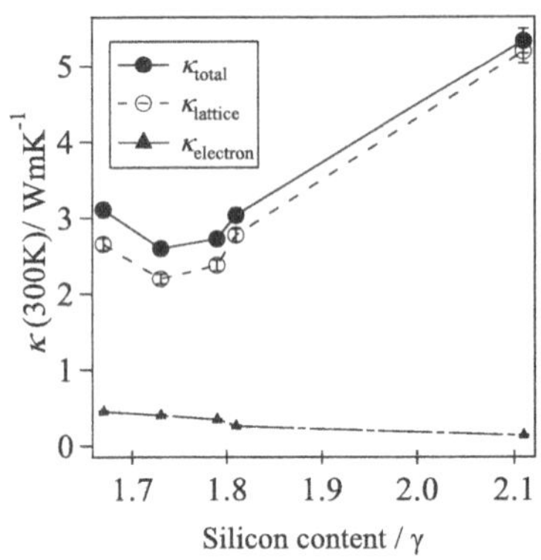

Fig. 6: The composition dependence of measured κ_{total}, with estimated $\kappa_{lattice}$, and $\kappa_{electron}$ for ($MnSi_{\gamma}$) SPS samples

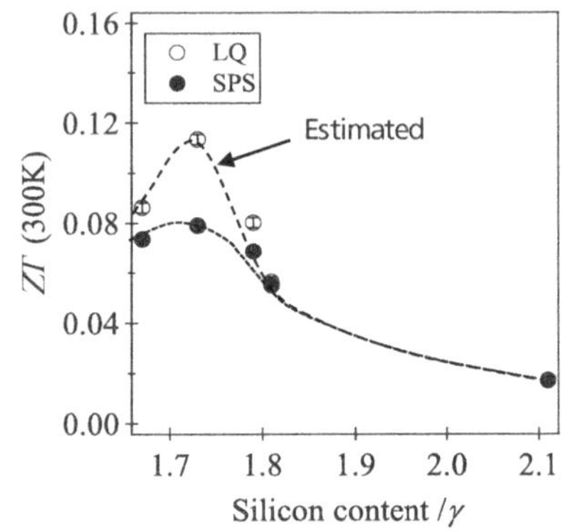

Fig. 7: The composition dependence of *ZT* for $MnSi_{\gamma}$ at room temperature.

ANOMALOUS TEMPERATURE GRADIENT IN NON-MAXWELLIAN GASES

George S. Levy, Entropic Power, 3980 Del Mar Meadows, San Diego, CA 92130, USA.

ABSTRACT

The theory presented in an earlier paper[1] is extended to explain anomalous data collected during very precise and careful experiments of the Seebeck effect. Despite meticulous procedures, researchers found irreducible temperature offsets and the temperature-voltage curve not passing through the origin. This paper shows that this phenomenon is quantum mechanical in nature. In semiconductors, the distribution of electrical carriers is the product of their density of states and the Fermi-Dirac distribution. A field applied to the carriers skews and de-normalizes their distribution. Renormalizing and comparing it with the original un-skewed one, indicates a shift in its first moment and a corresponding change in the carriers' temperature. The temperature offset is most significant and measurable when the semiconductor is high ZT and when the carriers are ballistic. The role of several distributions in the formation of temperature gradients is discussed. Particles following a non-Maxwellian distribution, such as Fermi-Dirac or Bose-Einstein statistics, can develop such gradients under certain conditions. However, classical particles following the Maxwell-Boltzmann distribution cannot. A calculator/graphing program illustrates the effect.

1. INTRODUCTION

This paper extends the theory presented in an earlier paper[1] explaining anomalous experimental data collected during very precise and careful experiments of the Seebeck effect in thermoelectric materials[23]. Despite meticulous efforts, researchers found irreducible temperature offsets and the temperature-voltage curve not passing through the origin.

It shall be shown that the experimental data is due to a quantum effect. Gas particles bound by quantum mechanics and complying with a non-Maxwellian distribution such as Fermi-Dirac statistics in a semiconductor, can develop a temperature gradient when they are subjected to a potential energy gradient. However, particles that follow classical physics and the Maxwell-Boltzmann distribution cannot produce this effect.

As shall be explained, the energy distribution of electrical carriers in a semiconductor is the product of their density of states and the Fermi-Dirac distribution. An electric field applied to the carriers skews and de-normalizes the carriers' distribution. Renormalizing to express it as a probability, and comparing it with the original un-skewed one, indicates a shift in its first moment and a corresponding change in the temperature of the carriers. This temperature offset is most significant and measurable when the semiconductor is high ZT (low coupling with heat phonons) and when the carriers are ballistic (non-interacting and thin geometry). This paper takes a general approach by exploring the role played by several distributions in the formation of temperature gradients when a force field is present. It shows that the Maxwell-Boltzmann distribution produces no temperature gradient. Paradoxically, the Fermi-Dirac and the Bose-Einstein distributions do generate gradients. A calculator/graphing program illustrates the effect. The following two cases shall highlight the role played by the statistical distribution of the gas.

1) Maxwellian gases
2) Non-Maxwellian gases.

2. MAXWELLIAN GASES

The result presented in this section is a direct consequence of the Second Law and therefore can be considered obvious. However going through the analysis is instructive as it highlights the influence of a distribution on the thermal behavior of a gas, and provides an analytical template to be used in the next section on non-Maxwellian distributions.

Consider a column of Maxwellian gas subjected to a vertical downward force field. As molecules rise between collisions, they convert kinetic energy to potential energy and become cooler. Vice versa, when they descend, they convert potential energy into kinetic energy and become warmer. Should a temperature gradient, hot at the bottom and cold at the top, spontaneously develop in such a column? Paradoxically, according to the Second Law, a gas column in thermal equilibrium must be isothermal with elevation. Two explanations vindicating the Second Law are provided:

1. A phenomenological description of the phenomenon at the micro-level.
2. A statistical analysis explaining the phenomenon at a more formal level in terms of the Maxwell-Boltzmann distribution of the gas.

2.1 Phenomenological Description

Two effects operating at the micro-level cancel each other out thereby ensuring that the gas column remains isothermal with elevation. The first one tends to produce a negative thermal gradient and the second one, a positive gradient, equal and opposite to the first.

Considering the first effect, let us assume an initial stable isothermal gas column subjected to a potential energy gradient such as gravity. At the microscopic scale, molecules are moving and colliding with each other. Between collisions, (i.e., along their mean free paths) a molecule remains adiabatic. As it rises, it loses kinetic energy and, on the average, becomes colder than its environment until it collides and exchanges energy with another molecule at a higher elevation. Similarly, a descending molecule gains energy and, on the average, becomes hotter than its environment until its next collision at a lower elevation when it can exchange energy. Heat is extracted at the top of a molecular path and deposited at the bottom. The net effect is a downward heat transfer. The temperature distribution ceases to be isothermal, prompting the column to shift from an isothermal to an adiabatic gradient. By itself this effect would cause heat to flow from cold to hot in direct violation of the Second Law.

This second effect is analogous to mass separation by a mass spectrometer. In this case, gravity separates molecules according to their energies allowing those with a higher total energy to move up and those with lower energy to stay down. Those with the higher total energy (kinetic + potential) are found at higher elevation simply because they start out with more total energy. Molecules get vertically separated according to their total energies. This second effect opposes the first. In a Maxwell-Boltzmann gas[4] these two effects cancel exactly, with the result that the column shifts towards an isothermal state in conformance with the Second Law.

2.2 Statistical Explanation

The following section provides a rigorous explanation for the temperature invariance of a gas column, in terms of its statistical distribution as a function of elevation. This section is most useful in establishing and validating an analytic template to be used in the next section which discusses gases following the Fermi-Dirac distribution. The analysis includes the following subsections:

1. Normalized energy distribution for a Maxwellian gas.
2. Unnormalized Maxwell-Boltzmann distribution with a force field.
3. Renormalization of distribution for any given elevation.
4. Temperature as function of elevation.

2.2.1 Normalized Energy Distribution for a Maxwellian gas

The normalized energy distribution in a Maxwellian gas is well known[4] and given by

$$f_{MB}(E_k) = 2\left(\frac{1}{k_B T}\right)^{3/2} \sqrt{\frac{E_k}{\pi}} \exp\left(\frac{-E_k}{k_B T}\right) \tag{1}$$

This equation does not include any potential energy term, and therefore does not *a-priori* indicate the effect of a force field on the distribution.

2.2.2 Unnormalized Maxwell-Boltzmann Distribution with a Force Field

The derivation below begins with an intermediate result[4] but also includes the inclusion of a force field. In a Maxwellian gas column subjected to a force field each component of the velocity vector has a normal distribution.

$$f_{MB}(v_x) = \left(\frac{m}{2\pi k_B T}\right)^{1/2} \exp\left(\frac{-mv_x^2}{2k_B T}\right) \tag{2}$$

$$f_{MB}(v_y) = \left(\frac{m}{2\pi k_B T}\right)^{1/2} \exp\left(\frac{-mv_y^2}{2k_B T}\right) \tag{3}$$

$$f_{MB}(v_z) = \left(\frac{m}{2\pi k_B T}\right)^{1/2} \exp\left(\frac{-mv_z^2}{2k_B T}\right) \tag{4}$$

Please note that since[5] $\int_{-\infty}^{\infty} \exp(x^2)dx = \sqrt{\pi}$ the distributions expressed in equations (2), (3) and (4) are normalized. If the gas is subjected to a vertical force such as gravity acting along the z axis, then kinetic energy associated with this axis must decrease with elevation, keeping the total energy $E_T = E_k + E_p = \frac{1}{2}mv_z^2 + mgz$ constant. Therefore $f_{MB}(v_z)$ needs to be modified to include a Boltzmann exponential factor as follows:

$$f_{MB}(v_z, z)_{unormalized} = \left(\frac{m}{2\pi k_B T}\right)^{1/2} \exp\left(\frac{-\frac{1}{2}mv_z^2 - mgz}{k_B T}\right). \tag{5}$$

The resulting distribution in equation (5) is normalized for $z = 0$ (i.e., ground level) and unnormalized for $z \neq 0$. Splitting the exponent into kinetic and potential terms yields,

$$\begin{aligned} f_{MB}(v_z, z)_{unormalized} &= \left(\frac{m}{2\pi k_B T}\right)^{1/2} \exp\left(\frac{-mv_z^2}{2k_B T}\right) \exp\left(\frac{-mgz}{k_B T}\right) \\ &= f_{MB}(v_z, z=0) \exp\left(\frac{-mgz}{k_B T}\right) \end{aligned} \tag{6}$$

where $f_{MB}(v_z, z=0)$ is normalized because it is at ground level as expressed by equation (4).

The probability distribution of the vector $\overline{v}$ can be seen as the joint probability of three independent probabilities $f_{MB}(v_x)$, $f_{MB}(v_y)$ and $f_{MB}(v_z,z)$ along each degree of freedom. Hence

$$f_{MB}(\overline{v},z) = f_{MB}(v_x,v_y,v_z,z) = f_{MB}(v_x)f_{MB}(v_y)f_{MB}(v_z,z) \tag{7}$$

Therefore, substituting equations (2) , (3) and (4) into (7) produces an unnormalized distribution because $f_{MB}(v_z,z)$ is generally unnormalized:

$$f_{MB}(\overline{v},z)_{unnormalized}\,dv_x dv_y dv_z = \left(\frac{m}{2\pi k_B T}\right)^{3/2} \exp\left(\frac{-m(v_x^2+v_y^2+v_z^2)}{2k_B T}\right)\exp\left(\frac{-mgz}{k_B T}\right)dv_x dv_y dv_z \tag{8}$$

The distribution is a function of elevation z. To obtain the distribution for the kinetic energy E_k, the velocity components are expressed in spherical coordinates. A velocity volume element is given by

$$dv_x dv_y dv_z = v^2 \sin\theta d\theta d\phi dv \tag{9}$$

Integrating θ and ϕ , this volume element becomes:

$$dv_x dv_y dv_z = v^2 dv \int_0^{2\pi} d\phi \int_0^{\pi} \sin\theta d\theta = 4\pi v^2 dv \tag{10}$$

Substituting equation (10) into equation (8) produces:

$$f_{MB}(v,z)_{unnormalized}\,dv = \left(\frac{m}{2\pi k_B T}\right)^{3/2} \exp\left(\frac{-m(v^2)}{2k_B T}\right)\exp\left(\frac{-mgz}{k_B T}\right)4\pi v^2 dv \tag{11}$$

Since $dE_k = mvdv = \sqrt{2mE_k}\,dv$ and defining the potential energy as $E_p = mgz$ results in

$$f_{MB}(E_k,E_p)_{unnormalized} = 2\left(\frac{1}{k_B T}\right)^{3/2}\sqrt{\frac{E_k}{\pi}}\exp\left(\frac{-E_k}{k_B T}\right)\exp\left(\frac{-E_p}{k_B T}\right) \tag{12}$$

which is unnormalized except at ground level $z=0$ where $E_p = 0$.

$$f_{MB}(E_k,E_p=0) = 2\left(\frac{1}{k_B T}\right)^{3/2}\sqrt{\frac{E_k}{\pi}}\exp\left(\frac{-E_k}{k_B T}\right) \tag{13}$$

This fact can easily be proven[6] by integrating equation (13) from 0 to ∞ and knowing that the Gamma function $\int_0^{\infty} x^{1/2}\exp(x)dx = \frac{\sqrt{\pi}}{2}$.

2.2.3 Renormalization of Distribution for any Given Elevation.

Substituting equation (13) into equation (12) $f_{MB}(E_k,E_p)_{unnormalized}$ can be expressed as

$$f_{MB}(E_k, E_p)_{unnormalized} = f_{MB}(E_k, E_p = 0)\exp\left(\frac{-E_p}{k_B T}\right) \tag{14}$$

The normalization quotient for equation (14) is

$$\int_0^\infty f_{MB}(E_k, E_p)_{unnormalized}\, dE_k = \int_0^\infty f_{MB}(E_k, E_p = 0)\exp\left(\frac{-E_p}{k_B T}\right) dE_k$$
$$= \exp\left(\frac{-E_p}{k_B T}\right)\int_0^\infty f_{MB}(E_k, E_p = 0) dE_k \tag{15}$$

and since $\int_0^\infty f(E_k, E_p = 0) dE_k = 1$ the normalization quotient is simply

$$\int_0^\infty f_{MB}(E_k, E_p)_{unnormalized}\, dE_k = \exp\left(\frac{-E_p}{k_B T}\right) \tag{16}$$

Therefore dividing equation (14) by $\exp\left(\frac{-E_p}{k_B T}\right)$ produces the normalized version

$$f_{MB}(E_k, E_p) = f_{MB}(E_k, E_p = 0) \tag{17}$$

The resulting renormalized energy distribution for any elevation is identical to the ground level distribution equation (13)

$$f_{MB}(E_k) = 2\left(\frac{1}{k_B T}\right)^{3/2}\sqrt{\frac{E_k}{\pi}}\exp\left(\frac{-E_k}{k_B T}\right) \tag{18}$$

Elevation invariance indicates that the Maxwellian gas column is isothermal as predicted by the Second Law. This invariance is due to the exponential nature of the Boltzmann factor $\exp\left(-\frac{E_p}{k_B T}\right)$. An energy term added to the exponent is identical to an amplitude factor. Renormalization eliminates this factor and the resulting distribution is the same as the original.

2.2.4 Temperature as Function of Elevation.

The expected value of the temperature $<T>$ can be calculated by first expressing the expected energy as

$$< E_k > = \int_0^\infty E_k f_{MB}(E_k)\, dE_k \tag{19}$$

Inserting (18) into (19) yields

$$< E_k > = \int_0^\infty 2\left(\frac{1}{k_B T}\right)^{3/2} \sqrt{\frac{1}{\pi}}\, E_k^{3/2} \exp\left(\frac{-E_k}{k_B T}\right) dE_k \tag{20}$$

which can be expressed as a gamma function[6]

$$< E_k > = k_B T \frac{2}{\sqrt{\pi}} \Gamma(5/2) \tag{21}$$

Since $\Gamma(5/2) = \frac{3}{4}\sqrt{\pi}$ the above equation unsurprisingly reduces to

$$< E_k > = \frac{3}{2} k_B T \tag{22}$$

and since $\frac{3}{2} k_B < T > \; = \; < E_k >$ the expected temperature is

$$< T > = T \tag{23}$$

which is trivial, yet when the same procedure is applied to a non-Maxwellian gas, the result is unexpected. The steps followed in the analysis presented above are illustrated in the following figure which shows 1) the original distribution in the absence of any force field, shown in (a) as the red-thick curve; 2) the unnormalized distribution after the application of a force field shown in (a) as the blue-thin curve; and 3) the normalized distribution after the application of the force field shown in (b) as the blue-thin curve. Noticeably, the renormalized distribution in the presence of a force field is identical to the original one without a force field, indicating that the gas is isothermal.

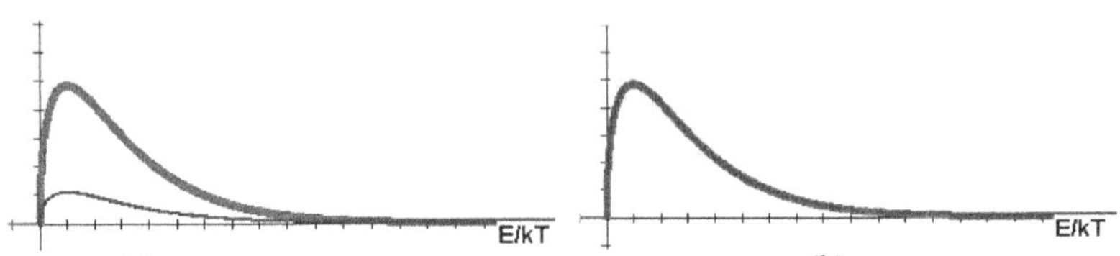

(a) (b)

Figure 1. Maxwell-Boltzmann distribution at two elevations before renormalization (a) and after renormalization (b). The distributions indicate that the gas is isothermal with elevation.

3. NON-MAXWELLIAN GASES

Non-Maxwellian particles in gas phase have energies distributed over a range of values as exemplified by the Fermi-Dirac or Bose-Einstein distributions. As shall be shown these distributions can be skewed if a force field is applied to the gas.

3.2 Fermi-Dirac Distribution

As discussed in the previous sections, Maxwellian gases in a potential energy gradient cannot develop a temperature gradient because of the exponential nature of the Boltzmann factor. This section will show that gases complying with Fermi-Dirac statistics can develop a temperature gradient. The same topics shall be addressed as in the previous sections.

1. Normalized energy distribution.
2. Unnormalized Fermi-Dirac distribution with a force field.
3. Renormalization of distribution for any given elevation.
4. Temperature as function of elevation.

3.2.1 Normalized energy distribution

Consider a Fermion gas embodied as electrical carriers confined to an n-doped semiconductor slab as shown in Figure 2.

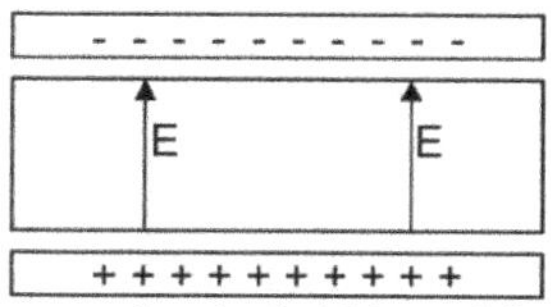

Figure 2. Electrical carriers are assumed to be held in a slab of high-Z thermoelectric material and subjected to an electrical field.

Carriers shall be considered to be electrons but the same reasoning presented here can be applied to holes. Furthermore, the slab is assumed to be thin enough that the carriers behave ballistically and that no significant space charge accumulates that would cancel the electric field.

The slab material is also assumed to have a high ZT factor such that the coupling between the carriers and the heat phonons in the crystal matrix is low, thereby allowing the electrons to behave mostly ballistically between the faces of the slab. As discussed by this author[1] this low coupling minimizes heat transfers between electrons and the solid matrix, thereby preventing thermal short circuits by heat phonons.

The heat capacity of electrical carriers determines to some extent the thermal properties of the material. When the carriers are electrons as in an n-doped semiconductor, the thermally significant electrons are those that happen to be in the conduction band and that are made available by the Fermi-Dirac distribution. Such electrons must have a total energy E_t greater than the potential energy of the band's bottom E_c, i.e. $E_t > E_c$. In the conduction band, the electron's density of states is given by[7,8]

$$g_c(E_t) = \frac{8\pi\sqrt{2}}{h^3} m^{3/2} \sqrt{E_t - E_c} \tag{24}$$

and the Fermi-Dirac distribution can be expressed as

$$F_{FD}(E_t) = \frac{1}{1+\exp\left(\dfrac{E_t - E_F}{k_B T}\right)} \tag{25}$$

In accordance with Landauer-Boltzmann formalism, the electron density as a function of energy $n(E_t)$ in the slab is given by the product of the carrier density of states and the Fermi distribution as shown in Figure 3. Keeping the same nomenclature as in the sections above, the Fermion distribution is defined as $f_{FD}(E_t)_{unnormalized} = n(E_t)$. Hence,

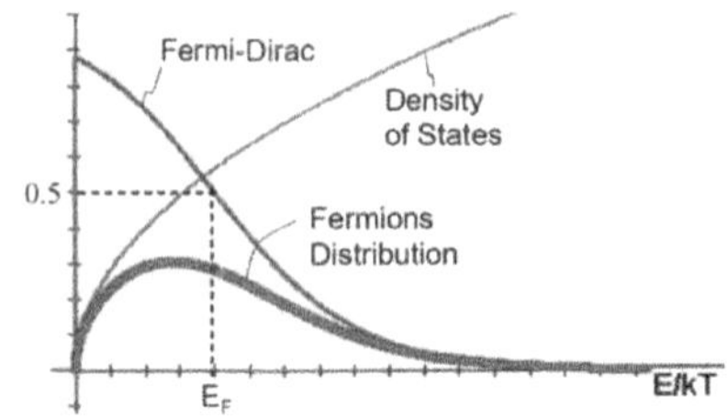

Figure 3. Fermion energy distribution is the product of the Fermi-Dirac distribution and the density of states.

$$f_{FD}(E_t)_{unnormalized} = n(E_t) = g_c(E_t)F_{FD}(E_t) \tag{26}$$

Substituting equation (24) and (25) into equation (26) yields

$$f_{FD}(E_t)_{unnormalized} = \frac{8\pi\sqrt{2}}{h^3} m^{3/2}\sqrt{E_t - E_c}\ \frac{1}{1+\exp\left(\frac{E_t - E_F}{k_B T}\right)} \tag{27}$$

Dividing equation (27) by a normalization quotient n_T yields the normalized distribution

$$f_{FD}(E_t) = \frac{f_{FD}(E_k)_{unnormalized}}{n_T} = \frac{1}{n_T}\frac{8\pi\sqrt{2}}{h^3} m^{3/2}\sqrt{E_t - E_c}\ \frac{1}{1+\exp\left(\frac{E_t - E_F}{k_B T}\right)} \tag{28}$$

where n_T is the total number of particles obtained by integrating $n(E_k)$ from $E_k = 0$ to $E_k = \infty$.

$$n_T = \int_0^\infty \frac{8\pi\sqrt{2}}{h^3} m^{3/2}\sqrt{E_t - E_c}\ \frac{1}{1+\exp\left(\frac{E_t - E_F}{k_B T}\right)} dE_t \tag{29}$$

Substituting equation (29) into (28) produces

$$f_{FD}(E_t) = \frac{\frac{8\pi\sqrt{2}}{h^3} m^{3/2}\sqrt{E_t - E_c}\ \frac{1}{1+\exp\left(\frac{E_t - E_F}{k_B T}\right)}}{\int_0^\infty \frac{8\pi\sqrt{2}}{h^3} m^{3/2}\sqrt{E_t - E_c}\ \frac{1}{1+\exp\left(\frac{E_t - E_F}{k_B T}\right)} dE_t} \tag{30}$$

Cancelling constant terms yields

$$f_{FD}(E_t) = \frac{\sqrt{E_t - E_c}\,\dfrac{1}{1+\exp\left(\dfrac{E_t - E_F}{k_B T}\right)}}{\int_0^\infty \sqrt{E_t - E_c}\,\dfrac{1}{1+\exp\left(\dfrac{E_t - E_F}{k_B T}\right)} dE_t} \tag{31}$$

Expressing kinetic energy as $E_k = E_t - E_c$ and substituting into equation (31) produces:

$$f_{FD}(E_k) = \frac{\sqrt{E_k}\,\dfrac{1}{1+\exp\left(\dfrac{E_k + E_c - E_F}{k_B T}\right)}}{\int_0^\infty \sqrt{E_k}\,\dfrac{1}{1+\exp\left(\dfrac{E_k + E_c - E_F}{k_B T}\right)} dE_k} \tag{32}$$

Defining a normalization constant A as

$$A = \left[\int_0^\infty \sqrt{E_k}\,\frac{1}{1+\exp\left(\dfrac{E_k + E_c - E_F}{k_B T}\right)} dE_k\right]^{-1} \tag{33}$$

and substituting in equation (31) generates

$$f_{FD}(E_k) = A\sqrt{E_k}\,\frac{1}{1+\exp\left(\dfrac{E_k + E_c - E_F}{k_B T}\right)} \tag{34}$$

3.2.2 Unnormalized Fermi-Dirac Distribution with a Force Field

Let an electrical field be applied across the slab. This field may be produced either internally e.g., as with a junction, or externally e.g., as with insulated capacitive plates as shown in Figure 2. The electrical potential energy induced by the field can be defined as a function of position z along the field, that is, across the slab. i.e., $(-e)V(z)$ and needs to be added to the potential energy term E_c in the Fermi exponential. Modifying equation (34) accordingly produces the unnormalized distribution as a function of E_k and V.

$$f_{FD}(E_k, V)_{unnormalized} = A\sqrt{E_k}\,\frac{1}{1+\exp\left(\dfrac{E_k + (-e)V + E_c - E_F}{k_B T}\right)} \tag{35}$$

As shown in Figure 4 the electrical field shifts the Fermi-Dirac distribution to the left.

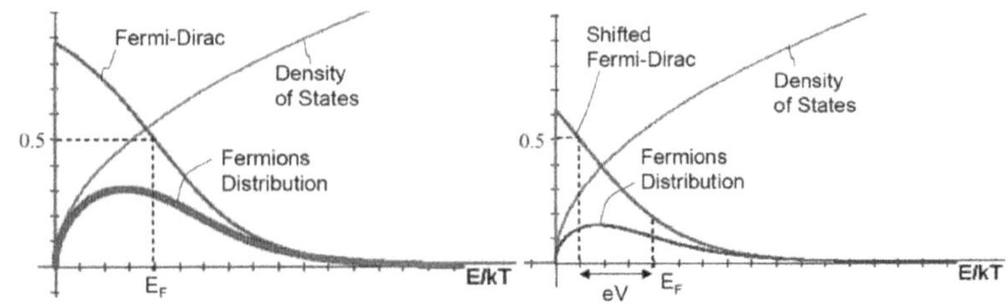

(a) (b)

Figure 4. Fermion energy distribution as a product of the Fermi-Dirac distribution and of the density of states, at low elevation (a) and at high elevation (b). The Fermi-Dirac distribution is shifted to the left by a potential $(-e)V = E_F$ thereby affecting the Fermion distribution.

3.2.3 Renormalization of distribution for any given elevation

The normalized distribution is obtained by multiplying equation (35) by a normalization factor. This factor is obtained by inserting the electrical potential $(-e)V$ into equation (33),

$$A(V) = \left[\int_0^\infty \sqrt{E_k} \frac{1}{1+\exp\left(\frac{E_k + (-e)V + E_c - E_F}{k_B T}\right)} dE_k \right]^{-1} \quad (36)$$

and the renormalized distribution is

$$f_{FD}(E_k, V) = A(V)\sqrt{E_k} \frac{1}{1+\exp\left(\frac{E_k + (-e)V + E_c - E_F}{k_B T}\right)} \quad (37)$$

which, unlike the Maxwell distribution, also expresses the non-isothermal property of the gas. There is no closed form expression for (36) and the best way to solve it is numerically. Figure 5 shows (a) the unnormalized distributions on the left, and (b) the normalized distributions on the right.

The similarities and differences between equations (18) and (37) are interesting. As E_k goes to infinity, the exponential Fermi-Dirac terms on the right of (37) becomes Maxwellian and the two equations become identical except for the normalization constant term in front. One must also note that potential energy $(-e)V$ is embedded inside the Fermi-Dirac expression and cannot be factored out and eliminated by renormalization as was done in equation (18). Therefore $f_{FD}(E_k, V)$ is not invariant with position along the potential gradient as was the case with the Maxwell-Boltzmann distribution.

3.2.4 Temperature Gradient

Temperature – like heat capacity – is determined by the active carriers whose distribution is shown in Figures 4 (a) and (b). This distribution is the product of the density of states and the Fermi distribution as expressed in equation (37). As can be seen from the figures, the temperature gradient is expected to change with the applied potential V. The expected temperature $<T(V)>$ is proportional to the first moment of the kinetic energy distribution of the active carriers. In other words, since $E_k = E_t - E_c = \frac{3}{2}k_B T$, the expected temperature is

$$<T(V)>=\frac{2}{3k_B}<E_k>=\frac{2}{3k_B}\int_0^\infty E_k\, f(E_k,V)dE_k \tag{38}$$

Substituting equation (37) into equation (38) and moving the $A(V)$ term out of the integral produces

$$<T(V)> =\frac{2A(V)}{3k_B}\int_0^\infty (E_k)^{3/2}\frac{1}{1+\exp\left(\frac{E_k+(-e)V+E_c-E_F}{k_B T}\right)}dE_k \tag{39}$$

Please note that T refers to the temperature of the Fermions at ground level (where the potential V is zero) and $<T(V)>$ is the expected temperature where the potential is V. There is no closed form expression for the above integral. The best way to solve this equation is numerically. Figure 5 corresponds to the steps followed in the analysis. The figure shows (red-thick curve) in (a) and (b) the original distribution in the absence of any force field. The figure also depicts (blue-thin curve) in (a) the unnormalized distribution after the application of a force field; in (b) the normalized distribution after the application of the force field. The renormalized distribution is different from the original one, indicating a variation in the temperature of the gas with elevation.

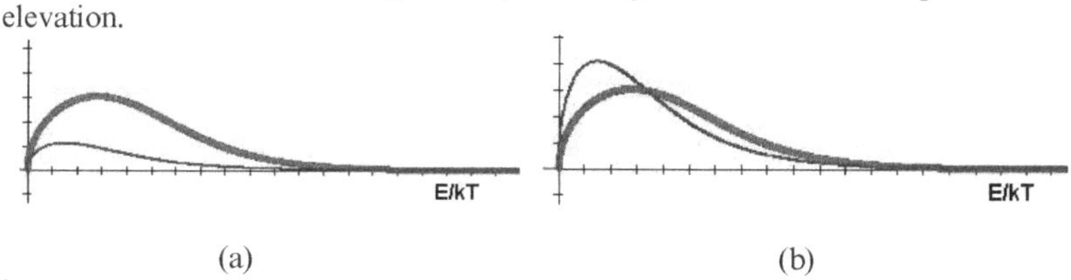

Figure 5. Fermion energy distribution at two elevations before renormalization (a) and after renormalization (b). The distributions indicate that the Fermions are not isothermal with elevation.

The temperature gradient produced in a non-Maxwellian gas (Fermions) column by a force field is further discussed by this author [9] in relation to a quantum game using Fermions in an energy well. In the game, the odds are shifted with respect to classical odds depending on the elevation of the Fermions in the well. Analogously, kinetic energy is shifted leading to a spontaneous temperature gradient in the column.

4. CONCLUSION

An explanation is proposed for mysterious voltage offsets observed by several researchers during meticulous thermoelectric experiments of high ZT materials showing the Seebeck curve not going through the origin.

The particle statistical distribution is identified as being responsible for this phenomenon. A Maxwellian gas remains isothermal with elevation because of the exponential dependence of the Boltzmann factor on potential energy and in conformance with the Second Law.

However, stepping outside of classical physics produces unexpected results. Fermions distribution is not purely exponential with potential energy. When renormalized, this distribution is not invariant with elevation and therefore indicates a temperature gradient. More generally, any non-Maxwellian gas in a force field is capable of producing a temperature gradient.

REFERENCES AND NOTES

1. Levy, G. Thermoelectric Effects under Adiabatic Conditions, *Entropy* **2013**, *15*, 4700-4715; doi:10.3390/e15114700 available on line at http://www.mdpi.com/1099-4300/15/11/4700, last checked on 10/29/2014.
2. Iwanaga, S.; Toberer, E.S.; LaLonde, A.; Snyder G.J. A high temperature apparatus for measurement of the Seebeck coefficient. *Rev. Sci. Instrum.* **2011**, *82*, pages. 063905-1 to 6
3. Snyder, J. Material Science, Caltech, Private communication on 2012/01/20 and 2012/04/09.
4. Maxwell-Boltzmann Distribution; available on line at http://en.wikipedia.org/wiki/Maxwell%E2%80%93Boltzmann_distribution (last accessed on 4 January 2015).
5. List of integrals of exponential functions; available on line at http://en.wikipedia.org/wiki/List_of_integrals_of_exponential_functions; last accessed on 4, January 2015.
6. Gamma Function; available on line at http://en.wikipedia.org/wiki/Gamma_function; last accessed on 4 January 2015.
7. Fermi-Dirac Statistics; available on line at http://en.wikipedia.org/wiki/Fermi%E2%80%93Dirac_statistics (last accessed on 4, January, 2014.
8. van Zeghbroeck, Bart; available on line at http://ecee.colorado.edu/~bart/book; last accessed on 4 January 2015.
9. Levy, G. Quantum Game Beats Classical Odds - Thermodynamics Implications, *Entropy* **2015**, *17*(11) 7645-7657; doi: 10.3390/e17117645.

THERMOPHYSICAL PROPERTY OF POLY-Si PHONONIC CRYSTALS FOR THERMOELECTRICS

Masahiro Nomura[1,2,3] and Oliver Paul[3]
[1]Institute of Industrial Science, the University of Tokyo
Tokyo, JAPAN
[2]Institute for Nano Quantum Information Electronics, the University of Tokyo
Tokyo, JAPAN
[3]Department of Microsystems Engineering (IMTEK), University of Freiburg
Freiburg, GERMANY

ABSTRACT

Thermal conduction and phonon transport in both single-crystalline and polycrystalline Si two-dimensional phononic crystal (PnC) nanostructures were investigated. The impact of phononic patterning on thermal conductivity was larger in polycrystalline Si PnCs than in single-crystalline Si PnCs. The difference in the impact is attributed to the difference in the thermal phonon mean free path (MFP) distribution induced by grain boundary scattering in the two materials. The grain boundaries and phononic patterning are efficient phonon scattering mechanisms for different MFP length scales. This multiscale phonon scattering structure covers a large part of the broad distribution of thermal phonon MFPs and thus efficiently reduces thermal conduction. Electrical and thermoelectric properties of polycrystalline Si 2D PnC nanostructures were also investigated at room temperature. The concept of phonon-glass and electron-crystal is valid in the investigated electron and phonon transport systems. The thermal conductivity is more sensitive than the electrical conductivity to the nanopatterning due to the longer mean free path of the thermal phonons than that of the charge carriers. The enhancement of the figure of merit *ZT* were factors of 2 and 4 for the p-doped and n-doped thin films, respectively, when the characteristic size of the phononic crystal nanostructure is below 100 nm.

INTRODUCTION

Thermal transport in nanostructured semiconductors has recently been investigated intensively because of its importance both in fundamental physics and practical applications[12]. Characteristic phonon transport phenomena have been reported in a variety of systems such as semiconductor super lattices[34], membranes containing nanoparticles[5], porous nanostructures[678], nanowires[9], and phononic crystal (PnC) structures[1011121314]. The phonon mean free path (MFP) is the most important physical parameter for determining the thermophysical PROPERTY of a system. The thermal conductivity of a crystal in a bulk material is reduced when the characteristic length of the system becomes comparable to or shorter than the phonon MFP in the bulk material. In such a system, the thermal property no longer follows the Fourier law but instead is described by semiballistic phonon transport. Phonon transport physics is especially interesting in PnC nanostructures because the wave nature of phonons appears[15] where coherence is preserved, and together with the incoherent phonon scattering process[16], determines phonon transport.

Thermoelectrics is an important application, in which phonon transport nanoengineering plays an active role. The figure of merit *ZT* is given by $S^2\sigma T/\kappa$, where T is temperature, S is the Seebeck coefficient, σ is electrical conductivity, and κ is thermal conductivity. The concept of "phonon-glass and electron-crystal" proposed by Slack yields the following strategy for increasing *ZT*: reduce thermal conduction while maintaining electrical conduction by taking advantage of different phonon and electron MFPs[17]. Many groups have demonstrated very low values of thermal conductivity by increasing phonon scattering in a variety of nanostructures1. However, for this purpose, it is important to be aware that thermal phonons are distributed over a broad range of frequencies[18,19], and thus there is a multiscale distribution of thermal phonon MFPs. Therefore, a phonon scattering mechanism that covers the whole MFP range is required in order to reduce thermal conductivity efficiently. Biswas *et al.* demonstrated the validity of this idea using an all-scale hierarchical architecture in which three different scattering mechanisms scattered phonon transport and covered different thermal phonon MFP ranges from the atomic scale to the mesoscale by using PbTe[20]. Si is considered one of the most promising candidates of a high performance thermoelectric material with a low environmental load. Therefore, it is important to investigate the validity of the strategy for Si and to obtain quantitative information for structural design.

In this study, we investigated the use of all-scale hierarchical architecture in Si for efficient reduction of thermal conduction. We fabricated two-dimensional (2D) PnC nanostructures using both single-crystalline and polycrystalline Si membranes and measured thermal conductivity at room temperature. We discuss the validity of multiscale phonon scattering in Si, using grain size analysis data. In addition, the electrical and thermal conductivities of p-doped and n-doped polycrystalline Si thin films with phononic nanopatterning were measured to investigate the enhancement of the thermoelectric performance. The enhancement of the thermoelectric performance is discussed from the viewpoint of nanoscale electron and phonon transport on the basis of the thermal and electrical conductivity data and the information on the grain size distribution. This data and discussion will be useful to enhance the thermoelectric performance of thin-film based thermoelectric power generators by nanopatterning.

EXPERIMENTAL

The single-crystalline Si samples were fabricated by using a commercially available (100) nominally boron-doped silicon-on-insulator wafer with a 145 nm-thick upper Si layer and a 1 μm-thick SiO_2 buried oxide layer. The poly-crystalline samples for electrical and thermal conductivity measurements were fabricated with polycrystalline Si thin films. The polycrystalline Si films were grown on a 1.5-μm-thick SiO_2 layer on a (100) Si wafer by low-pressure chemical vapor deposition. The dopants for the p-doped and n-doped thin films are phosphorus and boron, respectively. All the samples were annealed at 1050°C for 30 minutes in a nitrogen atmosphere to activate the dopant atoms and for better crystallization. The thickness of each film was measured by ellipsometry and it is about 143 nm. The PnC nanostructures were formed via electron beam lithography using a reactive-ion etching inductively coupled plasma system, with SF_6/O_2 gas as the etchant. For the samples for the electrical conductivity measurement, 250-nm-thick Al electrodes were deposited by electron beam deposition. The electrodes are used to connect the sample with a four probe measurement configuration. For the samples for the thermal conductivity measurement, the oxide layer under the Si thin film was removed with hydrofluoric acid after the nanopatterning in order to form the suspended structures. In the center of the structure, a 125-nm-thick Al layer was deposited to form a 4 μm × 4 μm pad on top of the Si layer, enabling thermoreflectance measurements. The dimensions of the micro- and nanostructures were measured with a scanning electron microscope (SEM).

Figure 1(a) shows an SEM image of the whole suspended structure with the 2D PnC structures. The Al pad in the center enables thermoreflectance measurements. The central Si island with the Al pad atop was then supported by two PnC structures. The width and the length of the structure were 5 μm and 25 μm, respectively. The 2D PnC structures were formed using circular holes periodically aligned as a square lattice, as shown in Fig. 1(b). PnC structures with a variety of radii were fabricated with a fixed period a = 300 nm.

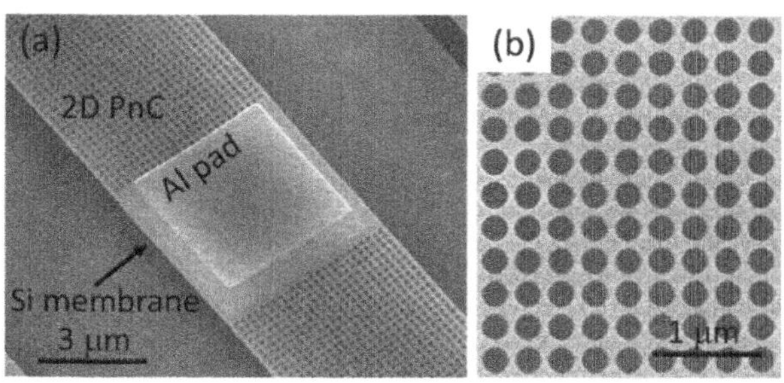

Figure 1. SEM images of (a) the whole suspended Si PnC nanostructure and (b) a 2D PnC nanostructure with circular holes aligned as a square lattice (a = 300 nm and r = 115 nm).

To measure in-plane thermal conductivities of the PnC nanostructures, a method using micro time-domain thermoreflectance (μ-TDTR) was developed, which has very high throughput as compared to electrical methods. The principle of this technique is the same as that for TDTR[21]; however, it can be applied to micrometer-sized systems. The Al pad was heated by a quasicontinuous laser beam (wavelength λ = 642 nm) for 500 ns, and the temporal evolution of the temperature of the Al pad (the TDTR signal) was monitored by a continuous-wave laser beam (λ = 785 nm). Both beams were collinearly focused on the Al pad, as shown in Fig. 2(a), by using a microscope objective lens with a numerical aperture of 0.65. The beam spot on the pad was approximately 700 nm in diameter.

The power of the pump pulse was set so the temperature increase was less than 5 K in the Al pad, and it was verified that the pump power did not change the thermal conductivity in the range of temperature increase. The details of the measurement can be found in our previous publication21. We roughly estimated the thermal radiation loss from the whole suspended structure by Stefan-Boltzmann law. The energy of the heating pulse, which causes temperature increase of 2 K, is about 60 pJ and the estimated radiant energy within the typical decay time of 10 μs is about 3 fJ, with an emissivity of 0.1. The radiant energy is on the order of 10^{-4} of the heating pulse energy and it is confirmed that the thermal radiation is not the main heat dissipation channel under the experimental condition. All the measurements were performed at room temperature and in a vacuum chamber to eliminate heat-convection loss. The PnC nanostructures that supported the central Si island were the only heat dissipation channels. This well-defined system made it possible to obtain a highly reliable value of the thermal conductivity by analyzing the temporal evolution of the TDTR signal. The temperature evolution in whole structure was simulated with exactly the same structure as shown in Fig. 2(b) via the finite element method (FEM) by using COMSOL Multiphysics®. All the structural parameters, the length and width of the suspended structure and the radius of the holes, were measured by SEM measurement. The TDTR signals were simulated by the thermal diffusion equation. In the simulation, κ is the only parameter for the unknown thermal conductivity of the PnC structures. The thermal conductivities were obtained by fitting the TDTR signal by a simulated temperature evolution using the least square method. Figure 2(b) shows the simulated temperature distribution, 15 μs after pulse heating, for a PnC structure with r = 115 nm.

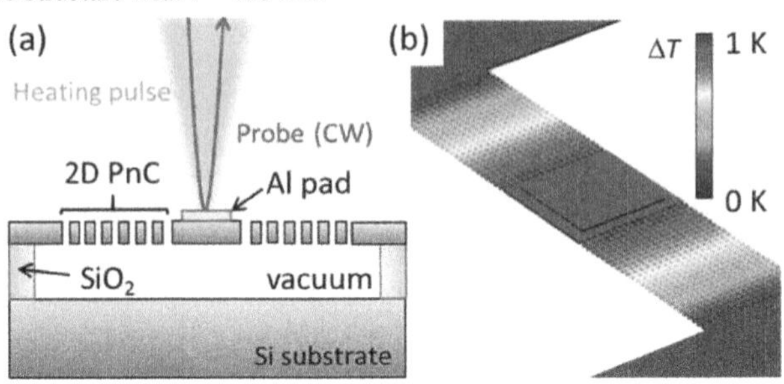

Figure 2. (a) Schematic of the -TDTR technique and cross section of the suspended structure. (b) Simulated temperature distribution at 15 μs after pulse heating a PnC structure with a = 300 nm and r = 115 nm.

RESULTS AND DISCUSSION

For both single-crystalline and polycrystalline samples, the thermal conductivity of the PnC nanostructure was measured by μ-TDTR at room temperature. Figure 3 shows TDTR signals for single-crystalline PnCs with radius of the holes r = 73 nm, 115 nm, and 126 nm, as well as for an unpatterned membrane. The dots and the solid lines represent experimental and simulated TDTR traces, respectively. The increase in the TDTR signal was caused by the temperature increase in the Al pad from the heating pulse, which was applied from time t = 0 to t = 500 ns. The heat was only dissipated through the PnC nanostructures, and the temperature of the Al gradually decreased on a timescale on the order of tens of microseconds. The best fitting curve for each PnC with r = 73, 115, and 126 nm, as well as for the unpatterned membrane, is given by thermal conductivities of 41 $Wm^{-1}K^{-1}$, 35 $Wm^{-1}K^{-1}$, 29 $Wm^{-1}K^{-1}$, and 75 $Wm^{-1}K^{-1}$, respectively.

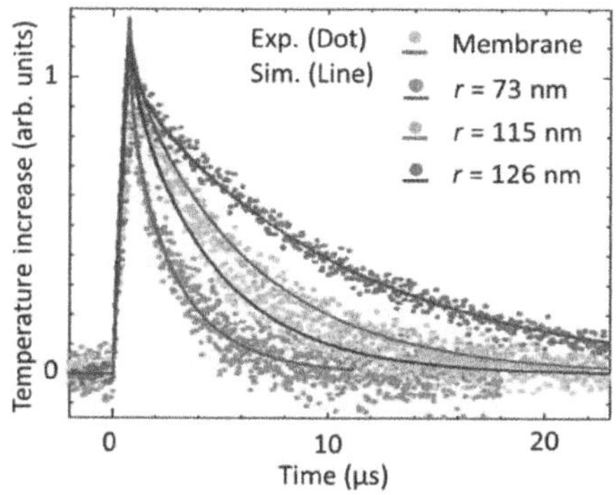

Figure 3. TDTR signals for the unpatterned membrane and the single-crystalline Si PnC nanostructures (r = 73 nm, 115 nm, and 126 nm). The dots and solid lines represent experimental and simulated TDTR signals, respectively.

Thermal conductivity was also measured in the polycrystalline PnC structures. Figure 4 summarizes the thermal conductivities obtained by FEM simulation for both types of crystalline Si PnC nanostructure, showing a decrease in thermal conductivity as the radius of the holes decreases. This indicates that the characteristic lengths of the PnC nanostructures were comparable to the MFPs of thermal phonons and that the system was in the semiballistic phonon transport regime. We found that forming holes, even when the value of r/a was small, dramatically reduced thermal conductivity. This result indicates that the side walls of the holes, which are perpendicular to the direction of in-plane phonon transport, backscatter phonons and largely reduce thermal conductivity. For single-crystalline Si, the thermal conductivity showed a steep reduction above r = 115 nm. In this regime, where the neck size, a − 2r, was smaller than the thickness (145 nm), the necking effect[7] became more effective and dramatically reduced thermal conductivity. This tendency had also been observed in our previous measurements for Si nanowires[22]. On the other hand, the polycrystalline sample does not show the steep reduction. This is because the grain boundary scattering is more frequent than surface scattering in the range of r. The steep reduction may be observed above r = 140 nm, where the neck size is similar to the grain size.

For polycrystalline PnC structures, the measured thermal conductivity for the membrane was 10.5 Wm−1 K−1. The radius dependence was similar to that of the single-crystalline samples below r

= 115 nm, but the impact of the phononic patterning was different because of the different thermal phonon MFP distributions. The magnitudes of the reduction in thermal conductivity for single-crystalline and polycrystalline Si PnCs can be compared by considering the data around r = 100 nm, where the necking effect is moderate. By interpolating the experimental plots, the thermal conductivities at r = 100 nm plots were estimated as 38 $Wm^{-1}K^{-1}$ and 4.3 $Wm^{-1}K^{-1}$ for the single-crystalline and the polycrystalline samples, respectively. The magnitudes of the reductions in membrane thermal conductivity were 49% and 59%, respectively. This indicates that the impact of the phononic patterning is larger for polycrystalline Si PnC nanostructures than for single-crystalline nanostructures.

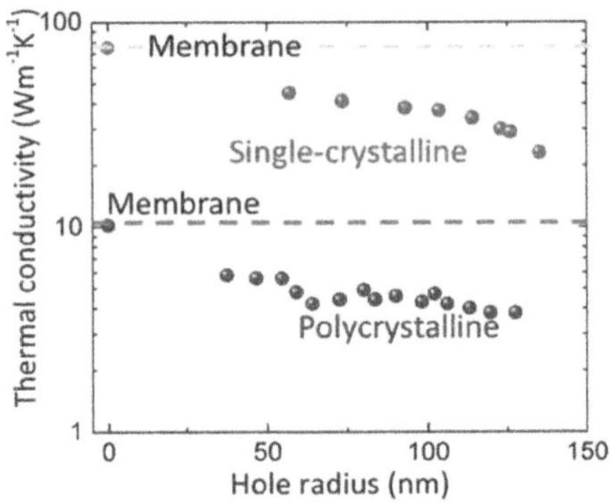

Figure 4. Measured thermal conductivities for a variety of radii in single- and polycrystalline Si PnC nanostructures.

The difference in the magnitudes of the reduction in thermal conductivity can be explained by examining nano- and mesoscopic phonon transport. Figures 5(a) shows the top image of the investigated undoped sample taken by a transmission electron microscope (TEM). The data shown in Figs. 5(b) and 5(c) for the p- and n-doped samples are used later. The grain size analysis on the surface is summarized in the histogram shown in Fig. 5(d). The grain sizes were mostly distributed around the 20 nm mark and with sizes on the order of a few nanometers to up to 60 nm. Thermal phonons, which have MFPs that are similar in size to the grains in single-crystalline Si (1–60 nm), are considered to be scattered more frequently by grain boundaries in polycrystalline Si than in single-crystalline Si. Therefore, thermal phonon MFPs are distributed over a larger MFP regime in polycrystalline Si than in single-crystalline Si. However, the characteristic lengths of the PnC structures, which are approximately the size of the neck (>50 nm), shorten the phonon MFPs to this length. The latter mechanism of phonon scattering is more efficient than the former for phonons with MFPs longer than the neck size. Therefore, the crystal grain and phononic patterning cover different ranges of phonon MFPs, which results in efficient phonon transport scattering. Thus, this multiscale phonon scattering is a Si analog of the all-scale hierarchical architecture demonstrated by Biswas *et al.* using PbTe (Ref.

20).

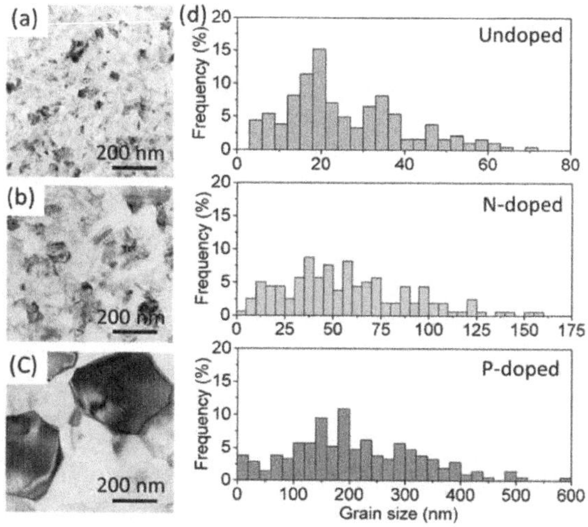

Figure 5. TEM images of top view for (a) undoped, (b) p-doped, and (c) n-doped polycrystalline Si thin films. (d) Histogram of the grain size measured on the surface for each thin film.

We investigated thermoelectric properties of the Si multiscale hierarchical architecture by measuring electric and thermal conductivity. The thermal conductivity of 2D PnC nanostructures were measured by the same method. For the electrical measurement, we fabricated 2D PnC on an SiO_2 layer. Figures 6(a) and 6(b) show an SEM image and a schematic cross section of the sample for the electrical conductivity measurement. The widths and lengths of the PnCs are about 3 μm and 9.5 μm, respectively. The 2D PnC nanostructures were formed using circular holes aligned periodically in a square lattic. The PnC structures with a variety of radii were fabricated with a fixed period of 300 nm. The electrical resistance of the PnCs were measured by the conventional four probe measurement technique. The electrical conductivities of the nanostructures were extracted from the resistance data by reproducing the measurement using a finite element model built with COMSOL Multiphysics®. Figure 6(c) shows the normalized simulated electric potential distribution under a static bias condition.

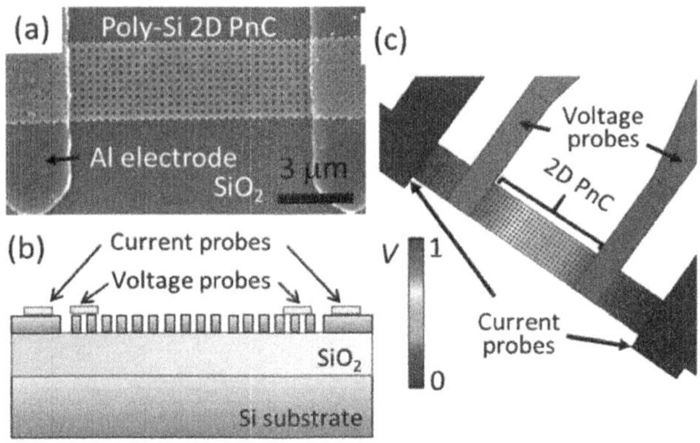

Figure 6. (a, b) SEM images and schematic cross section of the Si PnC nanostructures ($a = 300$ nm and radius $r = 110$ nm) for the four probe measurement (c) Normalized simulated electric potential distribution.

Figure 7(a) shows the obtained electrical conductivities for the PnC structures with different hole radii. The data points at $r = 0$ nm are for the unpatterned membranes. The value of σ is decreasing as the holes become larger due to the enhancement of surface scattering by the sidewalls of the holes. The creation of a depletion region in the vicinity of the holes may also contribute to the reduction. However, the contribution seems small in our 2D PnC nanostructures with neck size (defined as the shortest distance between two adjacent hole edges i.e. $a - 2r$), wider than 80 nm, because all the data points show a monotonic decrease between $r = 0$ to 110 nm. Figure 7(b) shows the obtained thermal conductivities for the PnC nanostructures with different hole radii. The thermal conductivity decreases as the neck size is decreased due to a stronger necking effect, which is observed in the semi-ballistic and ballistic phonon transport regimes. The reduction of thermal conductivity can be explained by incoherent phonon transport, since at room temperature the wave nature of phonons is not significant. The continuity between the unpatterned thin film and PnC nanostructure also results from the transport characteristics. In Fig. 7(a), the electrical conductivity shows an almost linear decrease, while the thermal conductivity in Fig. 7(b) shows large discontinuity. This data indicates that the charge carriers have much shorter MFPs than the characteristic size of the PnC nanostructures and their transport is diffusive, while the thermal phonon transport is semi-ballistic.

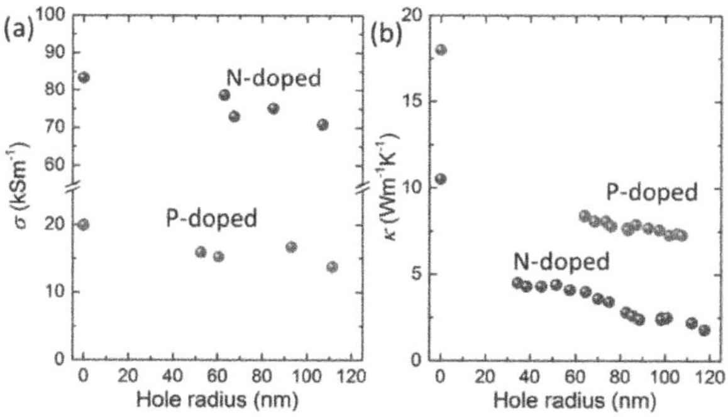

Figure 7. Measured electrical conductivities (a) and thermal conductivities (b) for the p-doped and n-doped polycrystalline Si PnC nanostructures with a variety of hole radii at 295 K.

Figure 8(a) shows the calculated values of power factor $S^2\sigma$ for the p-doped and n-doped unpatterned thin films and PnC nanostructures using the values of Seebeck coefficient S = 240 μV/K and -81 μV/K for the p-doped and n-doped thin films, which were obtained in our previous work[23]. We assumed that the phononic patterning does not change the value of S, which is the sum of a diffusion part and a phonon drag part, because the PnC nanostructure is not significantly small to change the electronic band structure of the crystal. The phonon drag part originates from the momentum transfer from the phonon system to the electron system by electron-phonon scattering. In our heavily doped samples, the phonon drag effect is negligible because of the high impurity and charge concentrations[24]. The power factor decreases more rapidly for the p-doped thin film as the radius of the holes increases, while the n-doped thin film maintains its value. Figure 8(b) shows the values of *ZT* for the p-doped and n-doped unpatterned thin films and PnC nanostructures. It was not possible to fabricate the PnC nanostructures with exactly the same hole radii for the electrical and thermal conductivity measurements. Therefore, the values of *ZT* were calculated for the four structures used in the electrical conductivity measurement with the estimated thermal conductivities from the densely recorded data shown in Fig. 7(b). The concept of 'phonon-glass and electron-crystal' is valid, and the values of *ZT* are successfully enhanced by factors of about 2 and 4 for the p-doped and n-doped samples, respectively, with phononic patterning.

For more detailed discussion on the observed enhancement of *ZT* by the nanopatterning, we analyzed the grain boundary distributions for the undoped, p-doped and n-doped thin films, using the surface images taken by a transmission electron microscope (TEM). The images are shown in Figs. 5(a–c) and the grain size distribution for each sample is shown in Fig. 5(d). The distributions for the undoped and p-doped samples are similar, but the sizes of the grains in the n-doped samples are much larger. This fact, and the lower thermal conductivity for the n-doped samples, indicates that the phonon transmission probability at the grain boundaries in the n-doped samples is much lower than that in the p-doped samples.

Thermal phonon transport in polycrystalline Si is complex, because many different phonon scattering mechanisms, such as phonon-phonon, surface, impurity, defect, and grain boundary scatterings, affect the thermal conduction. There are many previous studies on the thermal conduction in polycrystalline Si thin films, because of its importance in the field of MEMS[25,26,27]. In polycrystalline Si, the grain boundaries are very efficient phonon scattering interfaces, and the transmittance of phonons is quite difficult to estimate. Moreover, the transmittance depends on the frequency and vibrational mode of the phonons. The useful information for the discussion is that phosphorus segregates and migrates to the grain boundaries in the n-doped polycrystalline Si, while boron is uniformly distributed in the grains in the p-doped polycrystalline Si[28]. The observed large increase of σ/κ for the n-doped samples, as observed in Fig. 8(b), seems to result from the strong phonon scattering by the phosphorous atoms at the grain boundaries.

The nanopatterning with the neck size below 100 nm can enhance the value of *ZT* by a factor of 3, on average, and yields an estimated value *ZT* ~ 0.05 at room temperature. For the p-doped sample, the data point of σ at r = 93 nm is out of the monotonic decreasing trend in Fig. 7(a). Therefore, it is difficult to conclude that further enhancement can be expected or not, but it seems that the value of *ZT* is almost saturated. For the n-doped sample, the monotonically increased *ZT* value in Fig. 8(b) indicates that further enhancement is expected.

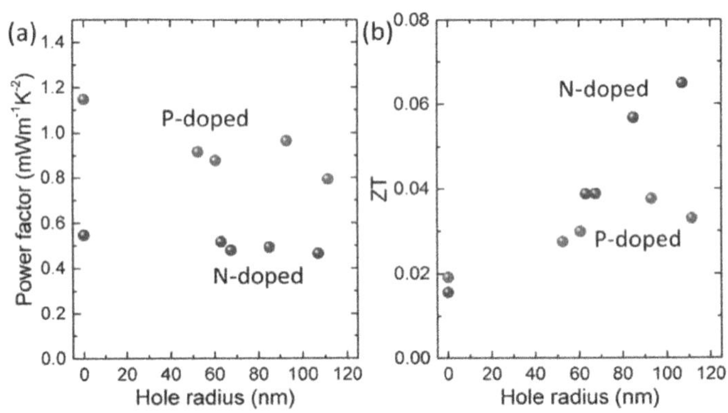

Figure 8. Power factor (a) and *ZT* (b) for the p-doped and n-doped polycrystalline Si PnC nanostructures with a variety of hole radii at 295 K.

CONCLUSIONS

In-plane thermal conductivity and phonon transport in single-crystalline and polycrystalline Si PnC nanostructures were investigated at room temperature. The impact of the phononic patterning was larger for polycrystalline membranes than for single-crystalline membranes. The multiscale phonon scattering described in this work is a Si analog of all-scale hierarchical architecture, which has been demonstrated to improve thermoelectric performance in PbTe. We conclude that it is important to scatter phonons over the whole range of thermal phonons MFPs by using appropriate materials and structural designs. The electrical and thermoelectric properties of 145-nm-thick polycrystalline Si thin films were also investigated. An enhancement of *ZT* by phononic patterning of a factor of about 3 was demonstrated at room temperature. It was found that both the p-doped and n-doped samples follow the principle of the phonon-glass and electron-crystal for PnC nanopatterning with neck size below 100 nm. These results provide useful information to enhance the thermoelectric property of thin-film based thermoelectric material by nanopatterning.

ACKNOWLEDGEMENTS

We thank J. Maire, Y. Kage, and D. Moser for the technical assistance, and J. Nakagawa, K. Yoshida and K. Hirakawa for the van der Pauw measurement. This work was supported by the Project for Developing Innovation Systems of the Ministry of Education, Culture, Sports, Science and Technology (MEXT), Japan, by KAKENHI (grant no. 25709090), by the Foundation for the Promotion of Industrial Science, and by Maeda Corporation.

REFERENCES

1. Vineis, C. J., Shakouri, A., Majumdar, A. & Kanatzidis, M. G. (2010). Nanostructured thermoelectrics: big efficiency gains from small features. Adv. Mater. **22,** 3970–80.
2. Shi, L. (2012) Thermal and Thermoelectric Transport in Nanostructures and Low-Dimensional Systems. Nanoscale Microscale Thermophys. Eng. **16,** 79–116.
3. Narayanamurti. (1979) Selective transmission of high-frequency phonons by a superlattice. Phys. Rev. Lett. **43,** 2012.
4. Luckyanova, M. N. et al. (2012) Coherent phonon heat conduction in superlattice. Science **338,** 936.
5. Rowe, D. M., Shukla, V. S. & Savvides, N. (1981) Phonon scattering at grain boundaries in heavily doped fine-grained silicon–germanium alloys. Nature **290,** 765–766.
6. Bera, C., Mingo, N. & Volz, S. (2010) Marked Effects of Alloying on the Thermal Conductivity of Nanoporous Materials. Phys. Rev. Lett. **104,** 115502.
7. Tang, J. et al. Holey silicon as an efficient thermoelectric material. (2010) Nano Lett. **10,** 4279–83.
8. Kashiwagi, M. et al. Enhanced figure of merit of a porous thin film of bismuth antimony telluride. (2011) Appl. Phys. Lett. **98,** 023114.
9. Hochbaum, A. I. et al. (2008) Enhanced thermoelectric performance of rough silicon nanowires. Nature **451,** 163–7.
10. Yu, J.-K., Mitrovic, S., Tham, D., Varghese, J. & Heath, J. R. (2010) Reduction of thermal conductivity in phononic nanomesh structures. Nat. Nanotechnol. **5,** 718–21.

11. Hopkins, P. E. et al. (2011) Reduction in the thermal conductivity of single crystalline silicon by phononic crystal patterning. Nano Lett. **11,** 107–12.
12. Zen, N., Puurtinen, T. a., Isotalo, T. J., Chaudhuri, S. & Maasilta, I. J. (2014) Engineering thermal conductance using a two-dimensional phononic crystal. Nat. Commun. **5,** 1–9.
13. Nomura, M. et al. (2015) Thermal phonon transport in silicon nanowires and two-dimensional phononic crystal nanostructures. Appl. Phys. Lett. **106,** 143102.
14. Nomura, M. et al. (2015) Impeded thermal transport in Si multiscale hierarchical architectures with phononic crystal nanostructures. Phys. Rev. B **91,** 205422.
15. Olsson III, R. H. & El-Kady, I. (2009) Microfabricated phononic crystal devices and applications. Meas. Sci. Technol. **20,** 012002.
16. Dechaumphai, E. & Chen, R. (2012) Thermal transport in phononic crystals : The role of zone folding effect Thermal transport in phononic crystals : The role of zone folding effect. J. Appl. Phys. **111,** 073508.
17. G. A. Slack. (1995) CRC Handbook of Thermoelectrics, ed. D. M. Rowe (CRC Press, Boca Raton, Florida, 1995).
18. Minnich, A. et al. (2011) Thermal Conductivity Spectroscopy Technique to Measure Phonon Mean Free Paths. Phys. Rev. Lett. **107,** 1–4.
19. Regner, K. T. et al. (2013) Broadband phonon mean free path contributions to thermal conductivity measured using frequency domain thermoreflectance. Nat. Commun. **4,** 1640.
20. Biswas, K. et al. (2012) High-performance bulk thermoelectrics with all-scale hierarchical architectures. Nature **489,** 414–8.
21. Cahill, D. G. (2004) Analysis of heat flow in layered structures for time-domain thermoreflectance. Rev. Sci. Instrum. **75,** 5119–5122.
22. Maire, J. & Nomura, M. (2014) Reduced thermal conductivities of Si one-dimensional periodic structure and nanowire. Jpn. J. Appl. Phys. **53,** 06JE09.
23. Moser, D., Ilkaya, D., Kopp, D. & Paul, O. (2012) Determination of the thermoelectric figure of merit of doped polysilicon thin films by micromachined test structures. IEEE Sens. Conf. 989.
24. Weber, L. & Gmelin, E. (1991) Transport Properties of Silicon. Appl. Phys. A Mater. Sci. Process. **53,** 136–140.
25. Von Arx, M., Paul, O. & Baltes, (1998) H. Test structures to measure the heat capacity of CMOS layer sandwiches. IEEE Trans. Semicond. Manuf. **11,** 217–224.
26. Volklein, F. & Baltes, H. A. (1992) Microstructure For Measurement Of Thermal Conductivity Of Polysilicon Thin Films. J. Microelectromechanical Syst. **1,** 193.
27. McConnell, A. D., Uma, S. & Goodson, K. E. (2001) Thermal conductivity of doped polysilicon layers. J. Microelectromechanical Syst. **10,** 360–369.
28. Mandurah, M. M., Saraswat, K. C., Helms, C. R. & Kamins, T. I. (1980) Dopant segregation in polycrystalline silicon. J. Appl. Phys. **51,** 5755–5763.

THE POTENTIAL OF MAXIMAL *ZT*-VALUE FOR THERMOELECTRIC MATERIALS OF Mn11Si19 HMS PHASE BY CALCULATING ELECTRONIC STRUCTURE

Akio Yamamoto[1], Koichi Kitahara[1], Hidetoshi Miyazaki[2], Manabu Inukai[2] and Tsunehiro Takeuchi[1,3,4]

[1] *Toyota Technological Institute, Nagoya 468-8511, Japan*

[2] *Department of Frontier Materials, Nagoya Institute of Technology, Nagoya 466-855, Japan*

[3] *PRESTO, Japan Science and Technology Agency, Chiyoda, Tokyo 102-0076, Japan*

[4] *GREMO, Nagoya University, Nagoya 464-8603, Japan*

ABSTRACT

The band structure and electron transport properties of $Mn_{11}Si_{19}$ higher manganese silicide (HMS), which solely consists of cheap and nontoxic elements and possess high *ZT*-value exceeding 0.5 at around 1000 K, were calculated to reveal the characteristics in electronic structure and their roles in making the electron transport properties suitable for thermoelectric materials. The value of $A = S^2 T\rho^{-1}\kappa_{el}^{-1}$, which indicates the maximum *ZT* obtainable for a given electronic structure, reached 4.06 at 1000 K, suggesting that the magnitude of *ZT* could be increased up to ~ 2 if the lattice thermal conductivity is effectively reduced to a small value comparable to the electron thermal conductivity. Here, S, ρ, and κ_{el} represent the Seebeck coefficient, electrical conductivity, and electron thermal conductivity, respectively. This large A is realized in association with the wide band gap, the presence of multiple bands near the edge of valence band, and the corresponding sharp peaks in the electronic density of states near the chemical potential.

INTRODUCTION

Thermoelectric generator, which is capable of recovering waste heat into useful electricity, has attracted considerable interest as a key technology to reduce the consumption of fossil fuels and the emission of global warming gases. However, the poor efficiency of energy conversion η, relatively high cost of thermoelectric materials, and their toxicity have prevented us from utilizing thermoelectric power generators in a variety of practical applications[1,2,3].

The value of η is an increasing function of dimensionless of figure of merit $ZT = S^2\sigma T/(\kappa_{el} + \kappa_{lat})$.[4] Here, S, σ, κ_{el} and κ_{lat} represent the Seebeck coefficient, electrical conductivity, electron thermal conductivity, and lattice thermal conductivity, respectively. For promoting the use of thermoelectric generators in practical applications, we need to develop thermoelectric materials possessing a large value of *ZT*. Besides, it would be more preferable for the thermoelectric materials to solely consist of cheap, and ubiquitous elements because a huge number of thermoelectric generators should be required to efficiently recovering energy from the diffusively disappearing waste heat.

Recently, thermoelectric materials possessing a relatively large ZT were developed in alloy systems consisting solely of cheap, environmental friendly elements: higher manganese silicide (HMS, $MnSi_{\sim 1.75}$), $NaCo_2O_4$ [5,6], $Ca_3Co_4O_9$ [7], TiS_2 [8], half-Heusler ZrNniSn [9], Colusites $Cu_{26}V_2M_6S_{32}$ (M = Ge, Sn) [10], Tetrahedrite $Cu_{10.5}Ni_{1.5}Sb_4S_{13}$ [11], etc. Among these thermoelectric materials we have focused, in this study, on HMS, which was reported to possess a large Seebeck coefficient exceeding 150 μV/K and small values of electrical resistivity less than 2 mΩ cm at the temperature range from 300 K to 800 K[12]. Though the lattice thermal conductivity is slightly large at 2-3 $W^{-1}mK^{-1}$ [12], these characteristics of electron transport properties naturally lead to a large magnitude of $ZT = 0.42$ at 700 K[12].

To reveal the mechanism leading to the electron transport properties suitable for thermoelectric materials, we have calculated the electronic structure of HMS using the first principles band calculation and have evaluated the electron transport properties in the context of Boltzmann transport equation. To confirm the reliability of calculations, we compared the calculated thermoelectric properties with the measured ones of the sample prepared at the composition. We have also investigated the temperature dependence of chemical potential because it dominantly works as a driving force to increase the magnitude of the Seebeck coefficient.

EXPERIMENTAL AND COMPUTATIONAL PROCEDURES

Appropriate amount of pure element powders, 99.9 % Mn and 99.999 % Si, were mixed together for 30 minutes using alumina mortar and pestle so as the nominal composition of prepared sample to be Mn_4Si_7. Powders were pressed into a disk-shaped pellet (φ10×25 mm) at the pressure of 20 MPa. Mother ingot was prepared by melting the pellet using an arc furnace in pressurized Ar gas atmosphere. After annealing at 1200 K for 3 days in vacuum, the mother ingot was crushed into powders and sintered into bulk using pulsed current sintering in vacuum under the condition of 1223 K at 70 MPa for 20 minutes.

The phases involved in a sample were identified by conventional powder X-ray diffraction (XRD) measurements using a Cu-Kα radiation source and a Bragg-Brentano-type diffractometer equipped in BRUKER D8 Advance. The density was evaluated by the Archimedes method using pure water, and it was confirmed that the relative density of the sintered sample was high enough to be 98 % of theoretical one.

The Seebeck coefficient and electrical resistivity of the prepared HMS sample were measured in a wide temperature range from 300 to 1000 K by the steady-state method and four-probe method, respectively, using ULVAC Model ZEM-3 M8 under a low-pressure helium atmosphere. The thermal conductivity was measured from 300 to 1000 K by the laser-flash method using NETZSCH, LFA457 Microflash.

We also employed first-principles band calculations to obtain information about the electronic structure of HMS. The distributed package programs, WIEN2K [13], was employed in this study. It uses the full potential linearized augmented plane-wave (FLAPW) method, and the electron correlation was taken into account by the generalized gradient approximation (GGA). The energy dependence of $\sigma(\varepsilon,T)\ /\ \tau(\varepsilon,T)\ =\ (e^2/3)N(\varepsilon)v_G^2(\varepsilon)$ was also calculated using the distributed program named BoltzTraP.[14] Here, $N(\varepsilon)$, $v_G(\varepsilon)$, $\tau(\varepsilon)$ represent electronic density of states, group velocity, and relaxation time of electrons, respectively.

RESULTS

The experimental and calculated XRD patterns of HMS are shown Figure 1 (a). Several different chemical formulas, such as Mn_4Si_7, $Mn_{11}Si_{19}$, $Mn_{15}Si_{26}$, $Mn_{27}Si_{47}$, were proposed for HMS phases. The difference of crystal structures were characterized by different stacking periods in two sub-lattices consisting solely of Mn or Si. The Pearson symbols and space groups of Mn_4Si_7, $Mn_{11}Si_{19}$, $Mn_{15}Si_{26}$, and $Mn_{27}Si_{47}$ are (tP44 and $P\bar{4}c2$), (tP120 and $P\bar{4}n2$), (tP164 and $I\bar{4}2d$), and, (tP296 and $P\bar{4}n2$), respectively[15-18]. By comparing the calculated XRD patterns of each structure with that experimentally obtained, we realized that the some of peaks of Mn_4Si_7 in the low angle were not observed in the experimentally obtained XRD pattern. This fact suggests that the structure of Mn_4Si_7 should be eliminated from the structure model of present HMS. However, other three structures showed the very similar powder XRD patterns with the measured one, and therefore, we did not succeed in selecting most plausible one from the qualitative analysis on the powder XRD patterns.

A typical scanning electron microscopy (SEM) image taken for the present sample was shown in Figure 1 (b). The high density of prepared sample (98% of theoretical value), which was concluded from the density measurement, was further confirmed in the SEM image that possessed no evidence of voids. The chemical compositions were determined using energy dispersive x-ray spectroscopy (EDX) at the position from 1 to 4 were almost same values and we confirmed that prepared sample was highly homogeneous. The compositions estimated from the EDX measurement were summarized in Table 1. The averaged composition was very close to that of $Mn_{11}Si_{19}$. Therefore we selected this structure for the electronic structure calculation.

The ε-$\boldsymbol{k}$ dispersion for $Mn_{11}Si_{19}$ calculated by WIEN2K is shown in Figure 2. The calculated electronic structure clearly shows that the HMS possesses a wide band gap of 840 meV in width and the multiple bands existing near the band edge.

According to the linear response theory, the electrical resistivity $\sigma(T)$, Seebeck coefficient $S(T)$, and electron thermal conductivity $\kappa_{el}(T)$ are expressed by the following equations[19].

$$\sigma(T)=\int\sigma(\varepsilon,T)\left(-\frac{\partial f_{FD}(\varepsilon,T)}{\partial\varepsilon}\right)d\varepsilon\ , \qquad (1)$$

$$S(T) = -\frac{1}{|e|T}\frac{\int \sigma(\varepsilon,T)(\varepsilon-\mu)\left(-\frac{\partial f_{FD}(\varepsilon,T)}{\partial\varepsilon}\right)d\varepsilon}{\int \sigma(\varepsilon,T)\left(-\frac{\partial f_{FD}(\varepsilon,T)}{\partial\varepsilon}\right)d\varepsilon}, \quad (2)$$

$$\kappa_{el}(T) = \frac{1}{|e|^2 T}\int \sigma(\varepsilon,T)(\varepsilon-\mu)^2\left(-\frac{\partial f_{FD}(\varepsilon,T)}{\partial\varepsilon}\right)d\varepsilon - S^2\sigma T, \quad (3)$$

Here $\sigma(\varepsilon, T)$ represents spectral conductivity that is given by the following formula.

$$\sigma(\varepsilon,T) = \frac{|e|}{4\pi^3\hbar}\int_{S_\varepsilon}\frac{\vec{v}_k(\varepsilon)\cdot\vec{v}_k(\varepsilon)\tau(\varepsilon,T)}{v_\perp(\varepsilon)}dS_\varepsilon \quad (4)$$

$\vec{v}_k(\varepsilon)$, $v_\perp(\varepsilon)$, and $\tau(\varepsilon, T)$ represent group velocity of electron, vertical component of group velocity at equi-energy surface, and relaxation time of electron respectively. Equations (1)-(3) suggest that we have to obtain information both of temperature dependent chemical potential $\mu(T)$ and spectral conductivity $\sigma(\varepsilon, T)$ to estimate electrical resistivity, Seebeck coefficient, and electron thermal conductivity. We obtained such quantities using the FLAPW-GGA band calculation for the $Mn_{11}Si_{19}$ HMS.

The electronic density of states $N(\varepsilon)$ and the normalized spectral conductivity $\sigma_{normed}(\varepsilon) = \sigma(\varepsilon, T) / \tau$ were shown in Figure 3. The Fermi energy was expressed as $\varepsilon = 0$. Since the temperature dependence of chemical potential provides the non-homogenous distribution of conduction electrons under the presence of temperature gradient to increase Seebeck coefficient, we calculated the chemical potential of HMS and plotted it as a function of temperature in Figure 4. The chemical potential drastically increased with increasing temperature because of the drastically decreasing electron density of states with increasing energy at ε_F, and at 1000 K it eventually reaches 136 meV. This value means that the chemical potential moves inside of band gap at 720 K. This significant temperature dependence of chemical potential naturally leads to a large magnitude of Seebeck coefficient. The Seebeck coefficient calculated using $\sigma_{norm}(\varepsilon) = \sigma(\varepsilon, T) / \tau(T)$ is shown in Figure 5. The calculated $S_{\sigma norm}$ reproduced well the measured one at wide temperature range.

By considering that the calculated electronic structure is reliable from the good fit between calculated and measured *S*, we estimated temperature dependence of relaxation time using the calculated and measured electrical resistivity. The consequently obtained relaxation time shown in Fig.6 possesses the order of 10^{-14} seconds, which is definitely comparable with that of other metallic materials[20]. The relaxation time clearly shows decreasing tendency with increasing temperature, but its temperature dependence is getting weaker with increasing temperature and almost no variation was observable at high temperatures above 800 K. This fact presumably indicates that the conduction electrons stay in the strongest scattering limit known as Mott-Ioffe-Regel limit[21,22] at high temperature above 800 K.

We also estimated the electron thermal conductivity κ_{el}^{total} using the relaxation time. The calculated electron thermal conductivity were shown in Fig.7 (a) together with the components calculated with the first term κ_{el}^{1st} and second term κ_{el}^{2nd} of Eq. (3). The electron thermal conductivity possessed weak temperature dependence and small values less than 0.4 $Wm^{-1}K^{-1}$ over the whole temperature range of the present calculation. The measured thermal conductivity κ_{meas} and calculated lattice thermal conductivity $\kappa_{lat} = \kappa_{meas} - \kappa_{el}^{total}$ were shown in Fig.7 (b). The estimated κ_{lat} shows behaviors generally observable for many materials: it moderately decreases with increasing temperature and the temperature dependence was getting weaker at high temperature. The decreasing tendency of κ_{lat} with increasing temperature is naturally attributed to the reduction of phonon mean free path at high temperature.

Table I. Compositions obtained from the EDX measurements (unit: at.%) for SEM image in Fig.1 (b)

Position	Mn	Si
1	36.62	63.38
2	36.69	63.31
3	36.72	63.28
4	36.60	63.40
Averaged composition	36.66	63.34
Nominal composition	36.36	63.64

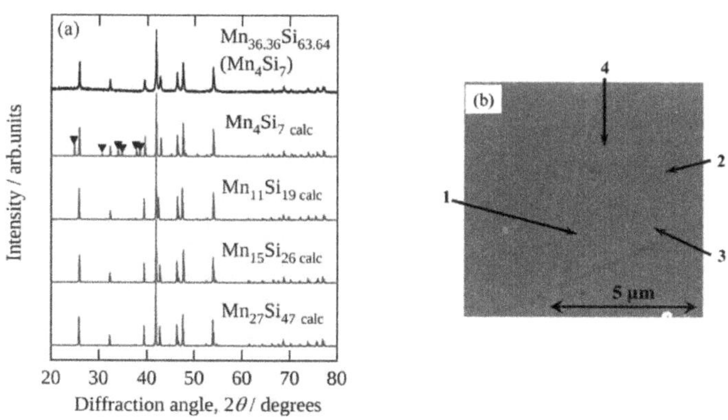

Figure 1. The experimental and calculated XRD patterns for HMS were shown in (a). (b) show SEM images obtained for sample prepared at the nominal compositions.

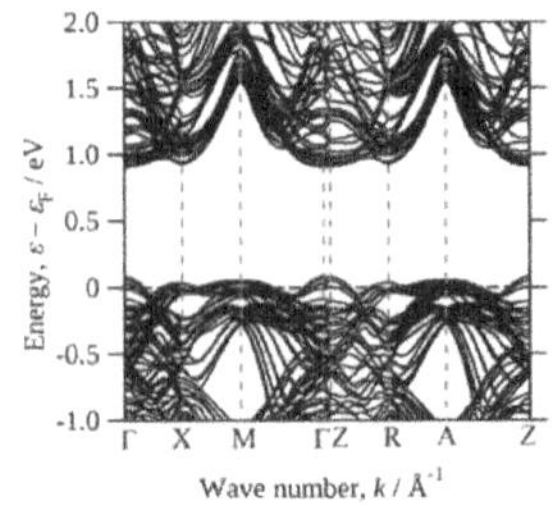

Figure 2. Band structure of $Mn_{11}Si_{19}$ (HMS). Several bands exist in the narrow energy range of a few hundred meV from both edge of the energy gap.

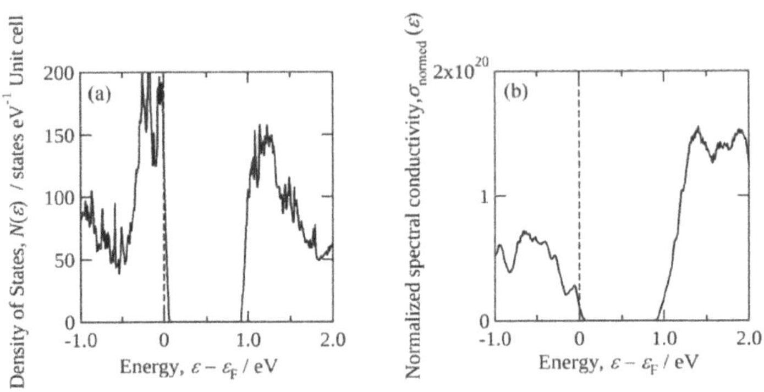

Figure 3. (a) and (b) represent Electronic density of states, $N(\varepsilon)$ and the normalized spectral conductivity, $\sigma_{\mathrm{normed}}(\varepsilon)$, for $Mn_{11}Si_{19}$ (HMS), respectively.

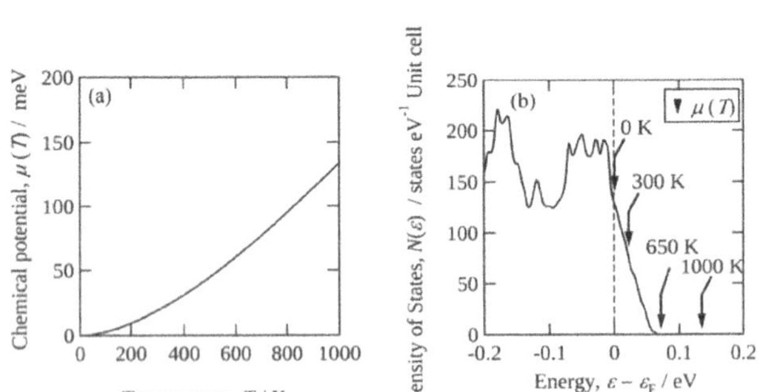

Figure 4. (a)Temperature dependence of chemical potential $\mu(T)$. (b) The electronic density of states $N(\varepsilon)$ together with $\mu(T)$ at $T=0$ K, 300 K, 650 K and 1000 K.

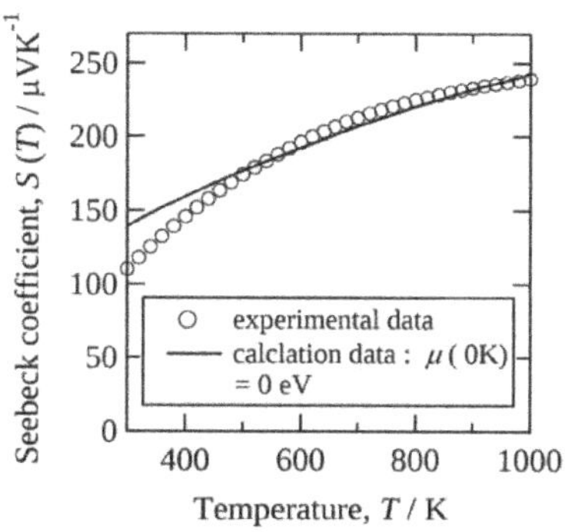

Figure 5. The experimental Seebeck coefficient at temperature range from 300 K to 1000 K together with calculated Seebeck coefficient at the energy of $\mu(0\ K) = 0$ meV.

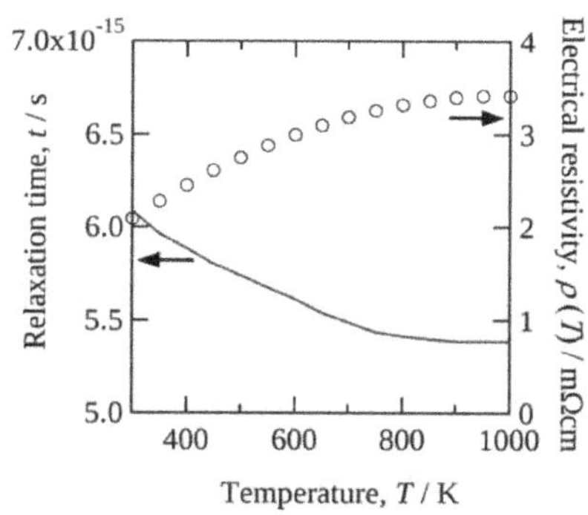

Figure 6. The experimental electrical resistivity and the calculated relaxation time of electron for HMS.

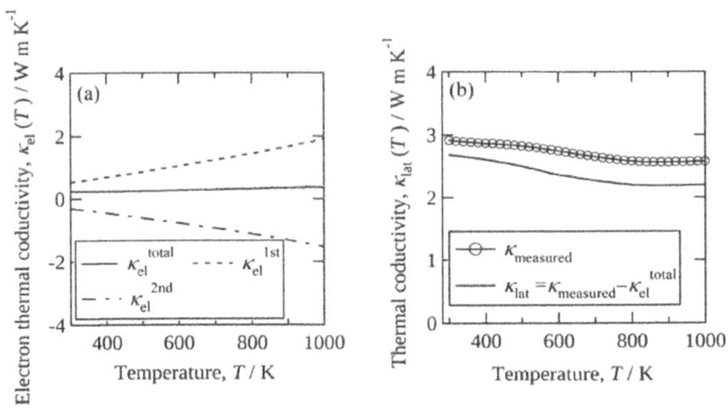

Figure 7. (a) The electron thermal conductivity κ_{el}^{total} were calculated by $N(\varepsilon)$ and $\sigma_{normed}(\varepsilon)$ and $\tau(T)$ together with the first and second term in Eq.(3). (b) The calculated lattice thermal conductivity were calculated by $\kappa_{measured}$ minus calculated κ_{el}.

DISCUSSION

We discuss the potential of $Mn_{11}Si_{19}$ (HMS) as a practical thermoelectric material. We evaluated $A = S^2\sigma T/\kappa_{el}$ and $B = (1+\kappa_{lat}/\kappa_{el})^{-1}$, and the resulting values are shown in Fig.8. The former indicates maximum ZT obtainable for a given electronic structure[19], and the latter is related to the concept of Electron Crystal & Phonon Glass[23]. T. Takeuchi[19] proposed three conditions leading to a large A-factor: (I) wide band gap, (II) multiple band existing near the band edge in the energy range where the electrons contribute the electrical properties, and (III) a large peak in electronic density of states staying at a few $k_B T$ apart from the edge of energy gap. The $Mn_{11}Si_{19}$ definitely possesses these characteristics. The estimated value of A-factor exceeded 4 at 1000 K. This value suggested that if the lattice thermal conductivity were reduced to be comparable with the electron thermal conductivity, we would have a large ZT-value exceeding 2.0. Unfortunately, however, the lattice thermal conductivity of the present sample was rather large at 2.2~2.7 WmK^{-1}, and these values were 6 ~ 11 times larger than that of electron thermal conductivity. This condition naturally leads to small value of B-factor less than 0.14, and hence the value of ZT is limited below 0.6. It is therefore of great importance to reduce its lattice thermal conductivity without affecting electronic structure for increasing ZT of HMS. We are now trying to do it and the result will be reported soon in elsewhere.

The present calculation strongly suggested that HMS certainly possesses characteristics of electronic structure leading to the electron transport properties suitable for practical thermoelectric materials. This fact suggests that we should find potential thermoelectric materials by carefully observing electronic structure near the chemical potential.

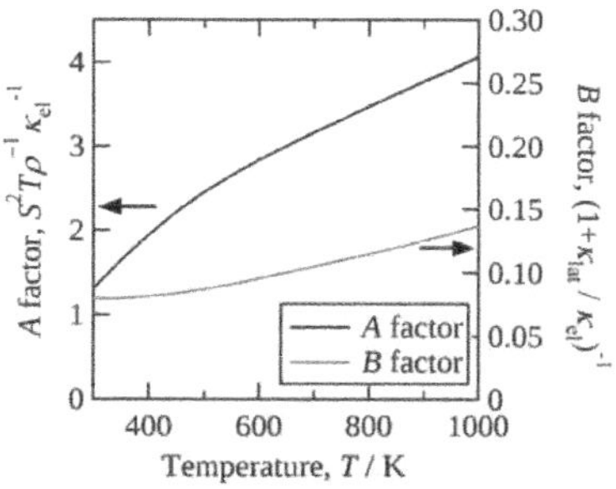

Figure 8. A-factor ($=S^2T\rho^{-1}\kappa_{el}^{-1}$) and B-factor [$=(1+\kappa_{lat}/\kappa_{el})^{-1}$] were calculated by $\sigma(\varepsilon, T)$ and $\mu(T)$.

CONCLUSION

In this work, we calculated the electronic structure and electron transport properties of $Mn_{11}Si_{19}$ HMS using FLAPW-GGA band calculation and Boltzmann transport equation to reveal the potential of $Mn_{11}Si_{19}$ as a practical thermoelectric material. The reliability of calculated electronic structure was confirmed from the good agreement between the calculated Seebeck coefficient and the measured one. Temperature dependence of relaxation time $\tau(T)$ was estimated from the electrical resistivity $\rho(T)$ using the calculated $\rho(T)\tau(T)$. The electron thermal conductivity $\kappa_{el}(T)$ and the lattice thermal conductivity $\kappa_{lat}(T)$ were also estimated from the measured thermal conductivity $\kappa(T)$, the calculated $\kappa_{el}(T)/\tau(T)$, and $\tau(T)$ determined from the electrical resistivity. These analyses suggested that the HMS is characterized by (a) large Seebeck coefficient in association with the strong temperature dependence of chemical potential, (b) the rather small electron thermal conductivity in association with the large power factor, and (c) weak but finite variation of relaxation time both of electrons and phonons. The estimated value of $A= S^2\sigma T/\kappa_{el}=4.0$ at 1000 K suggest that HMS has an electronic structure suitable for thermoelectric materials, but the slightly large magnitude of lattice thermal conductivity prevented us from obtaining large magnitude of ZT exceeding unity.

REFERENCES

[1] H. J. Goldsmid, A. R. Sheard, and D. A. Wright, *Brit. J. Appl. Phys.*, **9** 365 (1958).
[2] G.E. Smith, R Wolfe, *J. Appl. Phys.*, **33** 841-846 (1962).
[3] R. W. Fritts, *in Thermoelectric Materials and Devices*, 143-162 (1960).
[4] A. F. Ioffe, *Semiconductor Thermoelements and Thermoelectric cooling*, Infosearch limited,

London, 1-35 (1957).

[5] E. Maeda and M. Ohtaki, Trans. Mater. *Res. Soc. Jpn.*, **25** 237 (2000).

[6] M. Ito, T. Nagira, D. Furumoto, S. Katsuyama, and H.Nagai, *Scr. Mater.*, **48** 403 (2003).

[7] R. Funahashi, I. Matsubara, H. Ikuta, T. Takeuchi, U. Mizutani, and S. Sodeoka, *Jpn. J. Appl. Phys.*, **39** L1127 (2000).

[8] M. Ohta, S. Satoh, T. Kuzuya, S. Hirai, M. Kunii and A. Yamamoto Acta Materialia, **60** 20 pp. 7232-7240 (2012)

[9] S. Sakurada and N. Shutoh, *Appl. Phys. Lett.*, **86** 082105 (2005).

[10] K. Suekuni, F. S. Kim, H. Nishiate, M. Ohta, H. Tanaka and T. Takabatake, *Appl. Phys. Lett.*, **105** 132107 (2014)

[11] K. Suekuni, K. Tsuruta, M. Kunii, H. Nishitae, E. Nishibori, S. Maki, M. Ohta, A. Yamamoto and M. Koyano, *J. Appl. Phys.*, **113** 043712 (2013).

[12] Y. Miyazaki, Y. Saito, K. Hayashi, K. Yubuta and T. Kajitani, *Jpn. J. Appl. Phys.*, **50** 035804 (2011)

[13] P. Blaha, K. Schwarz, G. K. H. Madsen, D. Kvasnicka, J. Luitz, WIEN 2K, An Augmented Plane Wave + Local Orbitals Program for Calculating Crystal Properties, edited by Karlheinz Schwarz, Techn. Universitat, Wien, Austria, 2001. ISBN-3-9501031-1-2.

[14] G. K. H. Madsen and D. J. Singh, *Comput. Phys. Commun.*, **175** 67 (2006).

[15] U. Gottlieb, A. Supuce, B. Lambert-Andron and O Laborde, *J. Alloys Compd.*, **361** 13 (2003).

[16] O. Schwomma, H. Nowotny and A. Wittman, *Mantatsch. Chem.*, **94** 681 (1963).

[17] H. W. Knott, M. H. Mueller, and L. Heaton, *Acta Crystallogr.*, **23** 549 (1967).

[18] G. Zwilling and H. Nowotny, *Mantatsch. Chem.*, **104** 668 (1973).

[19] T. Takeuchi, *Mater. Trans.*, **50** 2359-2365 (2009).

[20] D. F. Zou, S. H. Xie, Y. Y. Liu, J. G. Lin, and J. Y. Li, *J. Appl. Phys.*, **113** 193705 (2013).

[21] A. F. Ioffe and A. R. Regel, *Prog. Semicond.*, **4** 237 (1960).

[22] A. Yamamoto, H. Miyazaki, and T. Takeuchi, *J. Appl. Phys.*, **115** 23708 (2014).

[23] G. A. Slack, *in Thermoelectric Handbook*, (ed. M. Rowe) 407-440 (CRC, Boca Raton, 1995).

Material Science and Technologies for Advanced Nuclear Fission and Fusion Energy

DEVELOPMENT OF Ga DOPED HOLLANDITES $Ba_xCs_y(Ga_{2x+y}Ti_{8-2x-y})O_{16}$ FOR Cs IMMOBILIZATION

Y. Xu[1], R. Grote[1], Y. Wen[2], L. Shuller-Nickles[2], K.S. Brinkman[1]

[1]Department of Materials Science and Engineering, Clemson University, Clemson SC 29634
[2]Department of Environmental Engineering and Earth Sciences, Clemson University, Clemson SC 29634

ABSTRACT

Among the many proposed hollandites compositions $(Ba_x,Cs_y)(M,Ti)_8O_{16}$, Ga substituted hollandite has been studied to a limited extent. By substituting Ti with Ga, hollandite can be synthesized with a standard sintering process producing dense ceramics at relatively low temperature of 1250°C. In the present work, Ga substituted hollandites containing different levels of Cs, from a pure Cs end-member to a pure Ba end-member were synthesized with a fixed A site occupancy level. X-ray diffraction patterns coupled with SEM-EDX mapping confirmed hollandite formation as a primary homogenous phase. Density functional theory calculations were performed on selected compositions to evaluate the energetics of the hollandite system. Computational results showed favorable formation of hollandite for these samples based on negative formation enthalpies.

INTRODUCTION

One of the major challenges confronting the nuclear power industry is to provide an enduring solution to the problem of high level waste disposal. Cs is one of the more problematic fission product radionuclides to immobilize because of its volatility at high temperature, and its tendency to form water soluble compounds. A number of ceramic materials are being investigated as hosts for immobilization of Cs. The goal of this work is to obtain dense single-phase hollandite ceramics which appear to be good candidates for Cs immobilization. The hollandite group of ceramics has the formula of $A_xB_8O_{16}$, where A site cations occupy a tunnel formed by the octahedrally-coordinated B site cations. In order to migrate along the tunnels, A site cations must overcome large energy barriers. This makes hollandite a good waste form for Cs immobilization.[1] By substituting the original Ti atoms with other transition metal ions with lower valence than 4, hollandite can be synthesized in an air atmosphere instead of a reducing atmosphere. The doped hollandite has a general formula of $(Ba_x,Cs_y)(M,Ti)_8O_{16}$ with $x + y < 2$ and M = divalent or trivalent cations.

The ability to incorporate Cs into the hollandite structure depends on the M cations radius and how the M and Ti cations arrange in the octahedral sites framework. Previous studies have shown that many transition metal ions such as Fe, Cr, Ga, Ti, and Ni have the ability to promote the Cs incorporation into hollandite in air atmosphere by standard ceramic sintering processes.[2] To the best of our knowledge, most of the work has been done on a relatively low Cs levels. Thus far, the composition of hollandite studied were close to $Ba_{1.04}Cs_{0.24}(M,Ti)_8O_{16}$.[3,4] Studies on the high Cs doped hollandites are limited. However, the possibility of forming high Cs loading waste form is of great importance. In this study, we first synthesized hollandite with different Cs levels from the pure Ba end-member to the pure Cs end-member.

Detailed studies on the composition and structure were performed integrating experimental measurements with atomistic calculations. The compositions were designed in a manner that fixed the A site occupancy in order to form single-phase hollandite. In a single hollandite unit cell, two A sites are available; DFT calculations were performed using three unit cells with varying Ba/Cs ratios resulting in a fixed site occupancy of 1.33 out of the 2 total sites available. The fixed occupancy on the A site was therefore chosen to limit computational expense. The Ga content was varied in order to maintain the charge balance induced by the Cs incorporation

EXPERIMENTAL

Synthesis of Hollandite Ceramic

(Ba,Cs,Ga)-hollandite ceramics of formula $Ba_xCs_yGa_{2x+y}Ti_{8-2x-y}O_{16}$ (x=1.33, 1.04, 0.667, 0; y=0, 0.24, 0.667, 1.33) were prepared by solid state reaction from reagent-grade oxide and carbonate powders: Ba_2CO_3, Ga_2O_3, TiO_2, Cs_2CO_3. Approximately 20 grams of dry oxide precursors were mixed stoichiometrically via ball milling and ground before making pellets. Pellets were calcined in air for 30h at 1150 °C. To reduce the risk of Cs vaporization during synthesis, the crucible was sealed with aluminum oxide cement. The synthesized hollandite were then ball milled and pressed into a pellet. The pallets were sintered in air at 1250 °C for 3 hours. All hollandite samples were characterized by X-ray diffraction (XRD) with Cu K wavelength (λ=1.54), scanning electron microscopy (SEM), energy dispersive X-ray analysis (EDX).

RESULTS AND DISCUSSION

X-ray diffraction confirmed that all samples exhibited primarily hollandite phases with the exception of the minor peak near 29 degrees 2θ (Figure 1.). This phase is currently un-identified, but SEM-EDS analysis has revealed that it is Ti rich and does not contain Cs. The shift in the (301) peak with increasing Cs composition indicates that the ring of the hollandite structure expands. Rietveld cell refinement also shows an increase in the *a* lattice parameters with increasing Cs concentration (Figure 2a). An increase of lattice parameter of *a* along with increasing Cs concentration is shown in Figure 2a, which is expected from the (301) peak shift. However, the *c* lattice parameter does not display a clear trend. This is in agreement with DFT results that that show a clear ring size expansion as Cs content increases. However, no clear trend was observed in the *c* lattice parameter. This is attributed to the competing effects on the lattice parameter due to dopants on the A sites and B sites.

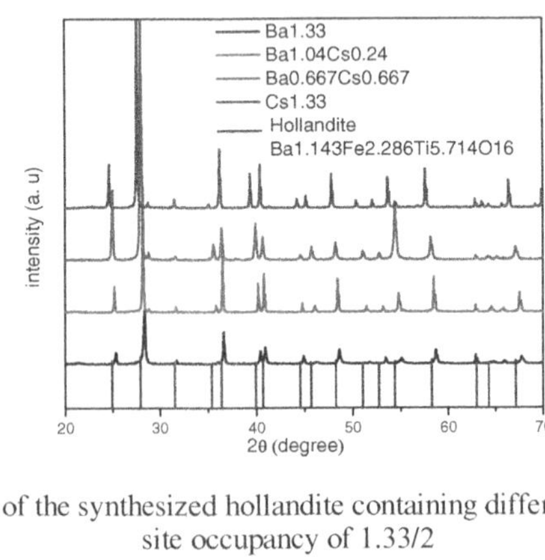

Figure. 1 XRD patterns of the synthesized hollandite containing different Cs levels with fixed A site occupancy of 1.33/2

In our study, as we increased the Cs content, the Ga concentration was decreased in order to maintain the charge balance of the crystal. At present, it appears that the Cs content affects the ring size more than the Ga dopant concentration, as the sample with the least Ga turns out to have the largest ring size. This trend will be examined more in future work with varying B site dopants.

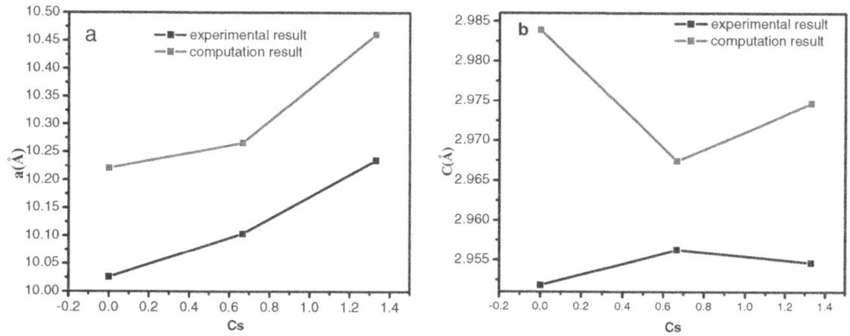

Figure 2. Evolution of lattice parameters versus Cs content; experimental results are plotted in black, computation results are plotted in red.

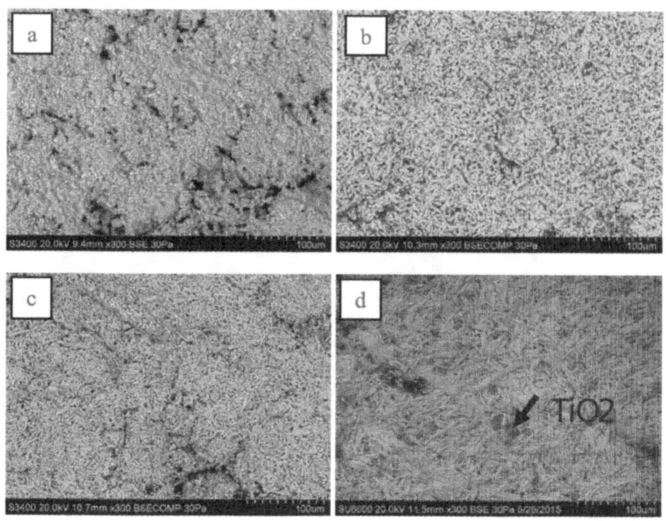

Figure 3. Back scattered images of the sintered pallets at magnification of 300, a) $Ba_{1.16}Cs_0Ga_{2.32}Ti_{5.68}O_{16}$; b) $Ba_{1.04}Cs_{0.24}Ga_{2.32}Ti_{5.68}O_{16}$; c) $Ba_{0.667}Cs_{0.667}Ga_2Ti_6O_{16}$. d) $Cs_{1.33}Ga_{1.33}Ti_{6.67}O_{16}$.

Back-scattered images were used to characterize the microstructure and morphology of the samples. The microstructures of the sintered pellets are displayed in Figure 3. The hollandite crystallized in a rod-like morphology, with the length of the rod like dimensions increasing as Cs content increases. Overall, the homogeneity of sintered pellets were confirmed by the SEM-backscattered images. For the high Cs loading sample ($Ba_0Cs_{1.33}$) in Figure 4d, we saw secondary phase as indicated by the dark areas. SEM-EDX mapping in Figure 4 demonstrated the homogeneous distribution of all elements except the highest Cs loading sample, which exhibited bright Ti hot spots. Semi-quantitative EDX elemental analysis results listed in Table 1 compared the target composition and the measured composition showing that Ga doped hollandites have very good Cs incorporation and retention ability during processing/crystallization. To compare with the reported hollandite, we also prepared a composition F2 (short ID for $Ba_{1.04}Cs_{0.24}Ga_{2.32}Ti_{5.68}O_{16}$) which is the same as reported in Aubin-Chevaldonnet's paper.[2] It is worth mentioning that we obtained a 95.8% Cs retention which is much higher than the 50% retention typically reported due to Cs volatility during high temperature sintering. The higher Cs retention in the present case is attributed to the use of a controlled atmosphere resulted in a high Cs vapor pressure in the system reducing the driving force for Cs loss from solid to vapor phase.

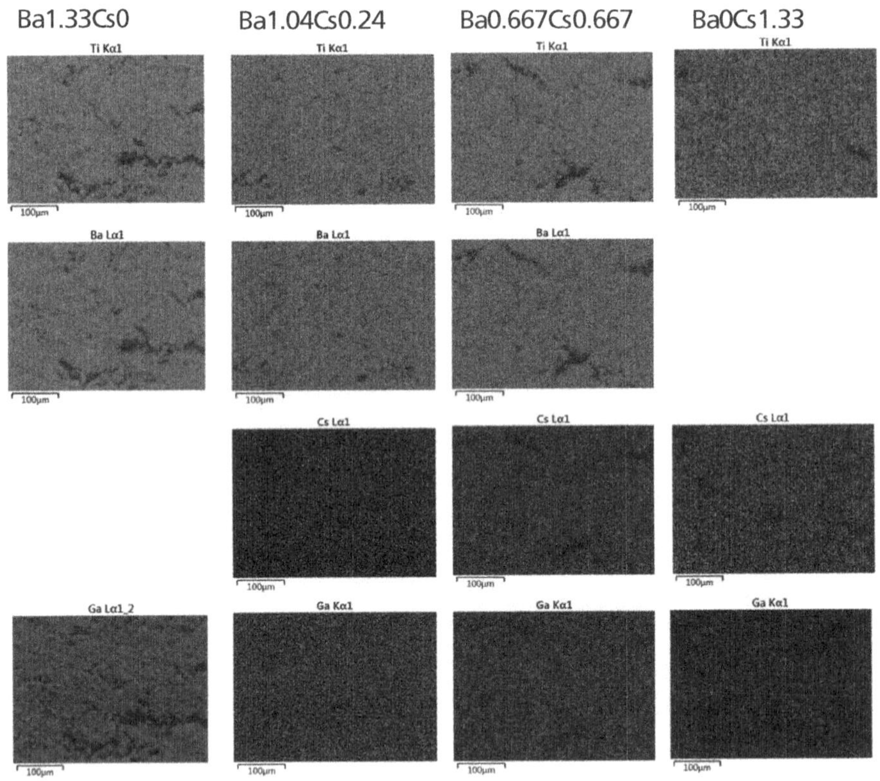

Figure 4. SEM-EDX elemental maps corresponding to the back-scattered images in Figure 3.

Higher magnification images and spectra were collected for high Cs loading sample F4 ($Ba_0Cs_{1.33}Ga_{1.33}Ti_{6.67}O_{16}$). A closer look and detailed EDX quantification indicated that the Ti hot spots shown in Table 2. The darker area in F4 was identified as the Ti rich phase, while the brighter area had a composition close to the target composition and was identified as hollandite. Low levels of Ti rich phases were therefore formed due to limited Cs loss at elevated temperature. However, from the determined composition, this Ti rich secondary phase does not contain Cs and is not expected to be detrimental to the leaching or durability of the waste form. In Aubin-Chevaldonnet's paper, it was reported that the Cs rich phase $CsGaSi_{0.4}Ti_{0.6}O_4$ which is similar to water soluble $CsAlTiO_4$ was favorable to form for Ga doped hollandite.[2] However, in the present work, no Cs rich phases were observed, even in samples containing the highest possible Cs loading. Also, no aluminum was detected from the EDX spectrum, even though aluminum crucibles were used.

The densities of sintered samples were measured by Archimedes method and are listed in Table 1. The relative density Da/Dt, where Da is the density measured by Archimedes methods and Da is the density calculated from XRD determined lattice parameter and molecular mass of the crystal. The solid-state reaction resulted in fairly good densification with relative densities approaching 90% in most samples. A decrease in the relative density as Cs level increases is related to the Cs evaporation during the sintering process resulting in increased porosity. The low density of F4 is a result of both Cs evaporation and existence of a low-density Ti phase. In summary, this initial work demonstrates that dense Ga doped hollandite ceramics incorporating Cs can be easily prepared by a solid-state reaction method.

Spot/Atom%					
1		2		3	
Cs	1.9	Cs	6.5	Cs	5.5
Ga	2.4	Ga	6.6	Ga	6.0
Ti	32.5	Ti	32.6	Ti	30.3
O	63.2	O	54.2	O	58.2

Table 1 Semi-quantitative EDS analysis of F4.

Table 2. Calculated hollandite compositions based on semi-quantitative measurement by SEM-EDX, normalized based on Ti.

Short ID	Target composition	EDX composition	Da	Dt	Da/Dt
F1	$Ba_{1.33}Cs_0Ga_{2.66}Ti_{5.34}O_{16}$	$Ba_{1.39}Cs_0Ga_{2.6}Ti_{5.68}O_{16}$	4.42	4.859	0.909
F2	$Ba_{1.04}Cs_{0.24}Ga_{2.32}Ti_{5.68}O_{16}$	$Ba_{1.19}Cs_{0.229}Ga_{2.6}Ti_{5.68}O_{16}$	4.31	4.744	0.908
F3	$Ba_{0.667}Cs_{0.667}Ga_2Ti_6O_{16}$	$Ba_{0.77}Cs_{0.5229}Ga_{2.06}Ti_6O_{16}$	4.15	4.745	0.874
F4	$Cs_{1.33}Ga_{1.33}Ti_{6.67}O_{16}$	$Cs_{1.22}Ga_{1.44}Ti_{6.67}O_{16}$	3.92	4.664	0.84

DENSITY FUNCTIONAL THEORY STUDY: ENTHALPY OF FORMATION

Enthalpies of formation were calculated based on the formation from the primary oxide components. For example, the formation reaction for $Ba_{1.33}Ga_{2.66}Ti_{5.34}O_{16}$ is:

$$1.33BaO+1.33Ga_2O_3+5.34TiO_2 \rightarrow Ba_{1.33}Ga_{2.66}Ti_{5.34}O_{16}$$

where the calculated formation enthalpy is the total energy of the product minus that of the reactants. The formation enthalpies of the selected compositions are listed in Table 3. All of the compositions exhibited negative formation enthalpies, indicating that are all thermodynamically favorable. This is in agreement with the experimental results.

Table 3. Calculated (CASTEP calculations with GGA-PBE and ultra-soft pseudopotentials) enthalpies of the selected compositions

Formulas	Formation Enthalpy (kJ/mol)
$Ba_{1.33}Ga_{2.66}Ti_{5.34}O_{16}$	-109.6
$Ba_{0.667}Cs_{0.667}Ga_2Ti_6O_{16}$	-133.9
$Cs_{1.33}Ga_{1.33}Ti_{6.67}O_{16}$	-140.8

CONCLUSIONS

This work examined the ability of Ga doped hollandites for Cs incorporation. Hollandites with different Cs loadings were prepared by a conventional ceramic sintering process. Ga doped hollandites showed good Cs retention during processing in controlled atmospheres. Dense, homogeneous pellets were successfully prepared at a relatively low sintering temperature of 1250°C. DFT calculations indicated that all of the compositions studied exhibited negative formation enthalpies, indicating that are all thermodynamically favorable. Further studies on the chemical durability/leaching resistance as a function of composition and structure is ongoing.

ACKNOWLEDGMENTS

The authors gratefully acknowledge financial support from the DOE-EPSCoR Project Number: DE-SC0012530, "Radionuclide Waste Disposal: Development of Multi-scale Experimental and Modeling Capabilities."

REFERENCES

[1] Ringwood, A. E.; Kesson, S. E.; Ware, N. G.; Hibberson, W. O.; Major, A. The SYNROC process: A geochemical approach to nuclear waste immobilization. *Geochem. J.* **1979**, *13*, 141-165.

[2] Aubin-Chevaldonnet, V.; Caurant, D.; Dannoux, A.; Gourier, D.; Charpentier, T.; Mazerolles, L.; Advocat, T. Preparation and characterization of $(Ba,Cs)(M,Ti)_8O_{16}$ (M;Al^{3+}, Fe^{3+}, Ga^{3+}, Cr^{3+}, Sc^{3+}, Mg^{2+}) hollandite ceramics developed for radioactive cesium immobilization. *Journal of Nuclear Materials* **2007**, *366*, 137-160.

[3] Amoroso, J.; Marra, J.; Conradson, S. D.; Tang, M.; Brinkman, K. Melt processed single phase hollandite waste forms for nuclear waste immobilization: $Ba_{1.0}Cs_{0.3}A_{2.3}Ti_{5.7}O_{16}$; Cr, Fe, Al. *Journal of Alloys and Compounds* **2014**, *584*, 590-599.

[4] Costa, G. C. C.; Xu, H.; Navrotsky, A. Thermochemistry of Barium Hollandites. *Journal of the American Ceramic Society* **2013**, *96*, 1554-1561.

ATOMISTIC SIMULATIONS OF CERAMIC MATERIALS RELEVANT FOR NUCLEAR WASTE MANAGEMENT: CASES OF MONAZITE AND PYROCHLORE

Y. Li[1,2], P. M. Kowalski[1,2], G. Beridze[1,2], A.Blanca-Romero[1,2] , Y. Ji, V. L. Vinograd[1,2,], J. Gale[3] & D. Bosbach[1,2]

[1]Institute of Energy and Climate Research (IEK-6), Forschungszentrum Jülich, Wilhelm-Johnen-Straβe, 52425 Jülich, Germany

[2]JARA High-Performance Computing, Schinkelstraβe 2, 52062 Aachen, Germany

[3] Nanochemistry Research Institute, Department of Chemistry, Curtin University, PO Box U1987, Perth, WA 6845, Australia

ABSTRACT

We performed systematic *ab initio* atomistic simulations of monazite ($\mathit{Ln}PO_4$) and pyrochlore ($A_2B_2O_7$)-type ceramics. The simulated properties include the thermodynamic parameters, the threshold displacement energies, the elastic constants and the energies of defects formation. We computed the excess effects of mixing in various monazite-type solid solutions and provided crucial information on their thermodynamic stability. To better understand the radiation damage resistance of the considered novel waste forms, we simulated the threshold displacement energies and the displacement probabilities of La cation in $LaPO_4$ monazite and computed defect formation energies in various pyrochlore compounds. The observed exothermic character of the anion Frenkel pair defects suggests that the ease of the defect formation correlates with instability of the pyrochlore structure relative to fluorite. Further calculations of activation barriers for the oxygen migration shed a light on the mechanism of the order-disorder transition in pyrochlore compounds.

INTRODUCTION

Large amounts of highly radioactive waste have been created worldwide mainly as a byproduct of electricity production[1]. Most of the waste is targeted for long-term storage and eventual geological disposal. Various suitable waste forms are currently discussed. Monazites and pyrochlores attract much of interest as suitable candidates for permanent immobilization of minor actinides (Np, Am and Cm) and Pu[2,3]. Naturally occurring analogues of these materials often contain significant amounts of radionuclides and despite the self-irradiation have been maintained their crystalline structure for millions of years despite the self-irradiation. Before these ceramic materials could be used in nuclear waste management, they should be well characterized.

The excess enthalpy of mixing, which is the heat absorbed or released upon mixing of the end members, is the major parameter that determines the thermodynamic stability of a solid solution[4]. Within the regular and subregular models, these excess effects are described with polynomials containing adjustable coefficients named the Margules parameters. It is usually very difficult to accurately measure the effects of mixing. For monazite-type phosphates experimental calorimetric data are available only for three $La_{1-x}\mathit{Ln}_xPO_4$ solid solutions, where Ln = Nd, Eu and Gd[5]. The uncertainties of the experimentally determined interaction parameters are often as large as 50% of the measured values. Previous research suggested that values of the interaction parameters in these solid solutions are related to the mismatch in volumes of the end members[5,6,7]. Such a relationship has also been confirmed in our study[8]. However, it is not yet clear whether this conclusion could be transferred to monazite-type solid solutions of other compositions including those with coupled substitutions. Four-valent actinides can be incorporated in the structure of the monazite by the coupled substitution mechanism:

$$M^{2+} + An^{4+} \rightarrow 2Ln^{3+} \qquad (1)$$

where M = Mg, Ca, Cd, Sr, Pb and Ba are di-valent cations. The excess enthalpies of the $(Ln_{1-2x}M_xAn_x)PO_4$ (M = Ca, Ln = La,Ce) solid solutions have been investigated experimentally[9], however, there are no relevant theoretical studies assessing stabilities of these solid solutions.

The radiation resistance of monazites has been studied by irradiating them with 800 keV Kr^{2+} ion beam in the temperature interval of 20 to 600 K[10]. The detected critical temperature for the $LaPO_4$ monazite, above which the material cannot be amorphized under irradiation, is ~330 K. The threshold displacement energy, E_d, is a crucial parameter determining the radiation tolerance of a material and the extent of the radiation damage. Due to difficulties in detecting the onset of defect formation by the experimental techniques, it is very hard to measure the value of E_d. On the other hand, with the aid of computer simulations one can determine the value of E_d through modeling of low energy displacement cascades with the aid of molecular dynamics[11]. In this contribution we present results of such simulations for $LaPO_4$ monazite.

Under irradiation caused by incorporation of actinides, two main point defects are formed in pyrochlore, the cation antisite (CA) and the anion Frenkel pair (AFP). An increase in the number of these defects leads to the order-disorder (O-D) phase transition from pyrochlore-type structure to the defect fluorite structure[12,13]. Despite significant of experimental effort[14,15], the mechanism of the O-D transition is not yet well understood. Previous force-field (FF) atomistic simulation studies have shown that the stability of pyrochlore vs. fluorite can be defined in the frame of defect formation energies (DFE). It has been demonstrated that CA DFEs are proportional to the cation radius ratio (r_A/r_B)[16,17]. However, the predicted energies are about 2 eV too large to introduce any disordering in the cation sublattice of the pyrochlores. On the other hand, for some pyrochlores, significantly lower values of the DFEs have been recently predicted by density functional theory (DFT) studies[18,19,20,21]. Therefore, we have recomputed the energies of the defect formation in different families of pyrochlore using DFT[22]. We have also applied DFT-based atomistic computer simulations to study the activation energy for oxygen ion migration in order to verify the results of previous force field calculations[23,24]. Here our results are compared to previous calculations and experimental estimates.

The paper is organized as follows. First, we briefly summarize the main results of our recent simulations of properties of monazite-type ceramics, namely the thermodynamics properties of monazite-type $(Ce_{1-x}Ln_x)PO_4$, $(Sm_{1-x}Ln_x)PO_4$ and $(Ln_{1-2x}Ca_xAn_x)PO_4$ solid solutions. Then we discuss the elastic constants and the threshold displacement energy for La cation in $LaPO_4$. Finally, in order to understand the oxygen migration mechanisms in pyrochlore, we discuss the activation energies for the oxygen migration in selected pyrochlore compounds.

COMPUTATIONAL METHODS

In this study, the Margules interaction parameters in $(Ce_{1-x}Ln_x)PO_4$, $(Sm_{1-x}Ln_x)PO_4$ and $(Ln_{1-2x}Mg_xAn_x)PO_4$ solid solutions were computed using a 2x2x2 supercell of monazite containing 192 atoms. The simulations of the defect formation energies and the activation energies for the oxygen migration in pyrochlore compounds, are performed with the unit cell containing 88 atoms.

All the properties except the threshold displacement energies were simulated *ab initio*. The first principles calculations were carried out using the plane-wave Quantum-ESPRESSO package[25]. We applied the PBEsol exchange correlation functional[26] for monazite, while the PBE functional was used for pyrochlore compounds[27]. The plane-wave energy cutoff was 50 Ryd. The ultrasoft pseudopotentials[28] were always employed. The $5s^2 5p^6 6s^2$ electrons of lanthanides and the $6s^2 6p^6 6d^1 7s^2$ electrons of actinides were treated explicitly. The $4f$ and $5f$ electrons of *Ln* and *An* were included into the core.

The excess properties of mixing of monazite-type solid solutions were studied with the aid of quasirandom structures. These structures were obtained via a search through randomly modified configurations of exchangeable atoms with a 2x2x2 supercell. Within this search the occurrence frequencies of pairs of dissimilar cations were counted at all possible distances within the supercell. The error function was defined as the total squared misfit between the observed frequencies and their theoretical values expected in a perfectly random solution. The structures with the minimum misfit were selected.

The elastic constants were computed *ab initio* from the stress-strain relationship[29]. Other mechanical constants such as bulk modulus, *B*, Young's modulus, *E*, and shear modulus, *G*, were calculated from the compliance tensor S ($S=C^{-1}$). The shear and bulk modules were determined using the Voigt–Reuss–Hill averaging scheme[30].

The threshold displacement energy, E_d, is defined as the minimum energy required to produce a permanent displacement of the primary knock-on atom. The calculations to determine the values of E_d through a large number of low-energy MD collision cascades was carried out using molecular dynamics code Lammps[31]. Interatomic interactions were modeled with a core-shell model employing the short-range Buckingham potential

$$\phi(r_{ij}) = A\exp(\frac{-r_{ij}}{\rho}) - \frac{C}{r_{ij}^6} \tag{2}$$

where the three parameters: A, ρ and C were determined by fitting to the structural parameters of $LaPO_4$ and some other phosphates computed *ab initio*.[32] The fitting was performed with the program GULP[33]. The obtained values of A, ρ and C are 17927.00 eV, 0.25934 Å, and 0.0 eV*Å^6 for La-O (shell), 877.30 eV, 0.3594 Å, and 0.0 eV*Å^6 for P-O (shell), 22764.30 eV, 0.1490 Å, and 27.879 eV*Å^6 for O (shell) – O (shell), the spring constant of the harmonic potential between the core and shell of oxygen is 74.92 eV/Å^2. The MD simulations were carried out at 300 K for a maximum of 5 ps, while the PKA energies ranged from 50 to 150 eV. 400 PKA directions of initial velocities were determined by a random spherical sampling.

We performed *ab initio* nugget elastic band (NEB) calculations of transition paths in order to calculate the activation energies of the oxygen migration in pyrochlore compounds. In the present study, the considered pyrochlore compositions (e.g. $Nd_2Zr_2O_7$, $Sm_2Zr_2O_7$, $Gd_2Zr_2O_7$, and $Er_2Ti_2O_7$) exhibited split vacancy behavior. Therefore, we calculated the transition barrier of the oxygen migration along the <111> direction. The transition path is a movement of an oxygen atom from one interstitial site to another interstitial site located along the <111> direction[23].

RESULTS

Thermodynamics Properties of Monazite-type Solid Solutions

We have performed systematic calculation of the $La_{1-x}Ln_xPO_4$ (*Ln* =Ce, ..., Tb) and $La_{1-x}Ln_xPO_4$ (*An*=Pu, Am and Cm) monazite-type solid solutions using the state-of-the-art *ab initio* methods of quantum chemistry8. The derived Margules interaction parameters for these solutions can be used for further geochemical modeling. Figure 1 (left), shows that the calculated interaction parameters in $(La_{1-x}Ln_x)PO_4$ increase monotonically as a function of the unit cell volume mismatch between the end members (ΔV, cm^3/mol). In the solid solution of $(La_{1-x}Ln_x)PO_4$, the increase in the parameter value was fitted to a quadratic function of ΔV, so that $W_{Ln}=0.618(\Delta V)^2$ kJ/mol, which is consistent with the results of previous numerical studies of other solid solutions [34, 35,36]. We note that the function, which describes the dependence of the interaction parameters of lanthanides on ΔV in $(Ln_{1-x}An_x)PO_4$ also describes relatively well the

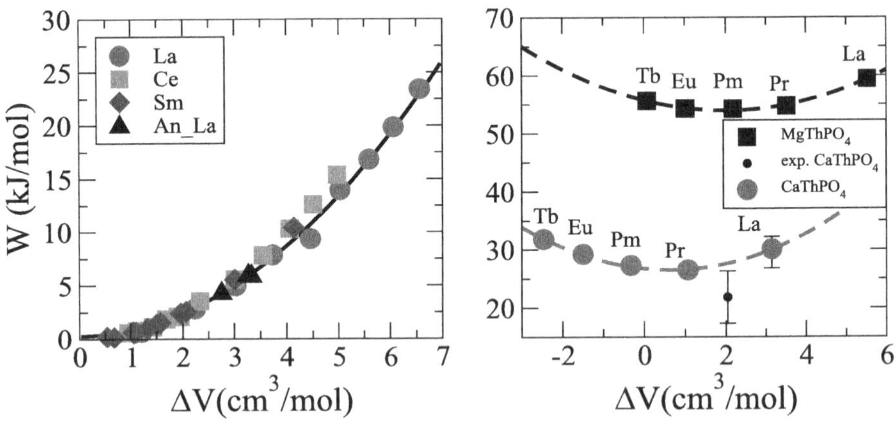

Figure 1. The Margules interaction parameters as a function of the endmembers volume mismatch for $Ln_{1-x}(Ln/An)_xPO_4$ (left) and MgTh*Ln*(PO_4) (right) solid solutions. In the left panel, circles and triangles represent the results for the solid solutions of $(La_{1-x}Ln_x)PO_4$ and $(La_{1-x}An_x)PO_4$, respectively. Square and diamond refer to the solid solutions of $(Ce_{1-x}Ln_x)PO_4$ and $(Sm_{1-x}Ln_x)PO_4$, respectively. In the right panel, squares and circles represent the results for $(Ln_{1-2x}Mg_xTh_x)PO_4$ and $(Ln_{1-2x}Ca_xTh_x)PO_4$, respectively, the experimental data is shown are the small black point. [37]

interaction parameters predicted here for solid solutions of actinides. In order to see whether the prediction still holds for the other monazite-type solid solutions, we also performed simulations of solid solutions of $(Ce_{1-x}Ln_x)PO_4$ and $(Sm_{1-x}Ln_x)PO_4$. Interestingly, for the solid solutions of $(Ce_{1-x}Ln_x)PO_4$ and $(Sm_{1-x}Ln_x)PO_4$, as shown in Figure 1 (left), the corresponding interaction parameters behave similarly to those of $(La_{1-x}Ln_x)PO_4$. A solid solution will remain stable in the whole compositional range at room temperature when $(W/K_B T) < 2$ (i.e. W < 5.0 kJ/mol)[38]. Therefore, we predict that the solid solutions with $\Delta V > 3$ cm^3/mol will tend to phase separate at ambient conditions.

Figure 1 (right) shows the Margules interaction parameters computed for the solid solutions with the coupled substitution, which have the composition of $(Ln_{1-2x}Mg_xTh_x)PO_4$ and $(Ln_{1-2x}Ca_xTh_x)PO_4$. The variation of the interaction parameter as a function of the end-member volume mismatch is quite different from the discussed above case of the single-site substitution. (Figure 1 (left)). This function also shows a minimum, which is located close to zero volume mismatch. However, the interaction parameter does not show a tendency to decrease to zero value with the decrease in the volume mismatch. In the case of the coupled substitution the interaction parameters appear to be much larger than those computed for the single substitution cases. The obvious reason for such a behavior is the chemical ordering of cations of different charge, which appears to be especially strong at x=1. The ordered alternation of different cations is thermodynamically supported, as it minimizes the electrostatic energy. At intermediate compositions this alternation is destroyed, causing the enthalpy of the mixed phase to be larger than that of a mechanical mixture of the end members. Thus, in solid solutions with coupled substitutions the volume mismatch is a relatively unimportant factor. The predicted values of the interaction parameters are consistent with the few available experimental data. Interestingly, the interaction parameters obtained for case with Ca/*An* doping are much smaller than in the case of Mg/*An*. We attribute this effect to a more ionic character of interatomic interactions in the Mg/*An* systems as compared to those in Ca/*An* systems. All studied systems with the coupled substitutions are predicted to phase separate at ambient conditions.

Threshold Displacement Energy in $LaPO_4$

An important parameter that describes the radiation damage in a material is the threshold displacement energy, E_d, which constitutes the minimal energy required to dislocate an atom from its original position inside the crystalline net. The threshold displacement energies may be used to simulate the extent of radiation damage in materials with codes such as SRIM.[39] The E_d of primary knock-out atom (PKA) is a property of the local structure of a material. It depends on the atom and on its local surrounding. The simulations of E_d of La in $LaPO_4$ monazite were performed by applying initial velocities to PKA lanthanum atom in 100-400 randomly chosen directions. Then NVE molecular dynamics runs were performed for 5 ps. The initial and final positions of PKA atoms were checked. The probability of a displacement as a function of different initial energy is illustrated in Figure 2. When the energy was smaller than 50 eV no displacements were observed indicating that this energy is too low to dislocate a La cation. On the other hand, at energies larger than 100 eV most of the events resulted in dislocations. We observed that a simple function [40]:

$$\mathrm{DFP}(E) = \begin{cases} 0 & \text{if} \le E_d \\ \frac{1}{\beta}(E^{\alpha}-(E_d)^{\alpha} & \text{If} > E_d \end{cases} \quad (3)$$

where α and β are fitting parameters and E is the energy of the PKA, describes the obtained statistics well. The obtained values of E_d, α, β are 51.85 eV, 9.48 and 0.57, respectively. Our

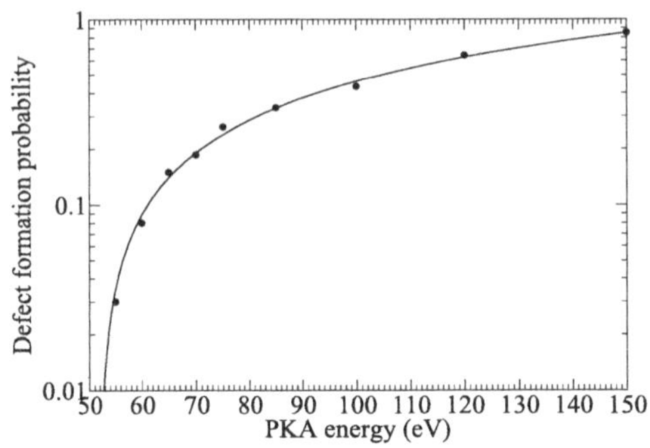

Figure 2 The probability of defect formation in $LaPO_4$ for La PKA. The solid line is the fit obtained using Eq. (3).

results thus show that E_d of lanthanum in La-monazite is ~52 eV, a relatively high value. This is comparable to E_d of Ti in rutile[41].

Elastic Moduli

	G (GPa)			E(GPa)		
$LnPO_4$	DFT	DFT+U	Exp.	DFT	DFT+U	Exp.
La	54	50	52	139	129	133 (140)
Nd	60	56	61	154	143	157
Eu	64	60	63	163	153	162(158)
Gd	65	60	67	165	158	173
M.A.E	2	4		11	7	

Table I. The shear (G) and Young's (E) moduli calculated with DFT and DFT+U methods and compared to the experiment. M.A.E refers to the mean average error.

The bulk and shear modules of $LnPO_4$ (Ln = La, Nd, Eu and Gd) calculated using DFT and DFT+U methods are listed in Table 1 together with the experimental data. The predicted values of the bulk and shear modules that fall in the ranges from 54 to 65 GPa and from 129 to 165 GPa, respectively, are in good agreement with the experiment. The DFT method with the f electrons modeled by the pseudopotential provides a slightly better agreement to the experimental values. relative to the f-in-core approach. The discussion of methods of computation of U values in $LnPO_4$ and a more detailed analysis of the structural and thermodynamic properties of these compounds can be found in our previous paper.[32]

Defect Formation Energy in Pyrochlore

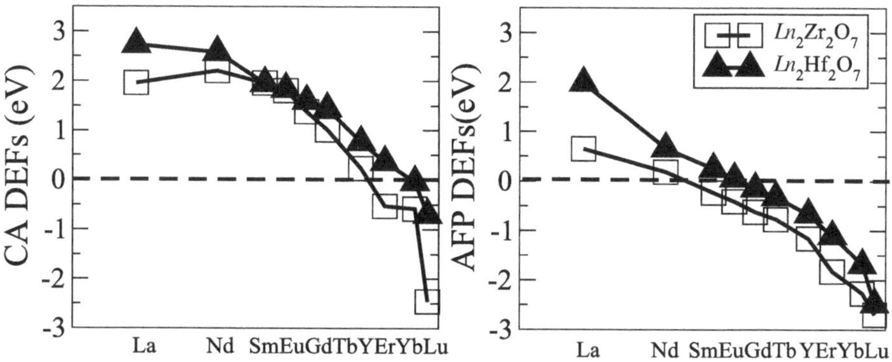

Figure 4. The calculated CA DEFs and AFP DEFs for $A_2Zr_2O_7$ and $A_2Hf_2O_7$ (A=La, Nd, Sm, Eu, Gd, Y, Er, Yb and Lu).

Here we discuss our recent DFT results on cation antisite (CA) and anion Frenkel pair (AFP) defect formation energies (DEFs) in zirconate- and hafnate-pyrochlores. Figure 4 (left panel) shows that for some $A_2Zr_2O_7$ pyrochlores with (A=Gd, Tb, Y, Er, Yb and Lu) the CA DFEs are smaller than 1 eV, while for A= Er, Yb and Lu these values are even negative. These low CA DFEs indicate the relative ease of disordering of the cation sublattice in these pyrochlores, what explains their tendency to crystallize as defect fluorite. However, for some compounds, that are known to form defect fluorite, the predicted CA DFEs are positive and relative large. This implies that the ease of the pyrochlore/fluorite transformation cannot be solely related to the ease of formation of CA defects. Figure 4 (right panel) shows that the values of AFP DFEs in pyrochlores are smaller than these of CA DFEs. The negative signs of AFP DFEs correlate very well with the data for the compositions (A = Y, Er, Yb and Lu). These compounds are known to crystallize as defect fluorite. $A_2Zr_2O_7$ pyrochlores with A = Sm, ..., Tb have relatively large CA DFEs, but are also typically found in the disordered form[42]. Interestingly, these compounds are characterized by negative values of AFP DFEs. It appears that the energy cost related to the accumulation of CA defects could be compensated by the energy gain due to the creation of the AFP defects. Consequently, our results suggest that the ease of AFP formation is essential for the disordering of the pyrochlore lattice. The similar conclusion can be also made based on the results for $A_2Hf_2O_7$. Our systematic simulations of CA and AFP DEFs of a wide range of pyrochlore compounds allowed us to correlate the defect formation energies with the stability field of pyrochlore.[22] Our results demonstrated that the negative AFP DEFs could be responsible for the instability of certain pyrochlore compounds.

Activation energy for the Oxygen Migration

To further investigate the role of oxygen defects in the order/disorder transition we computed transition barriers for the oxygen migration from the 48f site to a vacant 8a site from first principles. Our results compare well to the results of previous force field calculations (Table II) and to the available experimental data. All four considered pyrochlores, $Sm_2Zr_2O_7$, $Nd_2Zr_2O_7$, $Gd_2Zr_2O_7$ and $Er_2Ti_2O_7$, exhibit a split vacancy state (a configuration with O atom located between two 48f and 8a sites) as a final migration state. The overall motion of the oxygen ions occurs along <111> direction[23]. This conclusion is consistent with the previous force field calculations[24]. Generally, there is an acceptable agreement between our calculations and the experiment[43]. However, for $Er_2Ti_2O_7$ the predicted value of 0.34 eV is much smaller than the experimental value. Interestingly, the force field calculations seem to provide a better agreement with the experimental data. Evidently, further *ab initio* calculations of activation energies in different pyrochlore families are needed to clarify these issues.

	Calculated energy (*ab initio*)	Calculated energy (force field)	Experimental energy
$Sm_2Zr_2O_7$	1.17	0.67	0.69
$Nd_2Zr_2O_7$	0.71	0.78	0.76-0.84
$Gd_2Zr_2O_7$	1.20	0.58	0.73-0.90
$Er_2Ti_2O_7$	0.34	1.20	1.17

Table II. Predicted internal activation energies (eV) for oxygen vacancy migration in comparison with experiment data[43] and previous force field calculations[24].

CONCLUSIONS

Our study shows that the atomistic modeling can significantly complement the experimental effort aiming at characterization the ceramic materials that are proposed as matrixes for the disposal of nuclear waste. Indeed, many material properties cannot be easily studied by experimental methods. Based on atomistic simulations we calculated the Margules interaction parameters that describe the excess properties of monazite-type solid solutions. For $(La_{1-x}Ln_x)PO_4$ the interaction parameters are proportional to the squared mismatch between the volumes of end-members. This prediction also holds for $(Ce_{1-x}Ln_x)PO_4$ and $(Sm_{1-x}Ln_x)PO_4$. On the other hand, for monazite-type solid solutions of $(Ln_{1-2x}Mg_xAn_x)PO_4$ and $(Ln_{1-2x}Mg_xAn_x)PO_4$, the interaction parameters remain large (>15 kJ/mol) even when the volume mismatch is negligible. The elastic constants of $LaPO_4$, $EuPO_4$, $NdPO_4$ and $GdPO_4$ monazites were computed with the DFT+U approach. We also computed the threshold displacement energy of La cation in $LaPO_4$ monazite. The obtained elastic constants agree well with the available experimental data. The predicted large threshold displacement energy of ~52 eV for La in La-monazite possibly partially explains strong radiation damage resistance of monazites. For pyrochlore-type ceramics a new set of defect formation energies was computed *ab initio*. The instability of pyrochlore relative to defect fluorite correlates with the negative values of the formation energy of the anion Frenkel pair defect. We assume that the ease of the oxygen migration could play an important role in the mechanism of the order-disorder phase transition. To further investigate this phenomenon, we computed transition barriers for the oxygen migration in some pyrochlore compounds. Similar calculations are currently performed for monazite-type compounds and solid solutions.

ACKNOWLEDGEMENTS

We acknowledge the computing time on RWTH Aachen cluster awarded through Jülich-Aachen research alliance (JARA-HPC).

---

[1]Ewing, R. C. (2015). Long-term storage of spent nuclear fuel. *Nature Materials, Volume 14,* 252-257.

[2]Ewing, R. C., & Wang, L. (2002). Phosphates as Nuclear Waste Forms". Reviews in Mineralogy and Geochemistry 48, 673-699.

[3]Lumpkin, G. R. (2001). Alpha-decay damage and aqueous durability of actinide host phases in natural systems. *Journal. of Nuclear Materials, Volume 289,* 136-166.

[4]Prieto, M. Thermodynamics of solid solution-aqueous solution systems in: 19th Annual V M Goldschmidt Conference, Davos, Switzerland, June 19-21, 2009.

[5]Popa, K, Konings, R. J. M, & Geisler, T. (2007). High-temperature calorimetry of $(La_{1-x}Ln_x)PO_4$ solid solutions. *The Journal of Chemical Thermodyanmics, Volume 39,* 236-239.

[6] Schlenz, H, Heuser, H, Julia, Neumann, A., Schmitz, S., & Bosbach, D. (2013). Monazite as suitable actinide waste form. *Zeitschrift für Kristallographie-Crystalline materials, Volume 228*, 113-123.

[7]Konings, R. J. M, Walter, M., & Popa, K. (2008). Excess properties of the $Ln_{2-2x}Ca_xTh_x)(PO_4)_2$ (Ln = La, Ce) solid solutions. *The Journal of Chemical Thermodyanmics, Volume 40,* 1305-1308.

[8] Li, Y., Kowalski, P., Blanca-Romero, A., Vinograd, V. & Boshbach, D. (2014). Ab Initio Calculations of Excess Properties of $La_{1-x}(Ln,An)_xPO_4$. Journal of Solid State Chemistry, Volume 220, 137-141.

[9] Clavier, N., Podor, R., & Dacheus, N. (2011). Crystal Chemistry of the monazite structure, *Journal of the European Ceramic Society, Volume 31,* 941-976.

[10]Meldrun, A., Boatner, & L. A., Ewing, R. C. (1997). Displacive Radiation Effects in the Monazite- and Zircon-structure Orthophosphates. *Physical Review B, Volume 56,* 13805-13814.

[11]Robinson, M., Marks, N. A., Whittle, K. R., & G. R. Lumpkin, G. R. (2012). Systematic Calculation of Threshold Displacement Energies: Case Study in Rutile, *Physical Review B, Volume 85,* 104105-104115.

[12]Minervini, L., Grimes, R. W., & Sickafus, K. E. (2000). Disorder in Pyrochlore Oxides. *Journal of the American Ceramic Society, Volume 83,* 1873–1878.

[13]Wilde, P. J., & Catlow, C. R. A. (1998). *Defects and diffusion in pyrochlore structured oxides, Solid State Ionics, Volume 112,* 173–183.

[14]Zhang, F.X, Wang, J.W, Lian, J., Lang, M. K. , Becker, U., & Ewing, R. C. (2008). Phase Stability and Pressure Dependence of Defect Formation in $Gd_2Ti_2O_7$ and $Gd_2Zr_2O_7$ Pyrochlores" Physical Review Letter 100, 045503-045506.

[15]Lang, M., Zhang, F., Zhang, J., Wang, J., Schuster, B., Trautmann, C., Neumann, R., Becker, U., & Ewing, R. C. (2009). Nanoscale manipulation of the properties of solids at high pressure with relativistic heavy ions. *Nature Materials, Volume 8,* 793–797.

[16]Sickafus, K. E., Minervini, L., Grimes, R.W., Valdez, J. A., Ishimaru, M., Li, F., McClellan, K.

J., & Hartmann T. (2000). Radiation Tolerance of Complex Oxides. *Science, Volume 289*, 748–751.
[17]Sickafus, K. E., Grimes, R. W., Valdez, J. A., Cleave, A., Tang, M., Ishimaru, M., Corish, S. M., Stanek, C. R., Uberuaga, B. P. (2007). Radiation-induced amorphization resistance and radiation tolerance in structurally related oxides. *Nature Materials, Volume 6*, 217–223.
[18]Wang, J., Ewing, R. C., & Becker, U. (2014). Defect formation energy in pyrochlore: the effect of crystal size. *Materials Research Express, Volume 1*, 035501-035515.
[19]Chen, Z. J., Xiao, H. Y., Zu, X. T., & Gao, F. (2008) First-principles calculation of defect formation energies and electronic properties in stannate pyrochlores. *Journal of Applied Physics, Volume 104*, 093702-093707.
[20]Xiao, H. Y., Gao, F., & Weber, W. J. (2010). Threshold displacement energies and defect formation energies in $Y_2Ti_2O_7$. *Journal of Physics: Condensed Matter, Volume 22*, 415801-415809.
[21]Li, N., Xiao, H. Y., Zu, X. T., Wang, L. M., Ewing, R. C., Lian, J., & Gao, F. (2007).First-principles study of electronic properties of $La_2Hf_2O_7$ and $Gd_2Hf_2O_7$. *Journal of Applied Physics, Volume 102*, 063704-063709.
[22]Li, Y., Kowalski, Y., Beridze, G., Birnie, A., Finkeldei, S., & Bosbach, D. (2015) . Defect formation energies in $A_2B_2O_7$ pyrochlores. *Scripta Materialia*, in press.
[23]Pirzada, M., Grimes, R. W., Minervini, L., Maguire, J. F., & Sickafus, K. E. (2001). Oxygen migration in $A_2B_2O_7$ pyrochlore. *Solid State Ionics, Volume 140*, 201-208.
[24]Rushton, M. J. D., Stanek, C. R., Cleave, A. R., Uberuaga, B. P., Sickafus, K. E., & Grimes, R. W. (2007). Simulation of defects and defect processes in fluorite and fluorite related oxides: Implications for radiation tolerance. *Nuclear Instruments and methods in Physics Research B, Volume 255*, 151-157.
[25]Giannozzi, P., Baroni, S., Bonini, N., Calandra, M., Car, R., Cavazzoni, C., Ceresoli, D., Chiarotti, G. L., Cococcioni, M., Dabo, I., Corso, A. D., De Gironcoli, S., Fabris, S., Fratesi, G., Gebauer, R., Gerstmann, U., Gougoussis, C., Kokalj, A., Lazzeri, M., Martin-Samos, L., Marzari, N., Mauri, F., Mazzarello, R., Paolini, S., Pasquarello, A., Paulatto, L., Sbraccia, C., Scandolo, S., Sclauzero, G., Seitsonen, A. P., Smogunov, A., Umari, P., & Wentzcovitch, R. M. (2009). QUANTUM ESPRESSO: a modular and open-source software project for quantum simulations of materials. *Journal of Physics: Condensed Matter, Volume 21*, 395502-395520.
[26] Perdew, J. P., Ruzsinszky, A., Csonka, G. I., Vydrov, O. A., Scuseria, G. E., Constantin, L. A., Zou, X., & Burke, K. (2008). Restoring the Density-Gradient Expansion for Exchange in Solids and Surfaces, Physical Review Letter, Volume 100, 136406-136409.
[27]Perdew, J. P., Burke, K., & Ernzerhof, M. (1996). Generalized Gradient Approximation Made Simple. *Physical Review Letters, Volume 77*, 3865-3868.
[28] Vanderbild, D. (1990). Soft self-consistent pseudopotentials in a generalized eigenvalue formalism. *Physical Review B, Volume 41*, 7892-7895 (R).
[29] Wang, J., Zhou, Y., & Lin, Z. (2005). First-principles elastic stiffness of $LaPO_4$ monazite. Applied Physics Letters, Volume 87, 051902-051904.
[30] Hill, R. (1952). The Elastic Behaviour of a Crystalline Aggregate. *Proceedings of the Physical Society. Section A, Volume 69*, 349-354.

[31]Plimpton, S. (1995). Fast Parallel Algorithms for Short-Range Molecular Dynamics, *Journal of Computational Physics, Volume 117*, 1-19, http://lammps.sandia.gov

[32]Ariadna-Blanca, R., Kowalski, P., Beridze, G., Schlenz, H., & Bosbach, D. (2014).Performance of DFT+*U* Method for Prediction of Structural and Thermodynamic Parameters of Monazite-type Ceramics. *Journal of Computational Chemistry, Volume 35*, 1339-1346.
[33] Gale, J. D. (1997). GULP - a computer program for the symmetry adapted simulation of solids. *Journal of the Chemical Society, Faraday Transactions, Volume 93*, 629-637.
[34] Bosenick, A., Dove, M. T., Heine, V., & Geiger, C. A. (2001). Scaling of thermodynamic mixing properties in garnet solid solutions. *Physics and Chemistry of Minerals, Volume 28*, 177-187.
[35]Mohn, C. E., Lavrentiev, M. Y., Allan, N. L., Bakken, E., & Stølen, S. (2005). Size mismatch effects in oxide solid solutions using Monte Carlo and configurational averaging. *Physical Chemistry Chemical physics, Volume 7*, 1127-1135.
[36]Vinograd, V. L. , Brandt, F., Rozov, K., Klinkenberg, M., Refson, K., Winkler, B., & Bosbach, D. (2013). Solid–aqueous equilibrium in the BaSO4–RaSO4–H2O system: First-principles calculations and a thermodynamic assessment. *Geochimica et Cosmochimica Acta, Volume 122*, 398-417.
[37]Konings, R. J. M. Walter, M. and Popa, K. (2008). Excess properties of the $(Ln_{2-2x}Ca_xTh_x)(PO_4)_2$ (*Ln*=La,Ce). *J. Chem. Thermodyn. 40*, 1305-1308.
[38] Glynn, P. (2000). Solid-solution solubilities and thermodynamics: sulfates, carbonate and halides. *in Sulfate Minerals-crystallography, Gechemistry and Environmental Significance, Reviews in Mineralogy and Geochemistry, Volume 40*, 481-511.
[39] Ziegler, J. F. Zigler, M. D., Biersack, J. P. (2010). SRIM-the stopping and range of ions in matter, *Nucl. Intum. Methods B, Volume 268*, 1818-1823.
[40]Robinson, M., Marks, N. A., & Lumpkin, G. R. (2014). Structural dependence of threshold displacement energies in rutile, anatase and brookite TiO_2. *Materials Chemistry and Physics, Volume 147*, 311-318.
[41] Robinson, M., Marks, N. A., Whittle, K. R., & Lumpkin, G. R. (2012). Systematic Calculaiton of threshold displacement energies: Case Study in Rutile. *Physical Review B, Volume 85*, 104105-104115.
[42] Blanchard, P. E. R., Clements, R., Kennedy, B. J., Ling, C. D., & Reynolds, E. (2012). Does Local Disorder Occur in the Pyrochlore Zirconates?. *Inorganic Chemistry, Volume 51*, 13237-13244.
[43] Van Dijk, M. P., De Vries, K. J. & Burggraaf, A. J. (1983). Oxygen ion and Mixed Conductivity in Compounds with the Fluorite and Pyrochlore Structure. *Solid State Ionics, Volume 9-10*, 913-919.

DEVELOPMENT OF JOINING METHOD FOR ZIRCALOY AND SiC/SiC COMPOSITE TUBES BY USING FIBER LASER

Hisashi Serizawa
Joining and Welding Research Institute, Osaka University
11-1 Mihogaoka, Ibaraki, Osaka 567-0047, Japan

Yuuki Asakura and Joon-Soo Park
Organization of Advanced Sustainability Initiative for Energy System/Material (OASIS), Muroran Institute of Technology
27-1 Mizumoto, Muroran, Hokkaido 050-8585, Japan

Hirotatsu Kishimoto
Division of Engineering, Muroran Institute of Technology
27-1 Mizumoto, Muroran, Hokkaido 050-8585, Japan

Akira Kohyama
OASIS, Muroran Institute of Technology
27-1 Mizumoto, Muroran, Hokkaido 050-8585, Japan

ABSTRACT

As for the fuel cladding in the light-water reactor, silicon fiber reinforced silicon carbide composite (SiC/SiC composite) is expected as a potential replacement of Zircaloy due to its many superiorities, where it is necessary to develop the end-cap seal of SiC/SiC composite cladding. In this research, as one possible design, the applicability of fiber laser welding for joining between SiC/SiC composite and Zircaloy tubes was examined based on the caulking method. As the methods for fitting two tubes, two types of cutting, which were flat and screw cuttings, were conducted for the inner face of Zircaloy tube and the outer surface of SiC/SiC composite. The laser beam irradiation was circumferentially applied to the outer surface of Zircaloy tube. Regardless of types of cutting, the mechanical caulking joint of two tubes was successfully produced, but the airtightness was insufficient for the end-caps of SiC/SiC composite cladding. However, the screw cutting seems to be preferable because the screw threads of Zircaloy was melted by the laser irradiation and its shape changed so that Zircaloy adhered to SiC/SiC composite partially. Namely, it can be concluded that the circumferential laser irradiation would have a good potential for producing the caulking joint of Zircaloy and SiC/SiC composite tubes with the screw cutting.

INTRODUCTION

Silicon carbide fiber reinforced silicon carbide matrix (SiC/SiC) composite is one of the promising candidates for various next generation nuclear components such as fuel clad for the fission reactor[1-3] and blanket structures for the nuclear fusion reactor[4,5] due to many superiorities[6] : high-temperature thermos-mechanical and chemical stability, radiation tolerance and so on. After the disaster at Fukushima, Japan in 2011, there have been many researches and developments on the accident-tolerant fuel (ATF) for light-water reactor and SiC/SiC composite is expected as one of the enhanced ATF claddings[1-3] instead of Zircaloy because it is stable at temperature higher than 2000 °C and would not melt under loss of coolant accident (LOCA) conditions. In spite of various studies on the enhanced ATF cladding, the end-cap seal of SiC/SiC composite cladding is still a key problem to be solved[1-3,7-10]. Although the final goal for ATF is to use SiC/SiC composite (or monolithic SiC) as the end-caps, SiC/SiC composite cladding with Zircaloy end-caps seems to be one possible design since the clear definition about severe accident may change potential fuel pin design.

As for the joining method of SiC/SiC composite and Zircaloy, the brazing was successfully conducted for SiC/SiC composite claddings[9,10] which have been employed for Halden Boiling Water Rector (HBWR) irradiation test at Halden. However, the melting temperature of brazing filler metal is generally much lower than those of target materials joined and there is a strong demand to establish the joint between SiC/SiC composite and Zircaloy without the brazing filler metal. Recently, as a results of research and development about the joining technology, the high-power fiber laser welding has been developed as one of the high beam quality heat source[11]. Because the wave length of fiber laser is 1.064 μm which is the same as that of YAG laser, the diameter of beam spot at focal position becomes much smaller than that of CO_2 laser and then the intensity of fiber laser is about 100 times higher than that of CO_2 laser. In addition, the conversion efficiency of fiber laser is about 5 times higher than that of YAG laser. Moreover, since the high-power fiber laser can produce MW/mm^2 class power density similar to electron beam, the fiber laser welding is expected to minimize the total input energy in the welding and to reduce the welding distortion and the residual stress in comparison with the ordinary arc welding[12]. In this study, the applicability of fiber laser welding for joining between SiC/SiC composite and Zircaloy tubes was examined by using 4kW fiber laser.

EXPERIMENTAL PROCEDURE

The materials used were NITE (Nano Infiltration and Transient Eutectic-phase Process) SiC/SiC composite[5] and Zircaloy-2 tubes, whose inner & outer diameters are 12.0 & 10.0 and 12.5 & 10.75 mm, respectively. The composition of matrix in SiC/SiC composite is almost stoichiometric and its structure is highly crystalline, while the chemical composition of Zircaly-2 is shown in Table 1. Since a butt joint of SiC/SiC composite and Zircaloy seems to be unrealistic due to the large differences of the melting temperature and material type, a caulking method was employed for joining these two tubes. Namely, because the coefficient of thermal expansion of Zircaloy is larger than that of SiC/SiC composite[6,13,14], the inner face of Zircaloy tube and the outer surface of SiC/SiC composite tube were machined so that SiC/SiC composite tube was fitted into Zircaloy tube. Where thickness of Zircaloy and SiC/SiC composite tubes was reduced to be about 0.5 and 0.275 mm, respectively. As shown in Figure 1, two types of cutting were conducted

Table 1 Chemical composition of Zircaloy-2 (unit : wt%)

	Cr	Fe	Ni	O	Sn	Zr
Zircaloy-2	0.104 - 0.107	0.168 - 0.179	0.068 - 0.072	0.121 - 0.125	1.38 - 1.45	Bal.

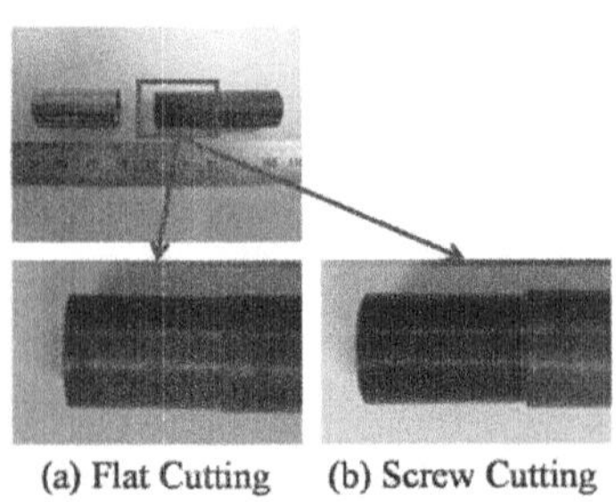

Figure 1 Photographs of Zircaloy and SiC/SiC composite tubes.

Figure 2 Photograph of experimental equipment for fiber laser irradiation.

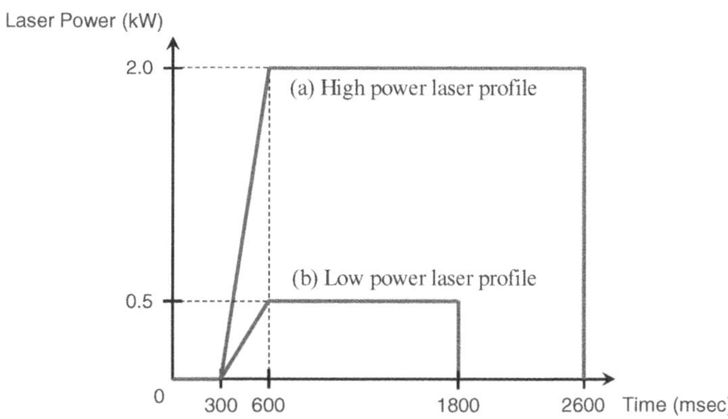

Figure 3 Laser profile for fiber laser irradiation.

for the inner face and outer surfaces. One was a flat cutting and another was a screw cutting. Then, the fiber laser beam was circumferentially irradiated on the outer surface of Zircaloy tube in order to caulk SiC/SiC composite thermally by using a special rotational apparatus. The maximum power of fiber laser is 4 kW and the diameter of beam spot at a focal position is 0.2 mm. Figure 2 shows a photograph of the experimental equipment for fiber laser irradiation. The positions of fiber laser and rotational apparatus were fixed, and Zircaloy and SiC/SiC tubes were rotated with a constant speed during the laser irradiation. The focal position was set to the outer surface of Zircaloy tube and an argon gas was used for the shielding gas whose flow was 40 L/min.

Although the final target value of airtightness for SiC/SiC composite cladding with Zircaloy end-caps is less than 2 x 10^{-10} $Pa \cdot m^3/s$, the simple bubble leak test was employed according to the following test method in order to identify the airtightness easily . The joint was dropped into the water and the air pressure, which is larger than water pressure, was applied into the tube by using the air compressor. Then, the airtightness was evaluated through the visual examination of bubbles.

WELDABILITY OF ZIRCALOY TUBE

In order to select the appropriate condition of fiber laser irradiation to Zircaloy tube, the power of laser and the rotational speed were set to be 2 kW and 50 rpm, respectively, based on the fiber laser welding condition for the stainless steel plate[15]. Because the outer diameter of Zircaloy tube was 12.5 mm, the irradiation speed was about 1.96 m/min and the time for one rotation was 1.2 s. The fiber laser profile at a first irradiation test was as shown in Figure 3(a), where the irradiation time during the maximum laser power was set to be longer than 1.2 s in order to overlap the circumferential irradiation line partially. After this irradiation test, Zircaloy tube was heated up to more than the melting temperature and the tube partially melted. So, in order to reduce the total heat input energy, the laser power and the irradiation time were changed to 0.5 kW and 1.2 s, respectively, where the laser profile became to be as shown in Figure 3(b) and the irradiation line slightly overlapped. Figure 4 shows the overview of Zircaloy tube after the second laser irradiation test. Although the diameter of laser beam spot is 0.2 mm, the width of penetration seems to be about 2 mm. However, there were not any cracks on the outer surface and the heat affected zone was also clearly observed on the inner face. Because the thickness of Zircaloy tube for the caulking joint was reduced to be 0.5 from 0.875 mm, the second laser irradiation condition seems to be the appropriate condition for caulking SiC/SiC composite tube.

ZIRCALOY & SiC/SiC COMPOSITE TUBES WITH FLAT CUTTING

The fiber laser irradiation test was conducted to Zircaloy & SiC/SiC composite joint. As shown in Figure 1(a), the inner face of Zircaloy and the outer surface of SiC/SiC composite were machined as the flat cutting and SiC/SiC composite tube was partially fitted into Zircaloy tube, where the length of overlapping part was about 12 mm. The laser beam was circumferentially irradiated at the middle of overlapping part under the condition as shown in Figure 3(b). Figure 5 shows the overview after the laser irradiation and any cracks were not observed. The width of penetration was about 2 mm, which is the same as the result shown in Figure 4. In addition, the mechanical caulking joint of Zircaloy and SiC/SiC composite tubes was successfully produced. However, the air leak test of this caulking joint indicated that this joint is insufficient for the end-

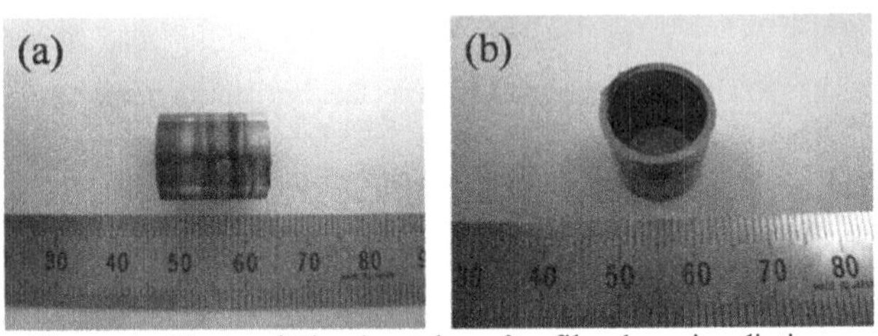

Figure 4 Photo of Zircaloy tubes after fiber laser irradiation.

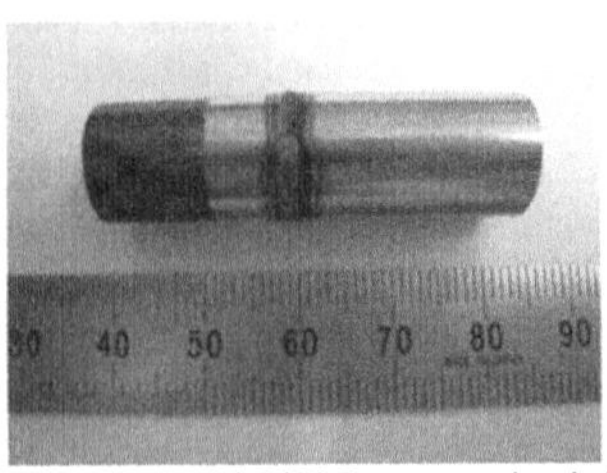

Figure 5 Photo of Zircaloy - SiC/SiC composite joint with flat cutting.

caps of SiC/SiC composite cladding. So, the microstructural observation was conducted by using scanning electron microscope (SEM) after cutting the joint along the joint length which is perpendicular to the irradiation line and Figure 6 shows the microstructure of only Zircaloy tube because the SiC/SiC composite tube was separated during the cutting process. As shown in this figure, Zircaloy tube was melted by the laser irradiation and the inner face of Zircaloy tube clearly dropped to the outer surface of SiC/SiC composite tube. In addition, as shown in Figure 6(b), the width of penetration was about 1.6 mm and a small fragment of SiC/SiC composite was attached to the inner face of Zircaloy tube. So, it is revealed that the fiber laser irradiation seems to be useful for producing the caulking joint of Zircaloy and SiC/SiC composite tubes. Then, in order to keep the tightness between Zircaloy and SiC/SiC composite during the laser irradiation, the screw cutting was employed as shown in Figure 1.

ZIRCALOY & SiC/SiC COMPOSITE TUBES WITH SCREW CUTTING

In order to improve the bonding property between Zircaloy and SiC/SiC composite during the laser irradiation, two circumferential irradiations were conducted in the overlapping part of two tubes, where the length of overlapping part was about 10 mm. In addition, the laser power was slightly increased to 0.6 kW because the thickness of Zircaloy tube at the screw thread was about 0.6 mm, which was thicker than 0.5 mm which was the thickness of Zircaloy tube with the flat cutting. Figure 7 shows the overview after the fiber laser irradiation test and the width of penetration was almost the same as the result shown in Figure 5, which was about 2 mm. There were not any cracks though the heat affected zones of two circumferential laser irradiation seem to overlap partially. In addition, although Zircaloy and SiC/SiC tubes were mechanically joined by the caulking method, the airtightness was insufficient for the end-caps of SiC/SiC composite cladding as same as the joint with the flat cutting. In order to examine the microstructural changes caused by laser irradiation, macroscopic observations were conducted by using a digital microscope. The caulking joint was cut to the half halves along the joint length and Figure 8

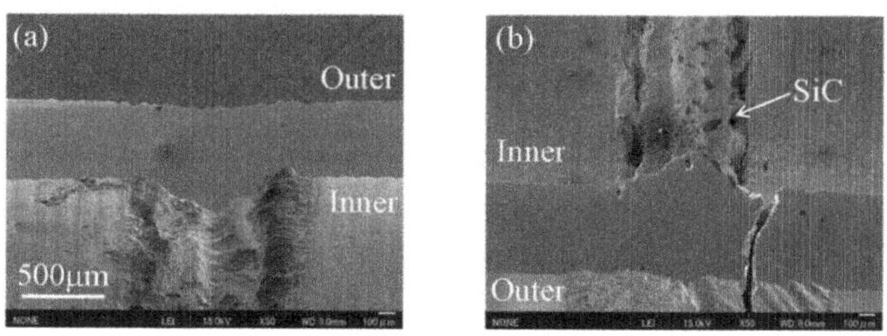

Figure 6 SEM images of Zircaloy - SiC/SiC composite joint with flat cutting.

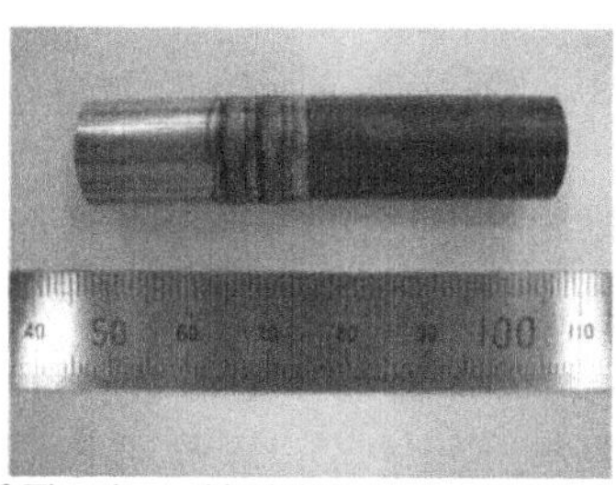

Figure 7 Photo of Zircaloy - SiC/SiC composite joint with screw cutting.

shows the macroscopic structures. As shown in this figure, there were many cracks in the inner face of SiC/SiC composite and the several parts seem to have peeled off. Figure 9 shows the cross sectional views of the overlapping part observed by using a digital microscope, where the red arrows indicate the damaged screw threads and the laser irradiation position was decided from the width of penetration. From this figure, it is found that the laser irradiation would not always induce the damaged screw threads. In other words, the occurrence of damaged screw threads seems to be independent of the irradiation position.

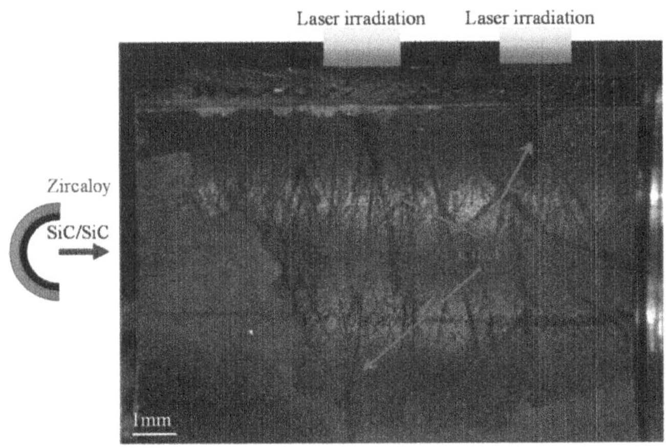

Figure 8 Digital microscope image of Zircaloy - SiC/SiC composite joint with screw cutting.

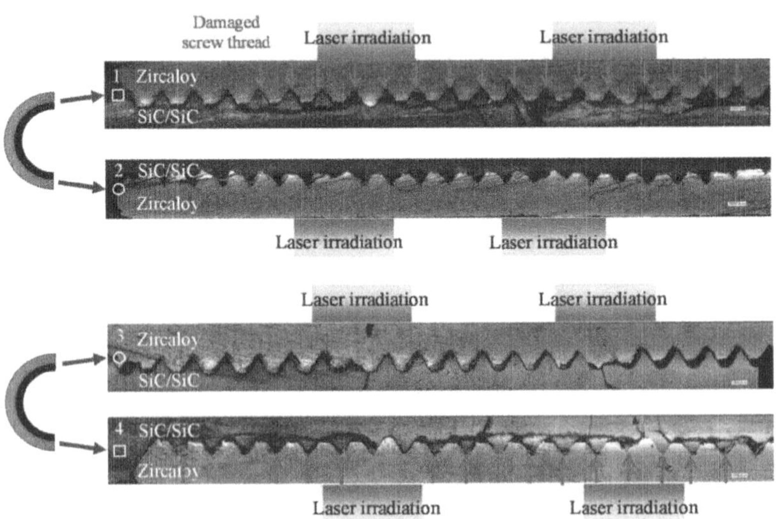

Figure 9 Digital microscope images of joint interface of Zircaloy - SiC/SiC composite joint with screw cutting.

In order to study the adhesion between Zircaloy and SiC/SiC composite, the microscopic observations near the screw threads were conducted by using SEM and the results are shown in Figure 10. Although the cracks were observed in the screw threads of SiC/SiC composite, there were not any cracks or pores in Zircaloy. In addition, the screw threads of Zircaloy was melted by the laser irradiation and its shape changed so that Zircaloy adhered to SiC/SiC composite partially. Namely, it can be concluded that the circumferential laser irradiation would have a good potential for producing the caulking joint of Zircaloy and SiC/SiC composite tubes with the screw cutting.

On the other hand, one possible reason for the damaged screw threads of SiC/SiC composite is the difference of the coefficient of thermal expansion between Zircaloy and SiC/SiC composite[6,13,14]. The coefficient of thermal expansions of Zircaloy and SiC/SiC composite were reported as 3.3 - 4.0 x 10^{-6} K^{-1} and 2.5 - 3.0 x 10^{-6} K^{-1}, respectively. Therefore, reducing the width of penetration may be effective in mitigating the thermal strain caused by the difference of thermal expansion and then the damaged screw threads of SiC/SiC composite might be prevented.

CONCLUSIONS

Based on the caulking method, the applicability of fiber laser welding for joining between SiC/SiC composite and Zircaloy tubes was examined, where a SiC/SiC composite tube was fitted into a Zircaloy tube. In addition, as the methods for fitting two tubes, two types of cutting, which were flat and screw cuttings, were conducted for the inner face of Zircaloy tube and the outer surface of SiC/SiC composite. The conclusions can be summarized as follows.

(1) By applying laser beam irradiation circumferentially to the outer surface of Zircaloy tube, the mechanical caulking joint of Zircaloy and SiC/SiC composite tubes was successfully produced regardless of types of cutting.

(2) Although airtightness of caulking joint with both flat and screw cuttings was insufficient for the end-caps of SiC/SiC composite cladding, the screw cutting seems to be preferable because the screw threads of Zircaloy was melted by the laser irradiation and its shape changed so that Zircaloy adhered to SiC/SiC composite partially.

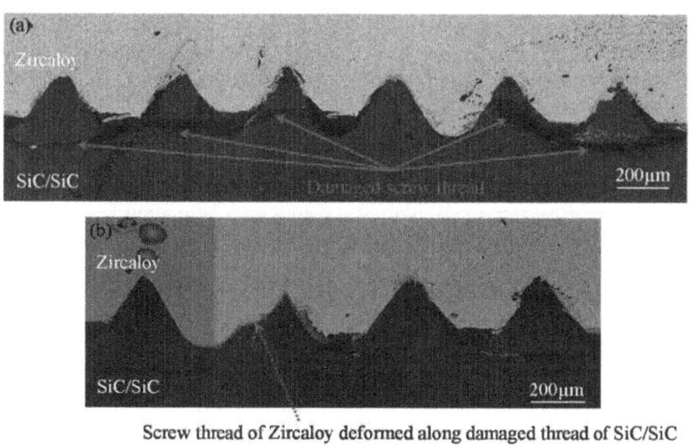

Figure 10 SEM images of joint interface of Zircaloy - SiC/SiC composite joint with screw cutting.

(3) By reducing the width of penetration produced by the laser irradiation, the thermal strain caused by the difference of thermal expansion between Zircaloy and SiC/SiC composite could decrease and then the damaged screw threads of SiC/SiC composite might be prevented.

REFERENCES

[1] E. D. Herderick, K. Cooper and N. Ames, "New Approach to Join SiC for Accidnet-Tolerant Nuclear Fuel Cladding", *Advanced Materials &Process*, Vol.170, No.1, January 2012 (2012) pp.24-27.

[2] S. B. Sitton, K. Barrett, I. Rooyen, D. Hurley and M. Khafizov, "Studying Silicon Carbide for Nuclear Fuel Cladding", *Nuclear Engineering International*, 19 April 2013 (2013). http://www.neimagazine.com/features/featurestudying-silicon-carbide-for-nuclear-fuel-cladding/

[3] S. B. Sitton, "Development of Advanced accident-tolerant fuels for commercial LWRs", *Nuclear News*, March 2014 (2014) pp.83-89.

[4] S. J. Zinkle and N. M. Ghoniem, "Operating Temperature Windows for Fusion Reactor Structural Materials", *Fusion Engineering and Design*, Vols.51-52 (2000), pp.55-71.

[5] Y. Katoh, A. Kohyama, T. Nozawa and M. Sato, "SiC/SiC Composites through Transient Eutectic-phase Route for Fusion Applications", *Journal of Nuclear Materials*, Vols.329-333 (2004) pp.587-591.

[6] S. J. Zinkle and L. L. Snead, "Thermophysical and Mechanical Properties of SiC/SiC Composites", *Fusion Materials, Semiannual Progress Report for Period Ending*, June 30 1998, DOE/ER-0313/24, Oak Ridge National Laboratory (1998) pp.93-100.

[7] M. Ferraris, M. Salvo, V. Casalegno, S. Han, Y. Katoh, H. C. Jung, T. Hinoki and A. Kohyama, "Joining of SiC-based Materials for Nuclear Energy Applications", *Journal of Nuclear Materials*, Vol.417 (2011) pp.379-382.

[8] Y. Katoh, J. O. Kiggans, C. Shih, T. Koyanagi, J. L. McDuffee and L. L. Snead, "Status of Silicon Carbide Joining and Irradiation Studies", ORNL/TM-2013/273, July 31 2013, Oak Ridge National Laboratory (2013).

[9] A. Kohyama and H. Kishimoto, "Irradiation Project of SiC/SiC Fuel Pin "INSPIRE": Status and Future Plan", *Structural Materials for Innovative Nuclear Systems (SMINS-3), Nuclear Science*, NEA/NSC/WPFC/DOC(2015)9 (2015) pp.215-222.

[10] H. Kishimoto and A. Kohyama, "SiC/SiC Fuel Cladding R&D Project "SCARLET": Status and Future Plan", *Structural Materials for Innovative Nuclear Systems (SMINS-3), Nuclear Science*, NEA/NSC/WPFC/DOC(2015)9 (2015) pp.233-240.

[11] Y. Kawahito, M. Mizutani and S. Katayama, "Elucidation of High-Power Fibre Laser Welding Phenomena of Stainless Steel and Effect of Factors on Weld Geometry", *Journal of Physics D: Applied Physics*, Vol.40 (2007) pp.5854-5859.

[12] H. Serizawa, S. Nakamura, M. Tanaka, Y. Kawahito, H. Tanigawa and S. Katayama, "Effect of Mechanical Restraint on Weldability of Reduced Activation Ferritic/Martensitic Steel Thick Plates", *Journal of Nuclear Materials*, Vol.417 (2011) pp.55-58.

[13] C. L. Whitmarsh, "Review of Zircaloy-2 and Zircaloy-4 Properties Relevant to N. S. Savannah Reactor Design", ORNL-3281, July 23 1962, Oak Ridge National Laboratory (1962).

[14] Q. Han, D. W. Kim, D. C. Kim, H. Lee, N. Kim, "Laser Pulsed Welding in Thin Sheets of Zircaloy-4", *Journal of Materials Processing Technology*, Vol.212 (2012) pp.1116-1122.

[15] J. Knaster, F. Arbeiter, P. Cara, P. Favuzza, T. Furukawa, F. Groechel, R. Heidinger, A. Ibarra, H. Matsumoto, A. Mosnier, H. Serizawa, M. Sugimoto, H. Suzuki and E. Wakai, "IFMIF: Overview of the Validation Activities", *Nuclear Fusion*, Vol.53 (2013) 116001(18pp).

Advanced Batteries and Supercapacitors for Energy Storage Applications

AN INVESTIGATION ON THE CYCLE PERFORMANCE OF $LiFePO_4$ POUCH CELLS BY A COMBINATION OF SYNCHROTRON BASED X-RAY DIFFRACTION AND ABSORPTION SPECTROSCOPY

G.T.K. Fey*[ac], Y.C. Lin[ac], K.P. Huang[ac], P.J. Wu[b], J.K. Chang[a] and H.M. Kao[a]
[a]Department of Chemical and Materials Engineering, National Central University, Chung-Li, 32054 Taiwan R.O.C
[b]National Synchrotron Radiation Research Center, Hsinchu , 30075 Taiwan R.O.C
[c]Bettery Energy Technology Inc., Chung-Li, 32063 Taiwan R.O.C.

ABSTRACT

The $LiFePO_4$ cathode with a Fe/P ratio of 0.99 has the best cyclic performance at the charge/discharge rate of 2 C/5 C by using a laminated pouch type 1.0 Ah full cell. The cycle life curves of the pouch cell show that the number of cycles increases to 2269 cycles with the capacity retention of 80% as the Fe/P ratio increases from 0.93 to 0.99. To investigate the effect of Fe concentration of stoichiometric $LiFePO_4$ on the aging mechanism of a pouch cell during long-term cycling tests, the structure of electrodes was characterized by a combination of electrochemical tests and synchrotron based X-ray diffraction and absorption spectroscopy.

To better understand the moisture effect on $LiFePO_4$/C cells under different humidity levels, we measured the water content, specific surface area and discharge capacity of commercial $LiFePO_4$/C cells over various periods of time. The water content and specific surface area increase with time and the change in surface area is irreversible. The discharge capacity of coin cells does not change with time, but the discharge capacity of pouch cells decreases due to delithiation caused by moisture.

INTRODUCTION

Lithium-ion batteries are widely developed for vehicle applications and energy storage systems owing to their high energy density, high power density, long cycle life, and lack of memory effect. The cycling stability of a Li-ion battery plays an important role in its practical application, especially for use in electric vehicles, which is generally affected by the types of electrode materials[1], binder properties[2], and electrolyte compositions[3].

The tendency for capacity degradation behavior of $LiFePO_4$ (abbreviated as LFP) cells is still seen as a problem for practical application. In recent years, as for the aging mechanisms of LFP-based cells, three main causes have been reported: (i) The lithium inventory loss resulted from side reactions at electrode/electrolyte interface relating to solid electrolyte interphase (SEI) growth and rearrangement[4]. (ii) The loss of active materials resulted from the dissolution of Fe^{3+} cations into electrolyte and/or particle isolation from the LFP electrode after long-term cycles[5]. (iii) Impedance rise associated with the internal resistance growth at the surface passivation film formed on both anode and cathode.

In this study, the capacity fading mechanism of laminated pouch type 1.0 Ah cell consisting of a graphite negative electrode and an olivine $LiFePO_4$ positive electrode during long-term cycling was studied by synchrotron X-ray diffraction (XRD) experiments.

Battery manufacturers are directing their focus to the performance of Li-ion batteries for next-generation electric vehicles (EVs) and hybrid electric vehicles (HEVs). It is known that impurities which cause capacity losses may be produced not only during powder sintering, but also during product storage in humid air[6, 7]. Therefore, we must find a method to evaluate how humid air impacts cathode materials.

X-ray absorption spectroscopy (XAS) has been widely used to study local structural aspects in a variety of systems. Although the application of synchrotron X-ray absorption near edge structure (XANES) techniques in the study of $LiFePO_4$ cathode materials has been reported by several researchers[8], the humid air effects on the cathode materials of pouch cells are still scarce. In this study, we characterized the electronic and chemical structures of dried and humid $LiFePO_4/C$ materials by X-ray absorption spectroscopy of O K-edge, Fe $L_{3,2}$-edge, and C K-edge. This might also be a useful method for controlling the quality of pouch cells.

EXPERIMENTAL

The $LiFePO_4/C$ powders used in this study were obtained from Bettery Energy Technology, Inc., Taiwan, R.O.C. To investigate the effect of Fe concentration of stoichiometric LFP on the aging mechanism during long-term cycling tests, the $LiFePO_4$ materials have the mole ratios of iron to phosphorus in the range of 0.93 to 0.99.

The pristine $LiFePO_4/C$ powder with the Fe/P ratio of 0.99 without any treatment was noted as sample A, which was dried at 245 ℃ under vacuum of 10^{-2} torr for 16 h. The dried powder was noted as sample B. Then, the dried powder was exposed in the air of 60% humidity for one day. The exposed powder was noted as sample C. The water content of samples A, B, and C is 5538, 1107, and 2625 ppm, respectively.

A laminated pouch type 1.0 Ah cell consisting of a graphite anode and an olivine $LiFePO_4$ cathode was made by Bettery Energy Technology, Inc., as shown in Figure 1. The discharge curves of pouch cells with samples A and B were measured at the charge/discharge rate of 1 C in the cut-off voltage range of 2.0–3.6 V. The pouch cells containing the dried $LiFePO_4$ cathode with different Fe/P ratios in the range of 0.93 to 0.99 were cycled at the charge/discharge rate of 2 C/5 C in the same voltage range.

Synchrotron X-ray diffraction (XRD) experiments were performed at the beamline 01-C1 of the National Synchrotron Radiation Research Center (NSRRC), Taiwan. All diffraction patterns in this study were collected using an imaging plate and a wavelength of 0.10332 nm selected for the measurements. Synchrotron X-ray absorption spectroscopy (XAS) was conducted by using the beamline 24-A1 in NSRRC and a Kratos Axis Ultra spectrometer using a focused monochromatized Al Kα radiation (hν = 1150 eV). The constant analyzer energy (CAE) mode was used with analyzer pass energy of 400 and 800 eV for preliminary wide scan. Synchrotron based XRD allowed us to access the structure information of impurities, and XAS was conducted to investigate the local atomic and electronic structure of these materials.

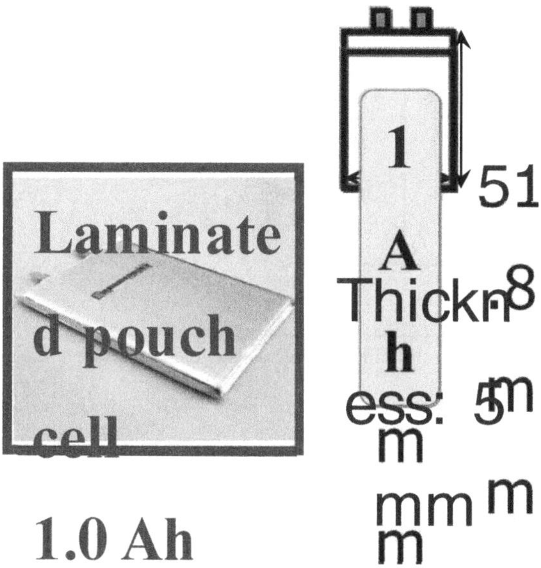

Figure 1: Schematic diagram of a laminated pouch type 1.0 Ah cell.

RESULTS AND DISCUSSION

The capacity fading mechanism of a laminated pouch type 1.0 Ah cell consisting of a graphite anode and an olivine $LiFePO_4$ cathode during long-term cycling tests was studied in this report. The $LiFePO_4$ cathodes have a mole ratio of iron to phosphorus in the range of 0.93 to 0.99. As shown in Figure 2, the LFP cell with the Fe/P ratio of 0.99 has the best cyclic performance at the charge/discharge rate of 2 C/5 C. The cycle life curves of the pouch cell display that the number of cycles increases to 2269 cycles with the capacity retention (C.R.) of 80% as the Fe/P ratio increases from 0.93 to 0.99.

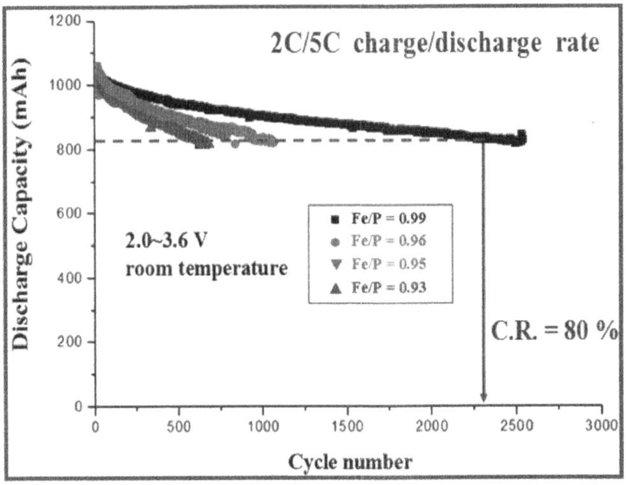

Figure 2. Capacity retention vs. cycle number for the LFP/graphite cells cycled at different mole ratios of Fe/P.

To investigate the aging mechanism of the $LiFePO_4$ cathode and graphite anode, the changes of the crystalline structure through the charge/discharge reactions were analysed by the synchrotron based X-ray diffraction (XRD) experiments. After long-term cycling tests, the LFP cathode with the 0.95 ratio shows well-crystallized diffraction peaks indexed to the delithiated orthorhombic $FePO_4$ in Figure 3(b), but the one with 0.99 ratio displays the two phase $LiFePO_4/FePO_4$ in Figure 3(d).

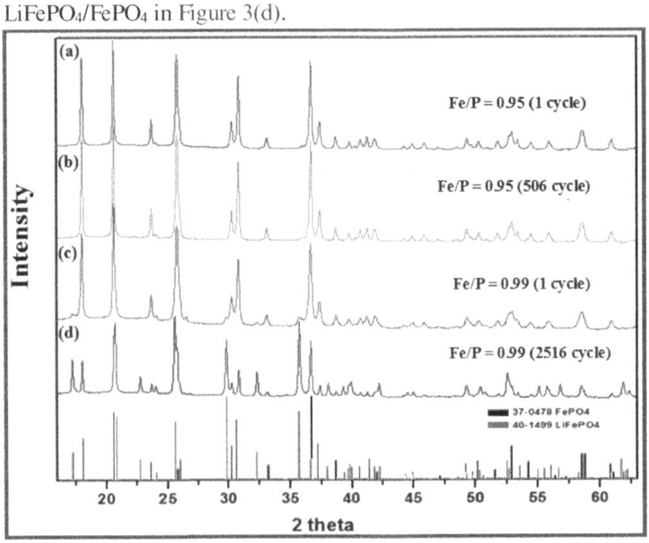

Figure 3. XRD patterns of the cycled $LiFePO_4$ cathodes with different Fe/P mole ratios were performed at the 2 C/5 C charge/discharge rate in the fully charge state.

Furthermore, the appearance of LiOH and LiOH · H_2O phases in the cycled anodes reveals that some Li ions released from the $LiFePO_4$ phase during the charge process did not come back to the $FePO_4$ phase, as shown in Figure 4. Due to the lack of lithium ions, some of the cathode materials lost its electrochemical activity and this brought down the reversible capacity significantly. In addition, the anode with the Fe/P ratio of 0.95 after 506 cycles has relatively stronger LiOH and LiOH · H_2O peaks than those of the other samples in Figure 4(b). It indicates that the decrease of the Fe concentration of stoichiometric $LiFePO_4$ resulted in a shortage of lithium ions for the cathode material. Consequently, the capacity fading rate increased.

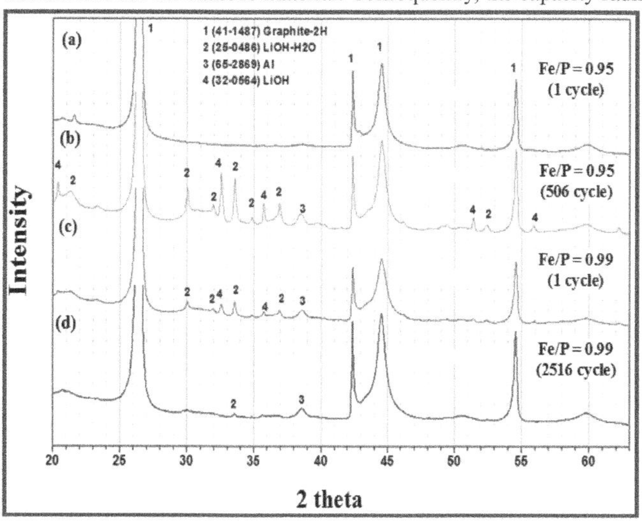

Figure 4. XRD patterns of the cycled graphite anodes corresponding to the LFP cathodes with different Fe/P mole ratios were performed at the 2 C/5 C charge/discharge rate in the fully charge state.

The water content and specific surface area of sample A increase with time up to 1000 hrs, after which it levels off, as shown in Figure 5. The surface area of samples A and B increases with time from Figure 6, meaning that the change in surface area of both samples is irreversible.

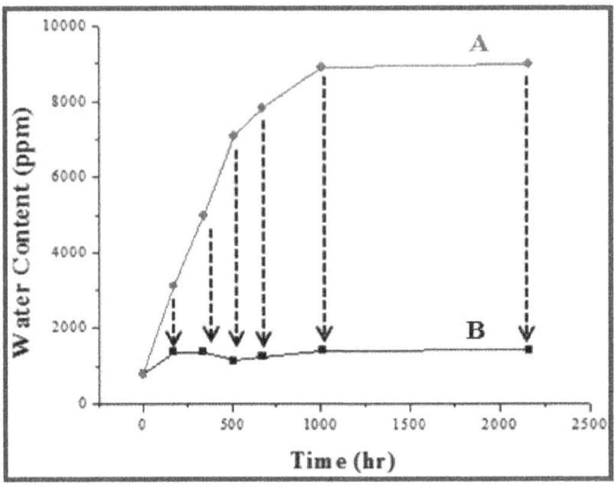

Figure 5. Water contents of samples A and B at different times in the air.

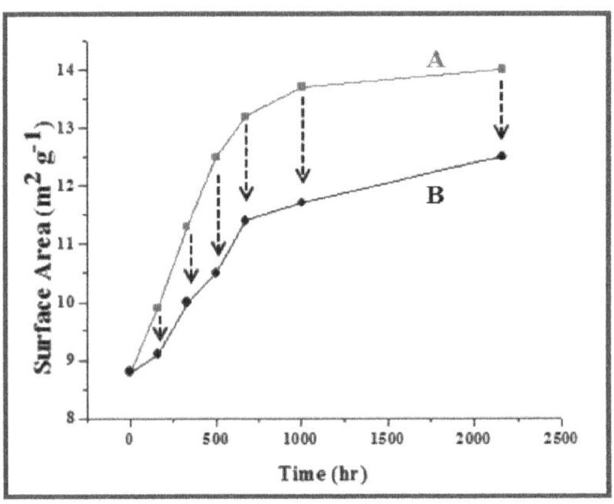

Figure 6. Specific surface areas of samples A and B at different times in the air.

The Fe $L_{3,2}$-edge XANES spectra of samples A, B, and C are displayed in Figure 7. The spectra exhibit two main peaks at around 707 and 710 eV, corresponding to Fe(II) and Fe(III), respectively[9]. The surface sensitive spectra taken by using auger electron yield (AEY) mode reveal a slightly larger Fe^{2+}/Fe^{3+} ratio of sample B with respective to the other two samples. However, the total electron yield data (deeper information) show a dramatic increase in Fe^{2+}/Fe^{3+} ratio of the dried sample B, but the ratio is lower for sample C exposed to humid air. This result indicates that the humid air can affect the Fe states in the deeper region of $LiFePO_4$/C cells, implying a delithiation process. This phase transition is reversible through a drying process. Thus it is conceivable that we can quantify humid air impacts based on the peak intensity of the Fe ions.

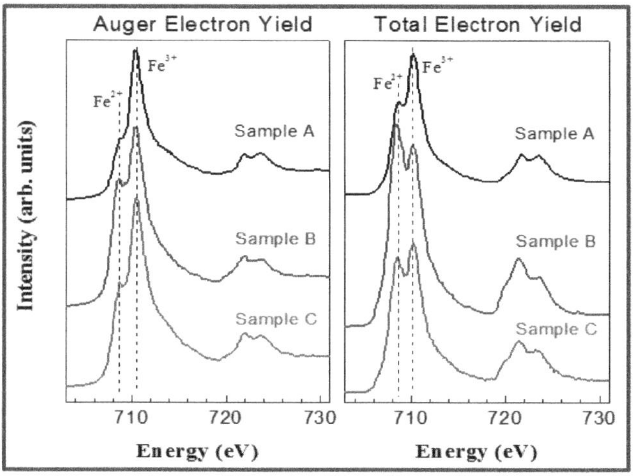

Figure 7. Fe $L_{3,2}$-edge spectra of samples A, B, and C.

The discharge curves of pouch cells (carbonaceous materials as anode) and 2032 coin cells (metallic lithium as anode) of samples A and B at 1 C are shown in Figure 8. Sample B delivers higher discharge capacities of the pouch cell than sample A at both 298 and 333 K, showing the lithium-deficient $LiFePO_4$/C cell will decrease the discharge capacity. However, it is clear both discharge capacities of coin cells are almost the same, showing humid air didn't affect the capacity of coin cells. It was reported humid air causes de-intercalation of lithium rather than the formation of an impurity if the timeframe in air is shorter, such as one day. The missing Li^+ of coin cells can be retrieved because all the available Li^+ in coin cells can be supplied by the anode (metallic lithium)[10].

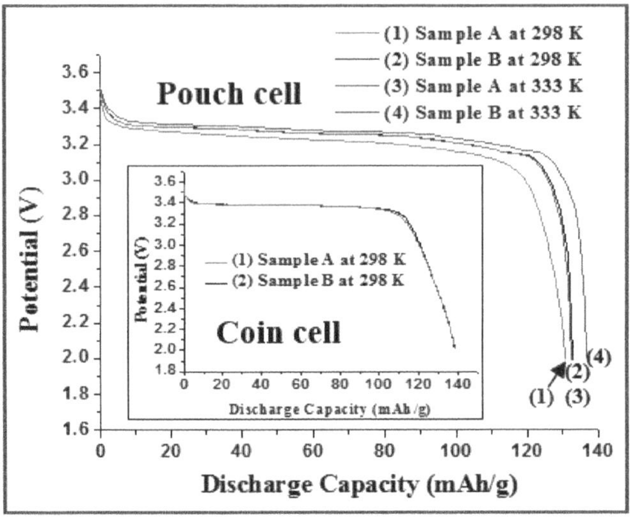

Figure 8. The discharge curves of samples A and B of pouch cells at 1 C. Insert: The discharge curves of 2032 coin cells at 1 C.

CONCLUSIONS

The decrease of the Fe concentration of stoichiometric $LiFePO_4$ material resulted in a shortage of lithium ion for the cathode material, leading to the deposits of lithium compounds on the graphite anode. Due to fewer lithium ions, some of the cathode material lost electrochemical activity and this brought down the reversible capacity significantly. Consequently, the capacity fading rate deteriorated rapidly.

The moisture effect on $LiFePO_4$/C cells under different humidity levels has been studied by XANES spectra. The results indicate that humid air can affect the Fe states in the deeper region of $LiFePO_4$/C cells, implying a delithiation process. This phase transition is reversible through a drying process. It is conceivable that we can quantify humid air effects based on the intensity of the Fe XANES spectra.

The discharge capacity of pouch cells was decreased by the delithiation process (missing Li^+). However, it can be retrieved in coin cells because all the available Li^+ in coin cells can be supplied by the Li metal anode.

ACKNOWLEDGEMENTS

The authors would like to thank the NCU, NSRRC, and BET Inc. for its support.

REFERENCES

[1] Han, X., Ouyang, M., Lu, L., Li, J., Zheng, Y., & Li, Z. (2014). A Comparative Study of

Commercial Lithium Ion Battery Cycle Life in Electrical Vehicle: Aging Mechanism Identification. Journal of Power Sources, Volume 251, 38-54.

[2] Zhang, Z., Zeng, T., Qu, C., Lu, H., Jia, M., Lai, Y., & Li, J. (2012). Cycle Performance Improvement of $LiFePO_4$ Cathode with Polyacrylic Acid as Binder. Electrochimica Acta, Volume 80, 440-444.

[3] Zhang, S.S. (2006). A Review on Electrolyte Additives for Lithium-ion Batteries. Journal of Power Sources, Volume 16, 1379-1394.

[4] Spotnitz, R. (2003). Simulation of Capacity Fade in Lithium-ion Batteries. Journal of Power Sources, Volume 113, 72.

[5] Tan, L., Zhang, L., Sun, Q., Shen, M., Qu, Q., & Zheng, H. (2013). Capacity Loss Induced by Lithium Deposition at Graphite Anode for $LiFePO_4$/graphite Cell Cycling at Different Temperatures. Electrochimica Acta, Volume 111, 802-808.

[6] Zaghib, K., Dontigny, M., Charest, P., Labrecque, J. F., Guerfi, A., Kopec, M., Mauger, A., Gendron, F., & Julien, C. M. (2008). Aging of $LiFePO_4$ upon Exposure to H_2O. Journal of Power Sources, Volume 185, 698.

[7] Julien, C. M., Mauger, A., Ait-Salah, A., Massot, M., Gendron, F., & Zaghib, K. (2007). Nanoscopic Scale Studies of $LiFePO_4$ as Cathode Material in Lithium-ion Batteries for HEV Application. Ionics, Volume 13, 395.

[8] Zhao, T., Chu, W., Zhao, H., Liang, X., Xu, W., Yu, M., Xia, D., & Wu, Z. (2010). XAS study of $LiFePO_4$ synthesized by solid state reactions and hydrothermal method. Nuclear Instruments and Methods in Physics Research, Volume A619, 122-127.

[9] Laffont, L., Delacourt, C., Gibot, P., Yue Wu, M., Kooyman, P., Masquelier, C., & Tarascon, M. (2006). Study of the $LiFePO_4/FePO_4$ Two-Phase System by High-Resolution Electron Energy Loss Spectroscopy. Journal of Chemistry Materials, Volume 18, 5520.

[10] Guo, X. F., Zuo, J., Zhan, H., & Zhou, Y. H. (2011). Storage Stability of $LiFePO_4$/C in Humid-hot Air. Chinese Journal of Chemistry, Volume 29, 2337.

THE INFLUENCE OF THE SYNTHESIS ROUTE ON ELECTROCHEMICAL PROPERTIES OF SPINEL TYPE HIGH-VOLTAGE CATHODE MATERIAL $LiNi_{0.5}Mn_{1.5}O_4$ FOR LITHIUM ION BATTERIES

M. Seidel[1], K. Nikolowski[2], M. Wolter[2], I. Kinski[3], A. Michaelis[1,2]
[1]TU Dresden, Institut für Werkstoffwissenschaften, Helmholtz. 7, 01069 Dresden, Germany
[2]Fraunhofer IKTS, Winterbergstr. 28, 01277 Dresden, Germany
[3]Fraunhofer IKTS, Michael-Faraday-Str. 1, 07629 Hermsdorf, Germany

ABSTRACT

The application of high-voltage cathode materials in lithium-ion batteries is of high interest to increase the batteries energy density. The cathode material $LiNi_{0.5}Mn_{1.5}O_4$ is a promising candidate, but very different electrochemical results are reported in literature. After all, this differences are highly depending on the specific synthesis route. But, for a final commercialization, a fundamental understanding is lacking, which is needed to optimize the particle morphology for slurry and electrode fabrication on one hand, and regarding the electrochemical properties on the other.

In this study, the synthesis of the high-voltage cathode material $LiNi_{0.5}Mn_{1.5}O_4$ by spray drying and its successful application in lithium-ion batteries are presented. The focus is on discussing the influence of two different synthesis routes, namely the application of different precursor compositions, on particle morphology and finally on electrochemical properties. It is shown that by using mixed precursors of acetates and nitrates a higher specific capacity is achieved compared to a route using only acetates. The deviating electrochemical behavior is connected with the different Mn^{3+} content observed in the samples. The relationship between electrochemical properties and particle morphology is consequently discussed.

INTRODUCTION

The foreseeable widespread use of lithium-ion batteries for e-mobility applications and the storage of renewable energy puts high demands on the used electrode materials: They have to ensure high energy and power density for batteries with a prolonged cycle life. Additionally, they should be cost-effective, reliable and environmental-friendly.

In order to meet these requirements, the properties of the well-known materials have to be further developed and optimized at different levels. To realize high energy density, a higher specific capacity as well as a high cell voltage are necessary in comparison to systems currently used. In order to achieve high power densities and a long life-time, it is necessary to tune exactly the particle size and morphology already during the synthesis process. This is the only way to create optimal conditions for ion and electrode transport in the solid and at the interfaces (e.g. to the electrolyte).

An interesting cathode material which allows higher cell voltages than the nowadays widely used materials is the high voltage material $LiNi_{0.5}Mn_{1.5}O_4$[1, 2, 3]. The material $LiNi_{0.5}Mn_{1.5}O_4$ was synthesized by partially substituting manganese with nickel in $LiMn_2O_4$ to improve the properties and maintaining the spinel type structure[4, 5, 6].

The electrochemical properties of $LiNi_{0.5}Mn_{1.5}O_4$ exhibit a high potential vs. metallic lithium of 4.7 V and a theoretical capacity of 147 mAh/g. $LiNi_{0.5}Mn_{1.5}O_4$ crystallizes in two different space groups, depending on synthesis conditions and ordering of cations: in the spinel-structure type with space group Fd-3m (number 227) and an ordered structure with space group $P4_332$ (number 212). Different approaches to improve the electrochemical properties were reported in literature[7, 8, 9, 10], including doping and coating the primary particles. Furthermore different synthesis routes have been applied. Nevertheless the $LiNi_{0.5}Mn_{1.5}O_4$ materials synthesized by different routes show deviating electrochemical properties.

To illustrate this effect literature results are compared in Table I. It is notable that although different routes using several precursor combinations have been applied, in all cases electrochemical active materials have been obtained. The capacities for the synthesized material vary roughly between 110 and 140 mAh/g. Nevertheless a precise comparison is hardly feasible, since parameters like electrode composition and preparation, electrochemical cell and test sequence differ between publications. Furthermore, the deviations cannot be attributed to single material properties like particle size, surface area or stoichiometry.

Table I. Comparison of starting materials, crystallographic structure and specific capacity of $LiNi_{0.5}Mn_{1.5}O_4$ synthesized using different synthesis routes.

Synthesis route	Starting materials	Structure and morphology	Specific capacity ($mAhg^{-1}$) at given C-rate
Spray drying	Li-, Ni- and Mn-acetates[11]	Fd-3m , octhahedra	142 (0.1 C)
	Li-, Ni- and Mn-acetates[12]	Coexistance of Fd-3m and $P4_332$, < 1 µm (granulates ~ 10 µm)	117 (0.1 C)* 129 (0.1 C)†
	Li-, Ni- and Mn-acetates[13]	Fd-3m	132 (0.1 C)
	Li-hydroxide, Ni-nitrate and Mn-acetate[14]	$P4_332$, agglomerated polyhedra	128 (3.5 C)
Precipitation	Li-hydroxide, Ni- and Mn-sulfate[15]	Fd-3m high crystallinity for T > 800 °C	>125 (1 C)
Sol-Gel	Li-, Ni- and Mn-acetates[16]	Fd-3m for > 750°C, high crystallinity for T > 800 °C	115-133 (0.1 C)
	Li-, Ni- and Mn-acetates[17]	$P4_332$ (600 °C), pseudo-octahedral shape 2 - 4 µm. (1000 °C)	133-107 (0.5 C)
Hydrothermal	b-MnO_2-NRs, Li-hydroxide, Ni-nitrate[18]	$P4_332$, rods	120 (20 C)
Solid-state	Li, Ni- and Mn-nitrate[19]	Fd-3m	~130 (0.5 C)
	Li, Ni- and Mn-nitrate[19]	$P4_332$	110 (0.2 C)

For this work spray drying has been chosen as synthesis route. In the following the influence of the precursors used for the synthesis of $LiNi_{0.5}Mn_{1.5}O_4$, namely mixtures of acetates

* as prepared
† after granulation

and/or nitrates is compared, and the different electrochemical properties for the synthesized powders in context of the particle morphology is discussed.

EXPERIMENTAL

Two different precursors where synthesizes by spray-drying. Precursor 1 was produced using $LiNO_3$ (99 %, abcr), $Ni(NO_3)_2 \cdot 6H_2O$ (99.9 %, abcr) and $Mn(CH_3COO)_2 \cdot 4H_2O$ (99 % Sigma-Aldrich) and precursor 2 using $CH_3COOLi \cdot 2\,H_2O$ (99 %, Sigma-Aldrich), $Ni(CH_3COO)_2 \cdot 4H_2O$ (99 %, Sigma-Aldrich) and $Mn(CH_3COO)_2 \cdot 4H_2O$ (99 % Sigma-Aldrich).

Stoichiometric amounts of the materials where dissolved in ethanol and the precursor was prepared via spray-drying (Büchi Mini Spray Dryer B-290). The precursors were heat treated at a temperature of 800 °C for 5 h under air, grinded and heat treated again at 800 °C for 5 h under air.

The SEM-images were taken using a Crossbeam NVISION 40 from Carl Zeiss SMT with an Everhart Thornley detector. The electrodes for the electrochemical measurements have been prepared by casting a slurry of 80 % active material ($LiNi_{0.5}Mn_{1.5}O_4$), 10 % binder (PVDF, Solef) and 10 % carbon black (Super P, Timcal) on Al-foil (Hydro) and drying for 20 h at a temperature of 60 °C. The casted electrodes had a loading of 8 mg_{LNMO}/cm^2 and a film thickness of about 40 µm. For characterization coin cells vs. metallic lithium have been assembled in an argon filled glove box. LP40 (1M $LiPF_6$ in EC:DEC 1:1 w/w, BASF) was used as electrolyte.

First, the cells were charged and discharged with different C-rates from 0.1 to 2 C for two times at each C-rate. Afterwards, a cycling test was applied for 60 cycles with a rate of 1 C. At the end the samples were charged and discharge again with 0.1 C for two times. The experiments were carried out using a Basytec CTS battery tester.

RESULTS AND DISCUSSION

After calcination the crystal shape of the two specimens varied for the samples derived from different precursors. Sample 1 (obtained by heat treatment of precursor 1) shows a mushroom-shaped form with a high volume which filled the whole crucible. In contrast, sample 2 (from precursor 2) shows a hard and solid product with a significant smaller volume compared to sample 1. This might be explained by the different release of acetate and nitrate during calcination.

(a) (b)

Figure 1: SEM-images of $LiNi_{0.5}Mn_{1.5}O_4$ synthesized using different precursors: (a) sample 1, (b) sample 2.

Figure 1 shows SEM-images of two samples synthesized from precursor 1 (Fig. 1(a)) and synthesized from precursor 2 (Fig. 1(b)). It can be seen that the samples synthesized from different precursor crystallizes in different morphologies. The sample 1 is shaped in inhomogeneous crystals and planes which can hardly be related to regular polyhedron. In figure 1(b) for sample 2 synthesized from the acetate precursor, octahedral shapes can be identified.

In figure 2 the performance test for sample 1 and 2 is shown. The capacity for sample 1 and 2 show different absolute values. The measured value for sample 1 is 131 mAh/g for a C-rate of 0.1 C. The capacity of sample 2 for the same C-rate is 119 mAh/g and therefore significantly lower. Both samples show a similar reduction in capacity for increased C-rates. At a C-rate of 2 C sample 1 still has a capacity of 119 mAh/g (9 % loss) and sample 2 of 110 mAh/g (7 % loss).

The discharge curve of the two samples (Fig. 3) indicates a different amount of Mn^{3+}. Here, as an example, the 2nd discharge curve is shown, because irreversible processes occur during the 1st cycle. This leads to an increase of capacity from the 1st cycle to the 2nd cycle, which can be seen in figure 2. This makes the 1st cycle capacity incomparable to the capacities measured during the following cycles. At a voltage of 4.0 V vs. metallic lithium sample 2 shows a more significant plateau then sample 1 and therefore quite likely[3] a higher amount of Mn^{3+}. The content of Mn^{3+} in $LiNi_{0.5}Mn_{1.5}O_4$ has an influence in the electrochemical properties and can therefore be a reason for the reduction of the capacity for sample 2. The reason for the different amount of Mn^{3+} can be assigned to the different shaped products during the calcination. In literature it has been reported that the amount of oxygen in contact with the precursor during the calcination has an effect on the Mn^{3+} content and the space group produced[20].

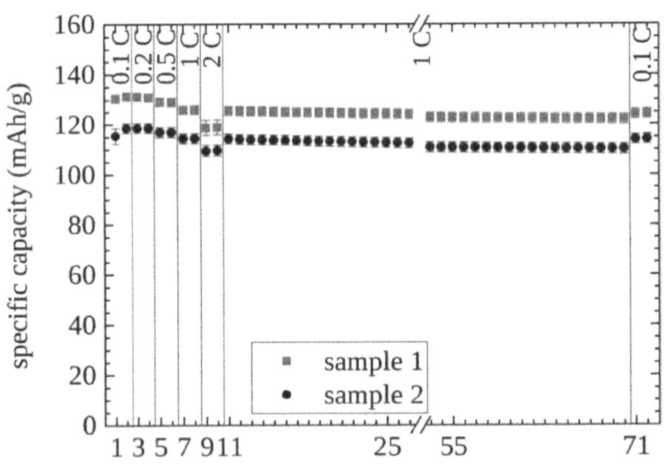

Figure 2: Discharge capacity with different C-rates for cells using $LiNi_{0.5}Mn_{1.5}O_4$ as cathode material and metallic lithium as anode material. The discharge C-rates are indicated in the figure. All cycles where charged with the same C-rate as the respective discharge.

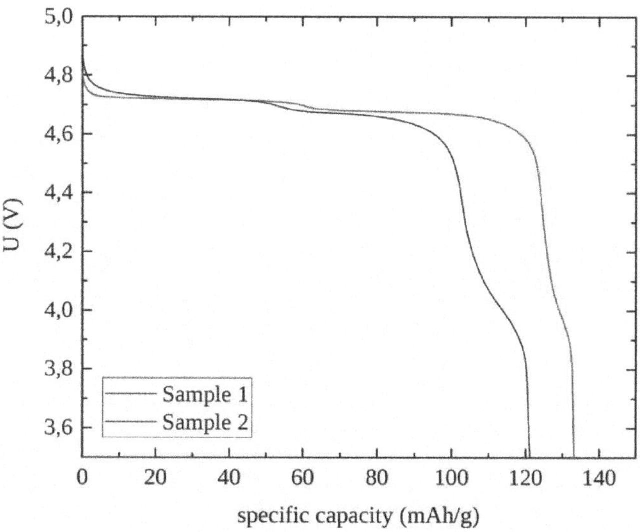

Figure 3: Second discharge curve of sample 1 and 2, respectively, at a C-rate of 0.1 C.

CONCLUSIONS

It was shown that samples crystallized from precursors consisting of different lithium, nickel and manganese compounds resulted in samples with different crystal morphology after calcination. The samples have different electrochemical properties, namely a different specific capacity, and the discharge curves indicate a different content of Mn^{3+}. The results indicate that a synthesis from a mixture of nitrates and acetate result in higher capacity and a lower amount of Mn^{3+} for the applied synthesis route and conditions. It therefore has been shown that the electrochemical performance of $LiNi_{0.5}Mn_{1.5}O_4$ can be tuned by adapting precursor chemistry, morphology and synthesis conditions.

ACKNOWLEDGEMENTS

The authors thank the German Federal Ministry of Education and Research BMBF (Bundesministerium für Bildung und Forschung) for financial support (project number 03X4637).

REFERENCES

[1] Julien, C. M.; Mauger, A. Review of 5-V electrodes for Li-ion batteries: status and trends. *Ionics* **2013**, *19*(7), 951–988. DOI: 10.1007/s11581-013-0913-2.

[2] Bhaskar, A.; Mikhailova, D.; Kiziltas-Yavuz, N.; Nikolowski, K.; Oswald, S.; Bramnik, N. N.; Ehrenberg, H. 3d-Transition metal doped spinels as high-voltage cathode materials for rechargeable lithium-ion batteries. *Progress in Solid State Chemistry* **2014**, *42*(4), 128–148. DOI: 10.1016/j.progsolidstchem.2014.04.007.

3 Liu Guoqiang. $LiNi_{0.5}Mn_{1.5}O_4$ Spinel and Its Derivatives as Cathodes for Li-Ion Batteries. In *Lithium Ion Batteries - New Developments*; Belharouak, I., Ed.; InTech, 2012; pp 83–100.

4 Guohua, L. The Spinel Phases $LiM_yMn_{2-y}O_4$ (M = Co, Cr, Ni) as the Cathode for Rechargeable Lithium Batteries. *J. Electrochem. Soc.* **1996**, *143* (1), 178. DOI: 10.1149/1.1836405.

5 Zhong Q., Bonakclarpour A., Zhang M., Gao Y. and Dahn J. R. Synthesis and Electrochemistry of $LiNi_xMn_{2-x}O_4$. *J. Electrochem. Soc.* **1997**, *144* (1), 205–213. DOI: 10.1149/1.1837386.

6 Tarascon, J. M. The Spinel Phase of $LiMn_2O_4$ as a Cathode in Secondary Lithium Cells. *J. Electrochem. Soc.* **1991**, *138* (10), 2859. DOI: 10.1149/1.2085330.

7 Li, S. R.; Chen, C. H.; Dahn, J. R. Studies of $LiNi_{0.5}Mn_{1.5}O_4$ as a Positive Electrode for Li-Ion Batteries: M3+ Doping (M = Al, Fe, Co and Cr), Electrolyte Salts and $LiNi_{0.5}Mn_{1.5}O_4/Li_4Ti_5O_{12}$ Cells. *J. Electrochem. Soc.* **2013**, *160* (11), A2166. DOI: 10.1149/2.075311jes.

8 Mo, M.; Hui, K.; Hong, X.; Guo, J.; Ye, C.; Li, A.; Hu, N.; Huang, Z.; Jiang, J.; Liang, J.; Chen, H. Improved cycling and rate performance of Sm-doped $LiNi_{0.5}Mn_{1.5}O_4$ cathode materials for 5V lithium ion batteries. *Applied Surface Science* **2014**, *290*, 412–418. DOI: 10.1016/j.apsusc.2013.11.094.

9 Lafont, U.; Locati, C.; Borghols, W.; Łasińska, A.; Dygas, J.; Chadwick, A.; Kelder, E. Nanosized high voltage cathode material $LiMg_{0.05}Ni_{0.45}Mn_{1.5}O_4$: Structural, electrochemical and in situ investigation. *Journal of Power Sources* **2009**, *189* (1), 179–184. DOI: 10.1016/j.jpowsour.2008.09.121.

10 Wu, H.; Belharouak, I.; Abouimrane, A.; Sun, Y.-K.; Amine, K. Surface modification of $LiNi_{0.5}Mn_{1.5}O_4$ by ZrP_2O_7 and ZrO_2 for lithium-ion batteries. *Journal of Power Sources* **2010**, *195* (9), 2909–2913. DOI: 10.1016/j.jpowsour.2009.11.029.

11 Wu, H.; Tu, J.; Chen, X.; Shi, D.; Zhao, X.; Cao, G. Synthesis and characterization of abundant Ni-doped $LiNi_xMn_{2-x}O_4$ (x=0.1–0.5) powders by spray-drying method. *Electrochimica Acta* **2006**, *51* (20), 4148–4152. DOI: 10.1016/j.electacta.2005.11.034.

12 Jing-kun Li, Meifeng Lu, Xiao-Zhen Liao, Zi-Feng Ma. Preparation of $LiNi_{0.5}Ni_{1.5}O_4$ cathode material through spray drying assisted annealing process and its electrochemical performance. Abstract #957. In *Pacific Rim Meeting on Electrochemical and Solid-State Sciences*; The Electrochemical Society, Ed., 2012.

13 Schroeder, M.; Glatthaar, S.; Geßwein, H.; Winkler, V.; Bruns, M.; Scherer, T.; Chakravadhanula, V. S. K.; Binder, J. R. Post-doping via spray-drying: a novel sol–gel process for the batch synthesis of doped $LiNi_{0.5}Mn_{1.5}O_4$ spinel material. *J Mater Sci* **2013**, *48* (9), 3404–3414. DOI: 10.1007/s10853-012-7127-2.

14 Li, D.; Ito, A.; Kobayakawa, K.; Noguchi, H.; Sato, Y. Electrochemical characteristics of $LiNi_{0.5}Mn_{1.5}O_4$ prepared by spray drying and post-annealing. *Electrochimica Acta* **2007**, *52* (5), 1919–1924. DOI: 10.1016/j.electacta.2006.07.056.

15 Fan, Y.; Wang, J.; Ye, X.; Zhang, J. Physical properties and electrochemical performance of $LiNi_{0.5}Mn_{1.5}O_4$ cathode material prepared by a coprecipitation method. *Materials Chemistry and Physics* **2007**, *103* (1), 19–23. DOI: 10.1016/j.matchemphys.2006.10.006.

16 Hwang, B.; Wu, Y.; Venkateswarlu, M.; Cheng, M.; Santhanam, R. Influence of synthesis conditions on electrochemical properties of high-voltage $Li_{1.02}Ni_{0.5}Mn_{1.5}O_4$ spinel cathode material. *Journal of Power Sources* **2009**, *193* (2), 828–833. DOI: 10.1016/j.jpowsour.2009.04.012.

17 Bhaskar, A. Investigations on $LiM_{0.5}Mn_{1.5}O_4$ (M = Fe, Co, Ni) Spinels as High-Volt Cathode Materials for Rechargeable Lithium-Ion Batteries. Dissertation, Technische Universität Darmstadt, Darmstadt, 2011.

[18] Yang, J.; Zhang, X.; Zhu, Z.; Cheng, F.; Chen, J. Ordered spinel $LiNi_{0.5}Mn_{1.5}O_4$ nanorods for high-rate lithium-ion batteries. *Journal of Electroanalytical Chemistry* **2013**, *688*, 113–117. DOI: 10.1016/j.jelechem.2012.09.042.
[19] Wang, L.; Li, H.; Huang, X.; Baudrin, E. A comparative study of Fd-3m and $P4_332$ "$LiNi_{0.5}Mn_{1.5}O_4$". *Solid State Ionics* **2011**, *193* (1), 32–38. DOI: 10.1016/j.ssi.2011.04.007.
[20] Pasero, D.; Reeves, N.; Pralong, V.; West, A. R. Oxygen Nonstoichiometry and Phase Transitions in $LiMn_{1.5}Ni_{0.5}O_{4-\delta}$. *J. Electrochem. Soc.* **2008**, *155* (4), A282. DOI: 10.1149/1.2832650.

Materials for Solar Thermal Energy Conversion and Storage

HIGH TEMPERATURE SOLAR RECEIVER WITH CERAMIC MATERIALS

Birgit Gobereit, Daniela Hofmann, Peter Schwarzbözl, Ralf Uhlig
German Aerospace Center (DLR)
Stuttgart/ Cologne, Germany

ABSTRACT:

Ceramic materials enable the development and design of advanced solar receivers for high working temperatures (> 600°C). Main advantage is that ceramic materials show very low creeping at high temperatures and the low thermal expansion coefficients, leading to the higher mechanical strength in comparison to metallic materials. In addition, these materials are highly resistant against corrosion or abrasion caused by heat transfer medias, such as solar salt, liquid metal or abrasive granular solids.

Currently, various receiver concepts for high temperature solar receivers with ceramic materials are under development at DLR. In open or closed type volumetric receivers a porous ceramic absorber is irradiated by the concentrated solar radiation. In this technology the heat is transferred to the fluid by taking advantage of the large surface area. Furthermore the heat must not be conducted through a wall as the absorbing surface is also the transfer surface. Another example are ceramic plate receivers, which can be used to transfer the solar radiation energy to pressurized air, which is flowing through channels inside ceramic plate modules.

Finally, in particle receivers solid granular ceramic particles are absorber, heat transfer and storage media.

Special conditions, such as the inhomogeneous solar radiation or the thermo-hydraulic boundaries from the power block, make the development of these solar receivers challenging. In this paper, the authors present an overview of the workflow and tools for receiver layout and receiver optimization for the different receiver types.

INTRODUCTION:

Concentrated solar power plants (CSP) are one of the most promising solutions to provide renewable electricity. The main advantage of CSP in comparison to other renewable technologies, such as PV or wind, is the possibility to integrate thermal storages. By this means, electricity can be provided on demand, which holds an important cost reduction potential, thanks to higher achievable power block load factor.

Figure 1 indicates the current and planned market share of CSP technologies in the market (based on data from [1]). While parabolic trough technology holds the highest market share today, the market share of power will increase significantly in the near future.

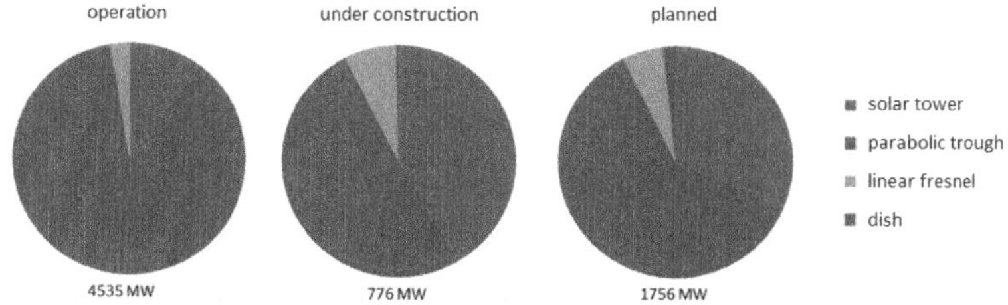

Figure 1: CSP technologies: actual market share and outlook, data from [1].

Solar power towers are point focusing systems. They have the ability to reach higher solar concentrations and consequently higher working temperatures. At the focal point a solar receiver absorbs the concentrated solar radiation and transfers the energy to a heat transfer media (HTM). Current studies aim at increasing the outlet temperature of the HTM in order to benefit from highly efficient power cycles, such as supercritical steam cycles with working temperatures in excess of 600°C or Brayton cycles. Especially in the case of steam cycles, the use of HTM at higher temperatures can enhance the thermal performance of the steam generator, while reducing the heat transfer surface. In this case, receiver outlet temperatures greater than 800°C are expected.

As a consequence, both the structural materials of the components and the HTM itself have to withstand high temperatures. Ceramic materials provide the required thermal stability. However their brittle nature must be considered very carefully during the engineering of the components. Solar receivers are particularly exposed to inhomogeneous and fluctuating solar radiation.

TECHNOLOGY OVERVIEW

In the following, some examples for solar receiver types with ceramic materials for high temperature applications and methods for the receiver development are presented and discussed.

Open Volumetric Receivers

The open volumetric receiver technology is used to heat ambient air to temperatures of about 700°C to fire the steam generator of a conventional Rankine power cycle (Figure 2). The thermal energy of the hot air can be stored as sensible heat in a parallel solid bed storage to overcome cloud passages or extend the production time in the evening.

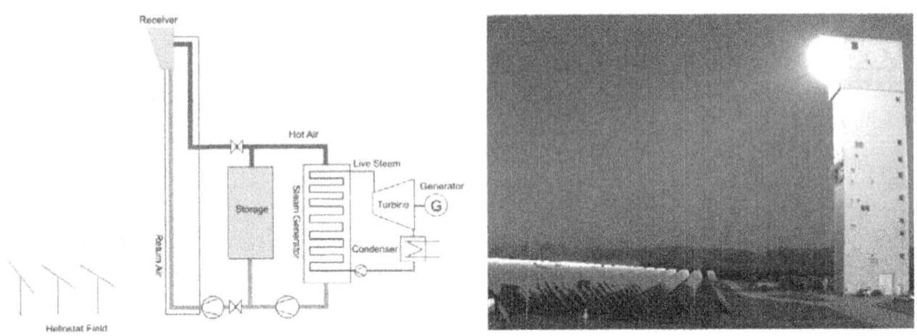

Figure 2: Solar tower power plant with open volumetric receiver technology: flow sheet (left) and demonstration plant Solar Tower Jülich in operation (right)

In the open volumetric receiver the concentrated radiation is absorbed inside a porous absorber structure and transferred to the air flow parallel to the direction of irradiation. This receiver is characterized by a large inner surface for absorption and heat transfer (hence the designation "volumetric") while it shows only a small front surface to the ambient (Figure 3). This allows using air as heat transfer fluid in combination with high radiation flux densities. The specific inner heat transfer surface per unit absorber volume usually takes values of 1000 m^2/m^3 and above [2].

Metal wire mesh, ceramic and metallic foams, metal sheets, printed ceramics and ceramic fiber mats were tested as absorber material [2, 3]. The current state of the art is an extruded

SiSiC honeycomb monolith structure of 80-200 cpsi, a spin-off from the automotive exhaust gas treatment development. The receiver technology with ceramic absorber modules of this kind was developed in several steps leading to the test of a modular 3 MWth receiver in 2003 with air outlet temperature of up to 750°C at an average flux density of up to 500 kW/m² and peak flux density of up to 900 kW/m². A complete solar tower power plant of 1.5 MWe nominal power with thermal storage is operated since 2009 as test and demonstration plant for the open volumetric receiver technology in Jülich, Germany, close to Cologne [4].

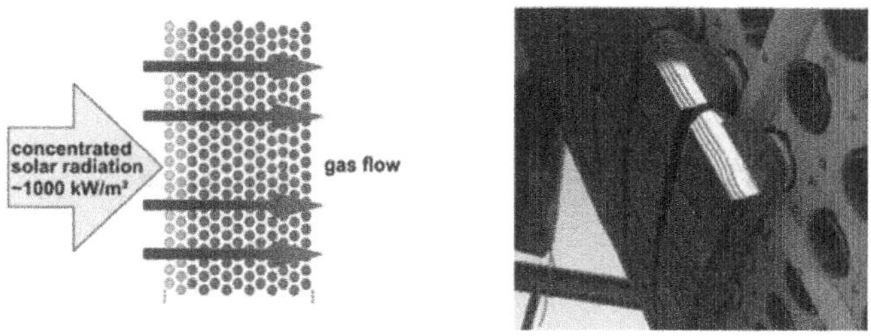

Figure 3: Principle of the volumetric absorber (left) and receiver surface covered with ceramic absorber modules (right).

Pressurized Volumetric Receivers

To solarize a power plant with a gas-turbine cycle, pressurized air from the turbine compressor is fed into solar receiver (Figure 4). After heating, the pressurized air can be heated up to higher temperatures by fossil burner and is the fed into the turbine. Integration of storage can be realized similar to the open volumetric system. Instead of storage, these plants can be operated for on demand electricity generation by hybridization (optional solar or fossil operation).

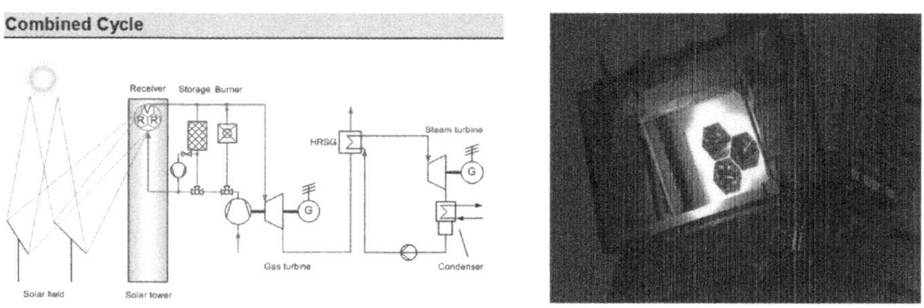

Figure 4: Schematic view of power plant with pressurized volumetric receiver in combination with a tube receiver (left) and cluster of three PVR integrated in CESA 1 tower on Plataforma Solar de Almería (right)

Pressurized volumetric receivers (PVR) consist of a pressure vessel with a quartz glass window in which metallic absorber grids and/or ceramic absorber foams are mounted in a way that cold, pressurized air can be heated up streaming through the solar flux absorber (Figure 4, left). The window has an ellipsoidal geometry in order to resist the pressure. This limits the vessel diameter. A cluster of receivers has to be used in order to cover the focal spot (see Figure 4, right schematic). Every receiver needs therefore a secondary concentrator which enables covering the gaps between them. A secondary concentrator is similar to a funnel concentrating the solar flux in the focal spot by guiding most of the flux power from the entrance aperture to the smaller exit aperture and finally into the receiver. The entrance aperture is hexagonal, in order to be able to cover the entire focal spot area.

Figure 4: Schematic of a PVR (left): pressure vessel (1), quartz window (2), inlet absorber (3), outlet absorber (4), secondary concentrator (5) and schematic of a receiver cluster (right)

The absorber consists of many plane porous absorber elements (foam) of a certain thickness which are mounted on a basket of ceramic composite material forming an elliptical dome. This geometry leads to a great surface, decreasing the solar fluxes on the absorber elements. Furthermore volumetric effects enhance efficiency: The hot absorbers emit constantly infrared rays which are mainly not lost since they are, due to the geometry, mostly absorbed by opposing absorber elements. As depicted by the arrows in Figure 4, the cold air streams alongside the inner vessel walls, is led through the (metallic) inlet absorber into the main receiver volume between window and ceramic absorber. Afterwards it streams through the (ceramic) outlet absorber and leaves the vessel. Former developments have been using metallic absorber grids solely. But with higher solar power leading to higher peak temperatures on the absorber surfaces the usage of ceramic foams becomes necessary.

Advantages of pressurized air receivers are:

- very high heat transfer due to large heat transfer area of the foam
- high working temperatures possible because of the use of ceramic foams as absorber
- volumetric effect leads to high efficiency
- useable as closed cycle for chemical reactors
- low pressure drop

Disadvantages:

- only gaseous fluids possible
- high costs (ceramics, quartz glass, secondary)
- limited size due to window
- secondary concentrator necessary (costs and additional reflection losses).
- safety risk due to brittle material under pressure (window)

Table 1: Data from actual receiver development

Pressure of fluid [bar]	10
Fluid inlet temperature [°C]	750
Fluid outlet temperature [°C]	1000
Max. metallic absorber temperature [°C]	915
Max. ceramic absorber temperature [°C]	1270
Absorbed thermal power [MW]	1.5
Receiver efficiency [%]	92
Receiver System efficiency with Secondary [%]	80
Vessel diameter [m]	1.5

Ceramic Plate Receivers

A new approach to pressurized solar receivers is the ceramic plate receiver. In this technology the solar radiation is absorbed at the surface of a ceramic module. Pressurized air flows through channels within the ceramic module and is heated up by taking advantage of the larger area of the channels (compared to the absorbing surface). Due to manufacturing limitations and to reduce failure risk a modular design was chosen for this kind of receiver. The modules (Figure 5, left) will be connected by distributor and collector pipes (Figure 5, right). Either these modules can be clustered in order to form a receiver, consisting only of ceramic modules (Figure 6, left). Or they can be combined with other receiver types, e.g. a metallic tube receiver (Figure 6, right).

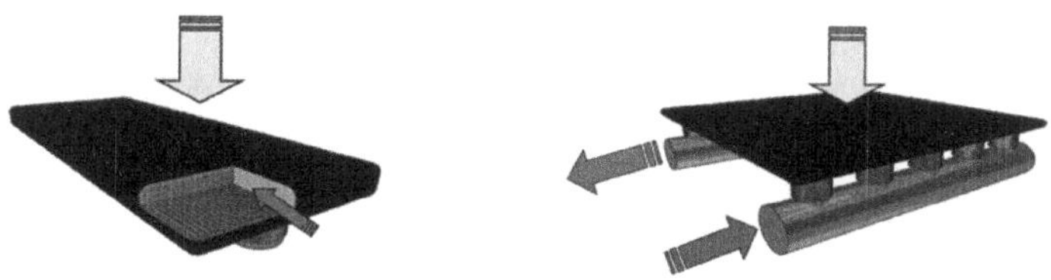

Figure 5: Single ceramic module (left), cluster of modules with distributor and collector

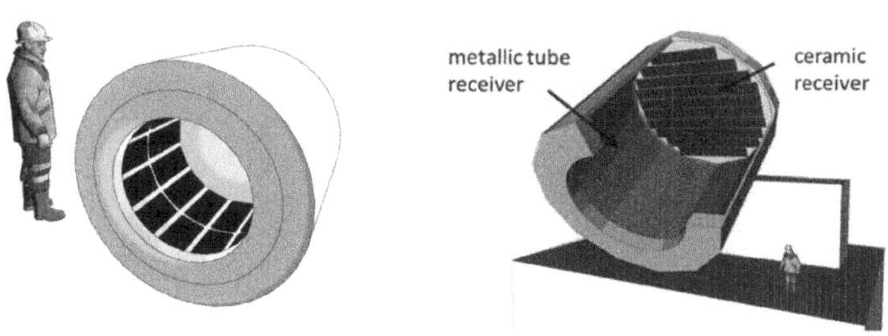

Figure 6: Examples for receiver layout; left: small scale, only ceramic modules, right: large scale, serial with metallic tube receiver

Advantages of a ceramic plate receiver are:
Possibility of high outlet temperatures due to good reliability of ceramic materials at high temperatures

- Low safety risk caused by low internal volume
- Freedom to design the inner structure (flow channels) for thermo-hydraulic optimum
- High thermal conductivity allows large heat transfer area while radiated area is small
- Modular approach allows geometric adaptation to heat flux distribution
- High stability against corrosive and abrasive heat transfer media (solar salt, liquid metal, granular material)
- closed loop (chemical processes, alternative HTM for advanced power cycles like CO2)

Disadvantages:

- sensitivity of brittle ceramic modules against tensile stresses which can be caused by the inhomogeneous and transient solar irradiation (low damage tolerance),
- necessity for a pressure-tight connection between the ceramic module and the metallic tubing.

Particle Receivers

In solid particle systems, ceramic particles are used as absorber, HTM and storage material. The solar radiation is focused on the receiver and direct absorbed by the particles. A schematic of a particle receiver plant is shown in Figure 7. Direct absorption of solar radiation in solid particles and the high temperature stability of ceramic particles are a key advantage of this concept. Also the availability of adequate ceramic particles in huge amounts and for an acceptable price is an important fact.

In comparison to the previously described air systems, the particles can be used directly as storage material. For this reason, no additional heat exchanger between HTM and storage material is necessary, thereby reducing both heat losses and costs.

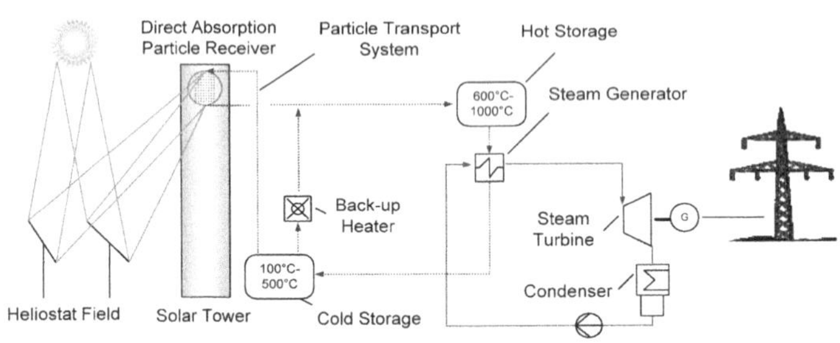

Figure 7: Schematic of particle receiver system

Two particle receiver concepts are currently under development at DLR: Falling particle and centrifugal receiver (Figure 8).

The falling particle receiver is a promising concept for large power tower systems (between 10 MW_{el} and 150 MW_{el},). Particles are released into the receiver via an inlet slit at the top and form a free falling curtain. The receiver geometry can be a cylindrical cavity with an aperture facing downwards.

Concentrated solar radiation from a surrounding heliostat field enters the aperture from below and is absorbed directly by the falling particles.

In a centrifugal receiver (CentRec), particles are introduced into a fast rotating cylinder, where centrifugal forces form an optically dense but thin layer of ceramic particles at the whole inner circumference of the receiver. The particles slide slowly along the wall in axial direction and are gradually heated by directly absorbing the incoming concentrated solar radiation. Depending on the load condition, the particle retention time in the receiver can be controlled by adjusting the rotation speed.

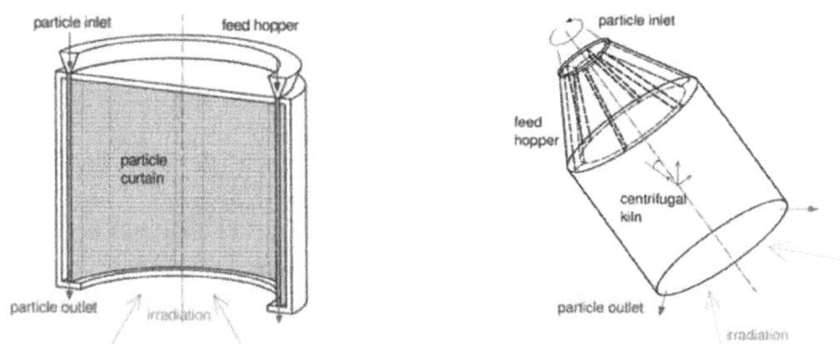

Figure 8: Schematic view of falling particle receiver (left) and centrifugal receiver (right).

LAYOUT, DESIGN AND OPTIMIZATION OF SOLAR RECEIVERS

In this section some general information about high temperature receiver development is given, followed by some specific aspect for the different receiver types. Only air receivers will be discussed in this paper. For the layout of particle receiver the authors would like to refer to [5, 6].

General Requirements for high temperature air receivers:

- Material temperature:
  - limited by maximum allowed temperatures for receiver materials
  - should be as low as possible (low thermal radiation losses, receiver life time)
- Costs:
  - as low as possible (system costs are relevant, not only receiver costs!)
  - high reliability (low O&M costs, low investment risk)
- Pressure drop
  - as low as possible (high system efficiency)
  - for pressurized air receiver: maximum pressure drop is defined by turbine specification

For all receiver layouts, solar boundaries have to be defined. Usually, this includes heliostat field layout and optimization (with simulation tools like HFLCAL [4]) and solar raytracing to determine the solar flux on the receiver surfaces (with simulation tools like SPRAY [5] or FEMRAY [9, 10]).

Based on the solar boundaries, a detailed receiver layout can be deduced. Receiver models can provide efficiency characteristics dependent on the solar load. These are important information to evaluate measures for receiver optimization and are one of the key inputs for the system analysis, which is needed to assess the influence of performance and cost on the economic potential, e.g. levelized cost of electricity (LCOE).

Open Volumetric Receivers

Unlike to tubular receiver technologies the heat transfer fluid flow in the open volumetric receiver is in parallel to the direction of radiation (Figure 3, left). Therefore, non-homogeneous radiation on the receiver surface cannot be compensated by adaptation of the flow path, but the fluid mass flow rate has to be adapted to the local radiation flux level. This done in two ways in the state of the art receiver technology:

First, the receiver is subdivided into small units, the so-called absorber modules (Figure 3, right), which are equipped with a fixed orifice of individual size to adjust the air mass flow rate. Second, the absorber modules are grouped to subreceivers, whose air mass flow can be controlled by adjustable flaps.

The usual simulation based layout procedure is to create an artificial design flux distribution representing the predominant flux situations (Figure 10, top, left). The orifice diameters are then designed to adapt the mass flow rate of the absorber modules accordingly. The individual mass flow rate is chosen to reach a uniform air outlet temperature level while keeping the material temperature limits of the absorber modules (Figure 10, top, right). In regions with very low radiation (edges and corners) the air outlet temperature is reduced to keep thermal efficiency and material temperatures at an acceptable level. Accordingly, this lower temperature and has to be compensated by higher air temperatures in the central regions (Figure 10, bottom, left). The air flaps are used during off-design operation to adapt the subreceiver mass flow rates to the shifted flux distribution (Figure 10, right).

At off-design situations with a flux density distribution very different from the design case the use of air flaps is inevitable to achieve the desired air outlet temperature and keep the material temperatures within limits ([7]).

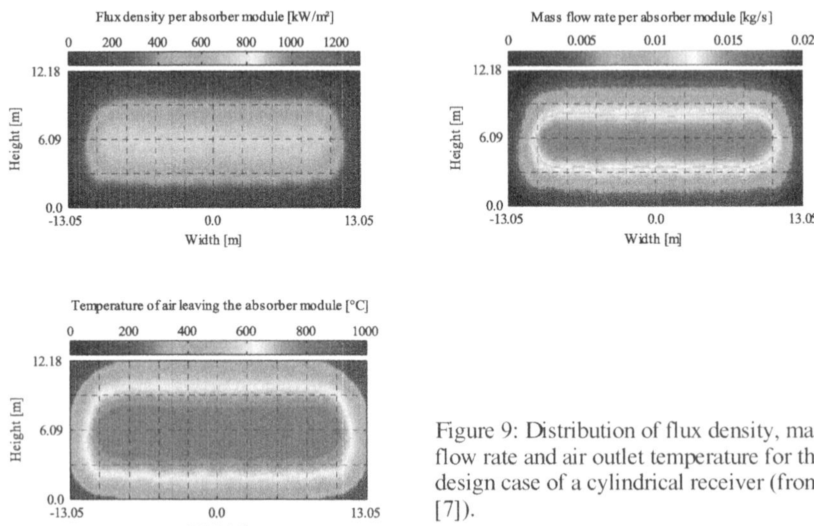

Figure 9: Distribution of flux density, mass flow rate and air outlet temperature for the design case of a cylindrical receiver (from [7]).

Pressurized Volumetric Receivers

Before the receiver layout can be computed, the secondary concentrator geometry layout has to be defined. For this purpose several secondary concentrator draft geometries are considered within an optimization progress using the simulation tool HFLCAL which identifies best field and tower properties (amount of heliostats and coordinates, tower height, receiver height and inclination angle).

The main effort in receiver design is the layout of the absorber structure. This is done within an optimization progress using the simulation tools SPRAY, SORSIM and VORECO.

SPRAY and SORSIM are ray-tracing tools. SPRAY simulates the sun rays reflected by the heliostat field reaching a defined aperture as the secondary concentrator entrance aperture or the receiver entrance aperture depending on site, date, time, heliostat field and tower properties. SORSIM is a sub tool of SPRAY, which simulates the ray traces starting from a source aperture (entrance aperture) within geometries, such as the secondary concentrator or the absorber cavity. Therefore it uses the results from SPRAY. The flux distribution on the absorber cavity (Figure 10) inside the pressure vessel is calculated by first identifying the exchange factors between the surfaces (absorber element – absorber element and window – absorber element) and then applying them to the incident flux distribution from the entrance aperture.

Figure 10: 2D and 3D plot of flux distribution on absorber, scale from 0 to 800 kW/m

VORECO is a simulation tool for the thermodynamic simulation of PVR and OVR. It identifies the temperatures of the absorber elements (Figure 11 left and middle) and the window. It needs as input the geometry data of the main receiver components (window, inlet and outlet absorber), metallic absorber data (wire diameter, grid mesh size, number of layers, ..) and/or ceramic absorber data (foam thickness, porosity, ..), fluid data (pressure, mass flow, fluid inlet temperature, ...) as well as the solar flux distribution on all components (obtained from ray-tracing simulation). Additionally for each absorber zone it is possible to define a mass flow regulation element (e.g. a perforated plate) in order to adapt the fluid flow in a way that fluid exit temperatures of the absorber elements are homogenized (see Figure 11 right).

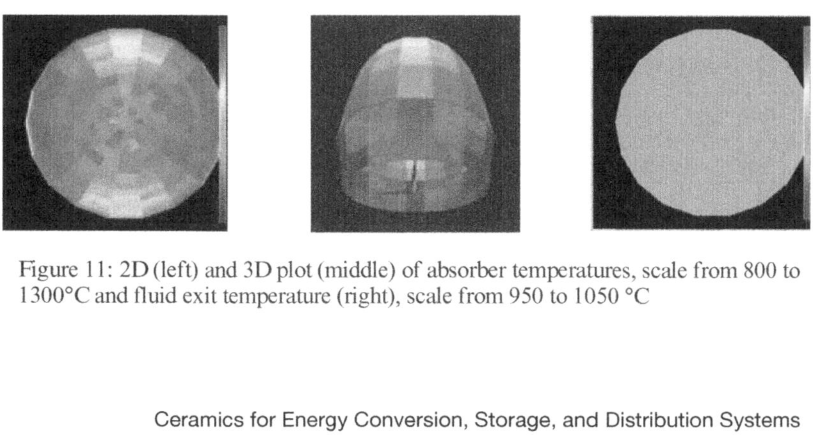

Figure 11: 2D (left) and 3D plot (middle) of absorber temperatures, scale from 800 to 1300°C and fluid exit temperature (right), scale from 950 to 1050 °C

Ceramic Plate Receivers

For the pre-design of a ceramic plate receiver a simplified parametric FE model was developed and a parameter study was performed. The workflow is showed schematically in Figure 12. The model assumes that a ceramic module consists of parallel rectangular channels with defined geometrical parameters (width w, height h, length l) and considers manufacturing constraints, such as maximum size of module or minimum size of channel.

In a first step, the pressure drop in one channel for a given geometry is determined with accepted correlations. The mass flow in one channel has to be adapted iteratively to meet the specified maximum pressure drop. Based on this, the number of channels, the irradiated area and the required net heat flux are determined. In a final step thermal stresses are determined using the thermal field and other mechanical boundaries (pressure).

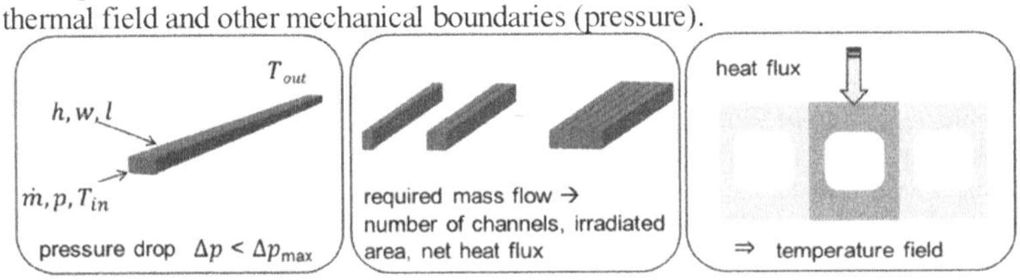

Figure 12: Workflow simple pre-design

Figure 13 displays typical results of a receiver design process: on the left side the temperature profile resulting from the net heat flux into the ceramic module and into the heat transfer fluid is shown. Resulting deformation and stresses are depicted on the right side of Figure 13.

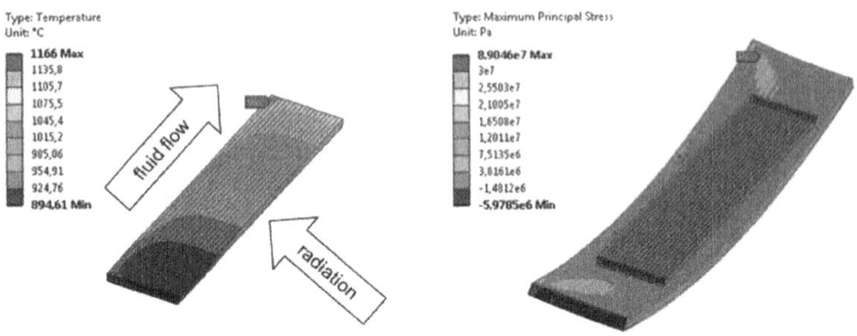

Figure 13: Typical result of simplified simulation of ceramic module; temperature field (left), deformation (not true scale) and stresses (right).

Results from a parametric study (variation of channel geometry and web width) show the dependency between maximum stress and irradiated area (Figure 14). The maximum stress is defined by the material properties. The irradiated area is an indicator for the costs. Hence, the pre-design process consists in minimizing the irradiation area while respecting the maximum stress boundary condition imposed by the material.

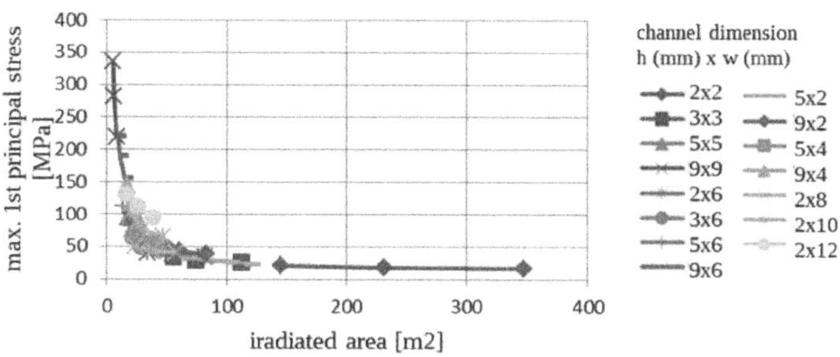

Figure 14: Example result from parametric study

CONCLUSION AND OUTLOOK

One very promising approach to minimize the LCOE of concentrated solar power systems is to increase the temperature of the heat transfer media. Ceramic materials were identified as suitable for a variety of high temperature applications. The application of ceramic materials has been studied for many years at DLR during several projects, analyzing different receiver concepts using ceramic materials. In parallel a multitude of methods and tools has been developed in order to design solar receivers for point focusing systems using ceramic materials. These tools both address solar boundary conditions, as well as specific material properties.

After defining the to be analyzed application (electricity generation, process heat, chemical reaction) ,the corresponding working fluid and the location of the plant, the receiver layout can be deduced using the above described tools. By these means the key parameters for further research focuses can be identified using sensitivity studies.

ACKNOWLEDGEMENTS

The authors would like to thank the German ministries for environment and economics and the EU for founding various research projects.

REFERENCES

[1] "Data from CSP Today, CSP Tracker," 2015.

[2] T. Fend, B. Hoffschmidt, R. Pitz-Paal and O. a. R. P. Reutter, "Porous materials as open volumetric solar receivers: Experimental determination of thermophysical and heat transfer properties," *Energy*, vol. 29, 2004.

[3] A. Ávila-Marín, "Volumetric receivers in Solar Thermal Power Plants with Central Receiver System technology: A review," *Solar Energy*, vol. 85, 2011.

[4] G. Koll, P. Schwarzboezl, K. Hennecke, T. Hartz and M. a. H. B. Schmitz, "The Solar Tower Jülich - A Research and Demonstration Plant for Central Receiver Systems.," in *Proc. of SolarPACES*, Berlin, 2009.

[5] P. Schwarzboezl, R. Pitz-Paal and M. Schmitz, "Visual HFLCAL – a software tool for layout and optimisation of heliostat fields.," in *Proc. of SolarPACES 2009*, Berlin, 2009.

[6] R. Buck and E. Teufel, "Comparison and optimization of heliostat," *J. Sol. Energy Eng.*, vol. 131, 2009.

[7] N. Ahlbrink, J. Andersson and M. a. P.-P. R. Diehl, "Optimization oft he Mass Flow Rate Distribution of an Open Volumetric Air Receiver.," *Journal of Solar Energy Engineering,* vol. 135, 2013.

[8] Y. C. Taide Tan, "Review of study on solid particle solar receivers," *Renewable and Sustainalbe Energy Reviews,* vol. 14, pp. pp 265-276, 2010.

DETERMINATION OF PARAMETERS FOR IMPROVED EFFICIENCY IN THERMAL ENERGY STORAGE USING ENCAPSULATED PHASE CHANGE MATERIALS

Laura Solomon [a], Alparslan Oztekin*[a], Sudhakar Neti [a], Himanshu Jain [b]

[a] Mechanical Engineering and Mechanics, P.C. Rossin College of Engineering and Applied Science, Lehigh University, Bethlehem, PA 18015, USA
[b] Materials Science and Engineering, P.C. Rossin College of Engineering and Applied Science, Lehigh University, Bethlehem, PA 18015, USA
* Corresponding author: *E-mail address:* alo2@lehigh.edu

ABSTRACT

Thermal energy storage has the potential to make electricity generation using solar energy possible any time of the day at competitive costs. Research toward improving the cost and efficacy of thermal energy storage systems is ongoing at several research centers. In particular, latent heat thermal energy storage systems offer the ability to store larger amounts of energy per unit mass than their sensible heat counterparts by taking advantage of the phase change of a material. Eutectic compositions in the sodium oxide - boric oxide binary system present potential novel materials for use in latent heat thermal energy storage. These materials have melting temperatures comparable to other studied phase change materials while having higher latent heat and sensible heat storage capabilities. The metal oxides considered here are very inexpensive and inert thus making good candidates for high temperature solar heat transfer fluids (HTF) as well as phase change materials (PCMs) for thermal energy storage (TES). Thus they have the potential to revolutionize the solar energy industry. In the present work, a heat transfer analysis is used to design an encapsulated phase change material capsule and to ensure crystallization of the material. The heat transfer analysis concludes that for the system considered, to ensure crystallization and related desired material behavior for the experimental methods used, the optimal capsule size is 50 mm in diameter with a capsule height of 50 mm.

NOMENCLATURE

h	Convective Heat Transfer Coefficient	[W/m^{2}K]
Fo	Fourier Number	[-]
Bi	Biot Number	[-]
J_n	Bessel Function of the first kind	[-]
k	Thermal Conductivity	[W/m K]
L	Latent Heat Of Fusion	[kJ/kg]
m	Mass	[kg]
Q	Enthalpy	[J]
r	Radius	[m]
t	Time	[s]
T_m	Melting Temperature	[K]
T_i	Initial Temperature	[K]
T_f	FinalTemperature	[K]
x	x Coordinate	[m]
c_p	Specific Heat	[J/kg K]

Greek symbols

θ	Dimensionless Temperature	[-]
α	Thermal Diffusivity	[m^2/s]
ρ	Density	[kg/m^3]

Abbreviations

CSP	Concentrated Solar Power
EPCMs	Encapsulated Phase Change Materials
HTF	Heat Transfer Fluid
PCMs	Phase Change Materials
TES	Thermal Energy Storage

INTRODUCTION

Research into replacements for current fossil fuel energy sources with renewable ones has long been an active area of research. In particular the use of solar radiation for electrical power generation using concentrating solar power (CSP) plants is promising. CSP plants produce electricity by directly using solar energy in the form of heat to drive a thermal power cycle such as the Brayton or Rankin cycle. However, while a CSP plant can produce electricity efficiently during periods of abundant solar radiation, it cannot meet the demand during periods of low radiation or at night. Currently thermal energy storage (TES) at a CSP plants is the only technology that allows for large amounts of excess energy to be stored during times of peak radiation for use at a later time, thus increasing the hours during which a CSP plant can efficiently generate electricity. Therefore, allowing the use of solar power at industrial power generation levels.

TES can be broken down into three categories; systems that use sensible heat, latent heat, and thermochemical energy storage, based on the principle of energy storage. The amount of energy stored in a material through sensible heat is dependent upon the specific heat capacity of the material and the temperature difference experienced by the material. In contrast to sensible heat storage, latent heat energy storage occurs at a nearly constant temperature during a phase change of the material. In a similar way thermochemical energy storage uses reversible chemical reactions to store and release energy. Currently, all CSP plants that have implemented thermal energy storage have done so through large sensible heat storage systems[1, 2]. By using a latent heat or thermochemical energy storage system the volume of material required could be drastically reduced. Additionally, a smaller temperature difference between the heat transfer fluid (HTF) and the storage material would be required. Thus both latent heat and thermochemical energy storage systems are a better, more efficient way to store thermal energy at a CSP plant. While thermochemical energy storage is an attractive means of TES, its development is still in the earlier stages. Therefore the present work focuses on latent heat TES.

Three different types of useful phase changes can occur in a material: solid-solid, solid-liquid, or liquid-gas. When the three phase transitions are compared for use in TES, the solid-liquid phase change is the most promising choice due to high heat of fusion and relatively small volumetric changes. Materials that have been studied for use as a phase change material (PCM) include organic materials such as paraffin and fatty acids, and inorganic salts such as sodium nitrate, potassium nitrate, and sodium chloride[1, 3, 4]. Notwithstanding, the use of PCMs for TES has its own difficulties, such as the low thermal conductivity of most currently studied PCMs. As the PCM solidifies it deposits solid PCM with lower heat transfer possibility on the surface of the heat exchanger, which due to a low thermal conductivity can increase the total time for energy transfer during the energy output from the system. In order to offset the low thermal conductivity of the PCMs, extensive studies have been conducted on ways to increase the thermal conductivity of the system, which include extending the heat transfer surface by fins or through encapsulation, adding a highly conductive material to the PCM, and employing multiple PCMs[5]. Encapsulation of a PCM increases the surface area over which heat transfer can occur, thereby decreasing the charging and discharging times and leading to better performance of the PCM[6]. Of particular interest here are encapsulated phase change materials (EPCMs). Nevertheless, encapsulation of PCMs raises additional problems, such as thermal compatibility as well as chemical and metallurgical interactions of the encapsulation material with the PCM, machinability of encapsulating materials, etc. Additionally, the encapsulation shell must allow for the volume change experienced by the PCM during the melting-solidification process while being able to withstand the thermal stresses in the encapsulation material itself[7]. This paper presents a detailed examination to determine an oxide glass for use as a PCM for TES at temperatures above 300 °C.

EVALUATION OF A METAL OXIDE FOR USE AS A PCM

Whereas numerous materials exist that have a melting temperature in almost any range required, this alone does not make them a viable PCM. The material must also exhibit other desirable thermo-physical, kinetic, and chemical properties such as a high latent heat of fusion, small volume change, sufficient crystallization rate, chemical stability, and be abundant at low cost[8]. The first consideration when choosing a potential PCM for TES is to ensure that the melting temperature of the material is within the operating range of the system. Initially, as part of the present work, twelve binary metal oxide systems, (Na_2O-SiO_2, Na_2O-B_2O_3, Na_2O-P_2O_5, CaO-SiO_2, CaO-B_2O_3, CaO-P_2O_5, Al_2O_3-SiO_2, Al_2O_3-B_2O_3, Al_2O_3-P_2O_5, K_2O-SiO_2, K_2O-B_2O_3, and K2O-P_2O_5) were considered as potential PCMs. The oxides considered here are very inexpensive and are very inert thus making good candidates for high temperature solar heat transfer fluids (HTF) as well as PCMs for TES. They have the potential to revolutionize solar energy industry. The number of systems was reduced to five (Na2O-B2O3, Na2O-P2O5, K2O-SiO2, K2O-B2O3, and K2O-P2O5) for a working temperature range of 300 °C to 1,000 °C based on the melting temperature of materials outside of the glass formation range[9, 10].

After determining which systems have eutectic materials with melting temperatures in the required range, a comparison of their latent heat of fusion was used to determine the material best suited for use as a PCM in a TES system. The compounds considered are $Na_2B_4O_7$, $NaBO_2$, $Na_4B_2O_5$, $K_2B_4O_7$, KBO_2, $Na_4P_2O_7$, KPO_3, and K_2SiO_3. The melting temperature, latent heat, and specific heats for the considered materials are listed in Table 1 along with other studied high temperature PCMs. As seen in Table 1 the borate systems have a higher latent heat than the phosphate systems and the values for the sodium oxide systems are higher than for the potassium oxide systems, but unfortunately some of the desired properties are yet to be determined. When compared to other studied PCMs for use in TES such as $NaNO_3$, sodium borates have comparable melting temperatures but have higher latent heat and specific heat values. For example the energy storage per unit mass over a 100 °C temperature range centered about the melting temperature $NaNO_3$ stores 324.4 kJ/kg whereas $Na_4B_2O_5$ stores 778.1 kJ/kg. This will results in a higher energy storage density of the system. Additionally, they should be less reactive with the encapsulation material than chloride and nitrate salts. Therefore the materials in the Na_2O-B_2O_3 are the most attractive for use as a PCM for TES.

Table I. Physical Properties of Potential PCMS

	Melting Temperature (°C)	Latent Heat (kJ/kg)	Solid Specific Heat (J/kg K)	Liquid Specific Heat (J/kg K)
$NaNO_3$	308[3]	162.5[3]	1,588[3]	1,650[3]
$MgCl_2$	714[11]	454[11]	798[4]	974[11]
NaCl	800[11]	481[11]	987[12]	1200[11]
Al	660[13]	397.3[14]	903[14]	1177[14]
$Na_2B_4O_7$	742[15]	403.5[15]	1174.3[15]	2213.3[15]
$NaBO_2$	967[15]	509.1[15]	1349.8[15]	2218.8[15]
$Na_4B_2O_5$	641[16]	617.3[15]	1166.4[15]	2048.8[15]
$K_2B_4O_7$	816[15]	446.4[15]	----	----
KBO_2	947[15]	383.4[15]	----	----
$Na_4P_2O_7$	970[17]	220.4[17]	----	----
KPO_3	810[17]	74.5[17]	----	----
K_2SiO_3	976[17]	325.4[17]	----	----

DESIGN OF $Na_4B_2O_5$ EPCM CAPSULES

Of the three eutectics in the non-glass forming region of the Na_2O-B_2O_3 system, $Na_4B_2O_5$ has the lowest melting temperature and highest latent heat, making it an ideal PCM candidate. To examine the performance of $Na_4B_2O_5$ as PCM, a sample EPCM capsule is designed for testing in an immersion calorimeter to establish the energy storage capabilities of the PCM. The design of the calorimeter has been reported in detail by Zheng et al.[3] and Zhao et. al[4]. Care must be taken as to the size of the capsule used in such calorimetry to avoid rapid cooling of the sample that would result in the formation of glass above certain temporal thermal gradients. Ota and Soga[18] conducted an extensive study into the critical cooling rates for the entire Na_2O-B_2O_3 system. Their work showed that in order to ensure crystallization of $Na_4B_2O_5$ it has to be cooled slower than 2 °C/s[18]. While the cooling rate in a TES system at a CSP plant is likely to be below this critical cooling rate, the cooling rates in immersion calorimeter experiments may not be so. To minimize the chance of glass formation (at high cooling rates in an immersion calorimeter), an analytical calculation was used to determine the size of the capsule.

The heat transfer within a 2-dimensional object can be calculated by superimposing the 1-dimensional solutions. In order to determine the radius and height of the calorimetry test samples, the solutions for a plane wall and an infinite cylinder are used together. The analytical solutions are listed below for the plane wall in equations 1 and 2 and for the infinite cylinder in equations 3 and 4[13].

$$\theta = \sum_{n=1}^{\infty} \frac{4\sin(\lambda_n)}{2\lambda_n + \sin(2\lambda_n)} e^{-\lambda_n^2 Fo} \cos\left(\lambda_n \frac{x}{L}\right) \tag{1}$$

$$\lambda_n \tan(\lambda_n) = Bi \tag{2}$$

$$\theta = \sum_{n=1}^{\infty} \frac{2}{\lambda_n} \frac{J_1(\lambda_n)}{J_0^2(\lambda_n) + J_1^2(\lambda_n)} e^{-\lambda_n^2 Fo} J_0\left(\lambda_n \frac{r}{r_0}\right) \tag{3}$$

$$\lambda_n \frac{J_1(\lambda_n)}{J_0(\lambda_n)} = Bi \tag{4}$$

$$Bi = \frac{hL}{k} \quad \text{or} \quad Bi = \frac{hL}{k} \tag{5}$$

$$Fo = \frac{\alpha t}{L^2} \quad \text{or} \quad Fo = \frac{\alpha t}{r_0^{\,2}} \tag{6}$$

In the above equations, θ is the non-dimensional temperature, Fo is the Fourier number, Bi is the Biot number, x is the distance from the plane wall centerline, L is the half thickness of the plane wall, λ_n is the nth solution to the characteristic equation, J_n is the Bessel function of the first kind, r is the distance from the axis, r_0 is the diameter of the cylinder, h is the outside heat transfer coefficient, k is the thermal conductivity, α is the thermal diffusivity, and t is time. A schematic of the capsule is given in Figure 1.

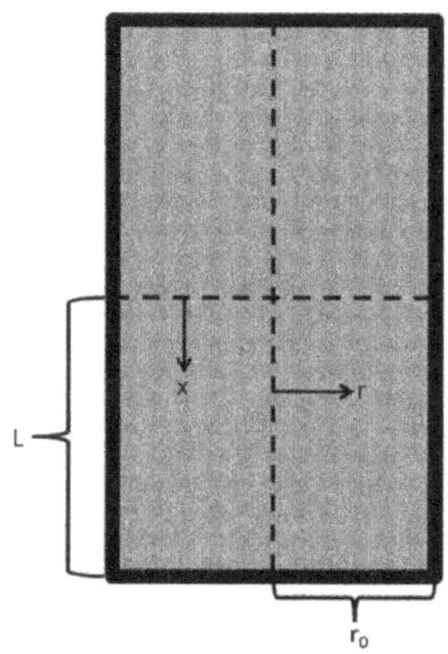

Figure 1. Schematic of EPCM Capsule

Using the above equations, the cooling rate for several sample sizes of varying heights and diameters was considered for an initial sample temperature 50 °C above the melting temperature of the PCM (641 °C). The results for a 25 mm or 50 mm diameter by 50 mm or 75 mm height capsule are presented in Figure 2. The critical cooling rate above which glass formation is feasible is also presented in the figure. From Figure 2 it can be seen that as the diameter and height of the capsule increase, the cooling rate decreases. The cooling rate within each of the capsule sizes considered is sufficiently slow that glass formation within the capsules is unlikely. It must be reiterated that the cooling considered here are only likely in the immersion calorimetry and not in CSP applications.

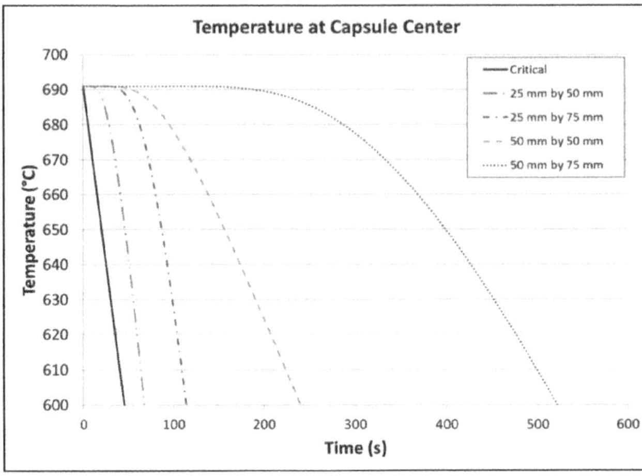

Figure 2. The temperature at the center of the EPCM capsule for a 25 mm by 50 mm (diameter by height), 25 mm by 75 mm, 50 mm by 50 mm, and 50 mm by 75 mm capsule size along with the critical cooling rate for $Na_4B_2O_5$

While the above analysis indicates using a capsule with as large a diameter and height as possible, the size of the capsule is limited by the overall temperature increase of the calorimeter. For the same capsules sizes used in the previous analysis, an energy balance was conducted to determine the maximum temperature rise of the calorimeter. The enthalpy in the capsule and the calorimeter are calculated using the following set of equations,

$$Q_{EPCM} = m_{cap}c_{p,cap}(T_f - T_i) + m_{PCM}c^l_{p,PCM}(T_f - T_m) + m_{PCM}L + m_{PCM}c^s_{p,PCM}(T_m - T_i) \qquad (7)$$

$$Q_{cal} = m_{cap}c_{p,cal}(T_{f,cal} - T_{i,cal}) \qquad (8)$$

where m_{cap} is the mass of capsule for PCM, $c_{p,cap}$ is the specific heat of the encapsulation, T_f is the final temperature of the EPCM capsule, T_i is the initial temperature of the capsule, m_{pcm} is the mass of the PCM, c_p^l is the liquid specific heat of the PCM, T_m is the melting temperature of the PCM, L is the latent heat of the PCM, c_p^s is the solid specific heat of the PCM, m_{cal} is the mass of the calorimeter, $c_{p,cal}$ is the specific heat of the calorimeter, $T_{f,cal}$ is the final temperature of the calorimeter, and $T_{i,cal}$ is the initial temperature of the calorimeter. The final temperature of the calorimeter can be determined by careful energy balance ($Q_{EPCM} = Q_{cal}$) and solving for $T_{f,cal}$.

Using the properties of the encapsulation material (stainless steel) and the solid and liquid specific heats and latent heat of $Na_4B_2O_5$ is listed in Table 1 the energy balances are calculated. The anticipated temperatures for the calorimeter fluid and heat transfer rate are calculated and the maximum temperature the calorimeter would reach for each capsule size is listed in Table 2. Based on maximum feasible parameters it is concluded that 50 mm by 50 mm or smaller EPCM can be used for the system considered here. The optimal size capsule for use to study $Na_4B_2O_5$ as a potential PCM for use is TES is a 50 mm diameter by 50 mm in height capsule.

Table II. Final Temperature of Calorimeter for Various Capsule Sizes

Capsule Size	Temperature
25 mm by 50 mm	33.70
25 mm by 75 mm	38.04
50 mm by 50 mm	59.60
50 mm by 75 mm	76.86

CONCLUSIONS

There are many factors that determine if a material is a good candidate for use as a PCM for TES, including having a suitable melting temperature, large latent heat of fusion, small volume change, low cost, corrosion and material behavior as well as abundant availability. Currently, the most studied PCM is sodium nitrate due to its worldwide availability and use at a range of solar power plants; however it has a low thermal conductivity. Suitable alternatives to sodium nitrate as a PCM include a range of oxides of eutectic composition, such as those in the range of 33% mol. to 66% mol Na_2O range of the Na_2O-B_2O_3 system. These materials have melting points comparable to other studied PCM but have higher latent heats of fusion, which would allow for smaller, more efficient storage systems. However, since these materials are able to form amorphous solids as well as crystalline solids care has to be taken in the design of the EPCM capsules to ensure sufficiently slow cooling. In the solar energy power applications this is rarely likely to be a problem since solar energy storage and discharge occur over many hours, resulting in low thermal temporal gradient systems. Using the theory of superposition of the one-dimensional conduction analysis in infinite cylinders and plane walls, the temperature within a two-dimensional cylindrical capsule was calculated for $Na_4B_2O_5$ to determine the optimal EPCM capsule size to be used in testing its TES capability. As the capsule diameter and height increase, the rate at which temperature changes within the capsules decreases. Additionally, for the four capsule sizes considered the cooling rate within the capsules was low enough that the molten PCM should crystalize easily. In general larger EPCM capsule sizes provide for much larger amounts of PCMs to be used for a given encapsulating material. This in turn provides for more PCM in the system unlike systems that use very small EPCM capsules. The final determination of the capsule size that may be used for testing is the temperature change in the calorimeter during testing. This is a self-imposed requirement based on the immersion calorimetry system that is to be used in the current work and is not a limitation on the actual TES use of these metal oxides in solar industry. The current immersion calorimetry system has a maximum operating temperature of 70 °C, which limits how large of a capsule can be used. Likewise a temperature increase greater than 10 °C is preferable to reduce error in the experimental measurements. Thus for the current applications it was determined that a capsule that is 50 mm in diameter with a height of 50 mm is deemed the optimal capsule size for calorimetry testing. Based on the properties of the three metal oxide eutectic (PCM) materials considered, a thermocline can be integrated into a CSP plant with the highest melting temperature material at the top (~ 1,240 K) and lowest melting temperature (~ 914 K) material at the bottom. The Na_2O-B_2O_3 system is an attractive material for use in TES; however additional testing and appreciation of the storage capability of the materials are required, which are underway.

REFERENCES

[1] A. Gil, M. Medrano, I. Martorell, A. Lazaro, P. Dolado, B. Zalba, and L. F. Cabeza, Renewable and Sustainable Energy Reviews **14**, 31 (2010).
[2] S. Kuravi, J. Trahan, D. Y. Goswami, M. M. Rahman, and E. K. Stefanakos, Progress in Energy and Combustion Science **39**, 285 (2013).
[3] Y. Zheng, W. Zhao, J. C. Sabol, K. Tuzla, S. Neti, A. Oztekin, and J. C. Chen, Solar Energy **87**, 117 (2013).
[4] W. Zhao, Y. Zheng, J. C. Sabol, K. Tuzla, S. Neti, A. Oztekin, and J. C. Chen, Renewable Energy **50**, 988 (2013).
[5] M. Liu, W. Saman, and F. Bruno, Renewable and Sustainable Energy Reviews **16**, 2118 (2012).
[6] D. P. Colvin, ASME-PUBLICATIONS-HTD **363**, 199 (1999).
[7] J. J. Blaney, S. Neti, W. Z. Misiolek, and A. Oztekin, Applied Thermal Engineering **50**, 555 (2013).
[8] A. Sharma, V. V. Tyagi, C. R. Chen, and D. Buddhi, Renewable and Sustainable energy reviews **13**, 318 (2009).
[9] E. M. Levin, R. S. Roth, C. R. Robbins, H. F. McMurdie, and A. C. Society, *Phase diagrams for ceramists* (American Ceramic Society, 1983).
[10] C. A. Harper, *Handbook of Ceramics, Glasses, and Diamonds* (McGraw-Hill, 2001).
[11] G. J. Janz, C. B. Allen, N. P. Bansal, R. M. Murphy, and R. P. T. Tomkins, NASA STI/Recon Technical Report N **80**, 10643 (1979).
[12] M. W. Chase and T. National Institute of Standards and, *NIST-JANAF thermochemical tables* (American Chemical Society ; American Institute of Physics for the National Institute of Standards and Technology, [Washington, D.C.]; Woodbury, N.Y., 1998).
[13] F. P. Incropera and D. P. DeWitt, *Fundamentals of Heat and Mass Transfer* (John Wiley & Sons, Inc., New York, 2002).
[14] R. A. McDonald, Journal of Chemical & Engineering Data **12**, 115 (1967).
[15] O. Kubaschewski, C. B. Alcock, and P. J. Spencer, *Materials thermochemistry* (Pergamon Press, 1993).
[16] T. M. Besmann and K. E. Spear, Journal of the American Ceramic Society **85**, 2887 (2002).
[17] F. D. Rossini, *Selected values of chemical thermodynamic properties* (U.S. Govt. Print. Off., Washington, 1952).
[18] R. Ota and N. Soga, Journal of the Ceramic Association, Japan **90**, 531 (1982).

TUNING THE SPECTRAL SELECTIVITY OF SIC-BASED VOLUMETRIC SOLAR RECEIVERS WITH ULTRA-HIGH TEMPERATURE CERAMIC COATINGS

Benoit Rousseau[1], Simon Guevelou[1], Jérôme Vicente[2], Cyril Caliot[3], Gilles Flamant[3]
[1] CNRS LTN UMR6607, Rue Christian Pauc, 44306 Nantes, France
[2] CNRS IUSTI UMR7343, Marseille, France
[3] CNRS PROMES UPR8521, Odeillo, France

ABSTRACT

Ultra-High Temperature Ceramics (UHTCs) are promising materials for developing high temperature solar air receivers (T~1300 K) with a high efficiency for converting concentrated solar energy into thermal energy. Indeed, when they are shaped as sintered pellets, they exhibit a high solar absorptivity and a low thermal emissivity. This optical behavior defines the so-called spectral selectivity, which has been intensively studied in the case of surface solar receivers. Practically, it can be envisaged to develop tubular receivers, based on UHTCs coatings. To go one step further, this work proposes to combine the good spectral selectivity of UTHCs and the volumetric effect observed on refractory ceramic open-cell foams, this effect conferring to porous volumetric receivers a higher conversion efficiency compared to tubular receivers. Starting from optical properties extracted on reflectivity spectra previously acquired on selected UHTCs single crystals (TaC, ZrC, ZrB_2), a Ray Tracing code is used to numerically compute, at T = 300 K, the radiative behaviors of open-cell foams coated with them. In particular, Ray Tracing calculations are applied within 3D digitalized images of the foams. For a given UHTC, the effect of the foam textures is discussed by comparing the normal reflectivity of single crystalline slab and the hemispherical reflectance of the foam. An analysis of the paths traveled by the rays is also proposed and allows to propose optimized solutions for coating the open-cell foams.

INTRODUCTION

Concentrated solar power (CSP) plants are hopeful technologies for electricity production with a low environmental impact, in the future.[1] Among all the developed forms, only solar power tower plants will be able to reach a solar-to-electricity conversion efficiency of 30-35% for a large scale production and with competitive costs. This expected conversion efficiency will be quite comparable to that of current fossil power plants.[2] A mandatory condition for achieving such a high overall efficiency is to work with a gaseous fluid (pressurized air or not) heated at temperatures above 1300 K in order to drive combined-cycle power blocks with a heat-to-electricity conversion efficiency of 60 %.[3] Several challenges must be overcome to bring air to such high temperatures including cheaper concentrators and high-temperature materials able to withstand high concentrated solar flux (~ 1000 suns). This latter requirement is thus crucial for designing solar air receivers which simultaneously ensure the absorption of the solar radiations and the heating of air, flowing through it, up to the targeted temperature.[4] A major attention is today paid for engineering solar air receivers with a high reliability. [5] Due to the so-called "volumetric effect" [2], porous solar receivers are tested since they offers a great potential of development.[6] In particular, research works aim to define the suitable texture of the porous material allowing both a good solar absorption[7] and a good air flow transport.[8] From all the volumetric receiver compared, those based on α-SiC open-cell foams are interesting because of their high solar absorptance, lightweight, high porosity for large penetration of the solar radiation through the volume of the receiver, high cell density to present a large surface area and high effective thermal conductivity to remove possible thermal spots. α-SiC open-cell foams are often obtained from the well-known replica[9] method which involves to burn a slurry composed of α-SiC powders deposited beforehand on a polyurethane foam pattern. However, commercial SiC powder contains impurities and its low cost elaboration processes generate electronic carriers which also favor strong absorption mechanisms in the infrared range spectral range. In other words, α-SiC ceramics show also a high thermal emittance. [10] Even if the volumetric effect decreases the temperature of front face exposed to the solar radiations, lowering the thermal emittance of the foam nearly zero, will reduce the thermal losses on the front face. To reach this objective, a way could be to deposit within α-SiC foams, layers composed of Ultra-High Temperatures Ceramics (UHTCs). Indeed, these refractory ceramics, exhibit a high solar absorbance and a low thermal emittance[11] added to their good thermal conductivity, thermo-mechanical and chemical properties up to the operating temperatures (T ~ 1400 K) required for CSP applications.

To appreciate the effect of the UHTCs coatings on the radiative properties of α-SiC open-cell foams, three single crystalline compounds are selected here: zirconium carbide[12] (ZrC), tantalum carbide[13] (TaC) and zirconium diboride[14] (ZrB_2). Their optical properties i.e. their index of refraction and their index of absorption where extracted from reflectivity spectra previously acquired at T = 300 K with the use of a suitable dielectric function.[15] Then a Ray Tracing code, called iMorphRad (C++) allowing the propagation of rays within 3D digitalized images of foams[16] will be apply to compute the radiative properties of the coated porous materials. Let us note, that UHTCs are often mixed with additive materials (SiC, $MoSi_2$[17]) to increase both their chemical and mechanical properties but the effects of the additive on the obtained radiative properties will not be investigated in this work. Finally, effects of the UHTCs on the radiative of coated and uncoated α-SiC open-cell foams will be discussed. In particular, by statically studying the paths travelled by rays within the foams practical law for predicting both the thermal emittance and the solar absorptance of foams coated with UHTCs, will be also investigated. Practical solution for covering the pores network defined by the foams will be furthermore considered.

NUMERICAL PART : MODELLING OF THE EMITTANCE

Identification of the optical properties from the knowledge specular reflectivities

With the objective of being used for CSP applications, the radiative properties of UHTCs, have been recently studied by performing solar absorptance and thermal emittance measurements up to T = 1400 K. [11, 18] Let us notify that radiative data have been acquired on dense sintered ceramics making that some precautions must be taken to directly extract the optical properties from experiments. In particular, surface light backscattering and grain size distribution are at the origin of extrinsic absorption mechanisms which can interfere with the intrinsic absorption mechanisms only due to the crystalline bulk. This explains why previous reflectivity spectra performed on single crystals at T = 300 K are preferentially studied in this work. The identified optical properties will constitutes a precious guideline to probe the spectral selectivity of the virtually coated foams.

The methodology applied for extracting the optical properties has been extensively described.[19] Briefly, once the normal spectral reflectivity, $R(\omega)$, is carefully measured on the optically polished surface of an oriented single crystal, the Fresnel's law is used to obtain the complex dielectric function, $\tilde{\varepsilon}(\omega)$ since:

$$R(\omega) = \left| \sqrt{\tilde{\varepsilon}(\omega)} - 1 \Big/ \sqrt{\tilde{\varepsilon}(\omega)} + 1 \right|^2 \qquad (1)$$

In this expression, ω stands for the wavenumber (cm^{-1}). As the optical behavior of the three selected UHTCs of this work, behaves as semi-metallic materials a Double Drude Damping complex dielectric function model is used to reproduce their optical responses. This model, fully detailed by Gervais in Ref [19], allows to reproduce the optical responses of mobile carriers known as plasmon as the one of localized carriers known as trapped polarons. Its expression is given now by :

$$\tilde{\varepsilon}(\omega)=\varepsilon_{\infty}\left(\prod_{j}\frac{\Omega_{jLO}^{2}-\omega^{2}-i\gamma_{jTO}\omega}{\Omega_{jTO}^{2}-\omega^{2}-i\gamma_{jTO}\omega}-\frac{\Omega_{P}^{2}+i(\gamma_{P}-\gamma_{0})\omega}{\omega(\omega+i\gamma_{P})}\right) \tag{2}$$

Since all the modelling will be performed as the same temperature, T, in the following , this variable is omitted from Eq. (2). Then, the product term stands for the factorized form of the dielectric function introduced by Kurosawa, Ω_{jTO} and Ω_{jLO}being respectively the transverse (TO) and longitudinal (LO) optical wave numbers of the jth oscillators and γ_{jTO} and γ_{jLO} the corresponding TO and LO dampings. This function reproduces the optical signature of the phonons and of trapped polarons. The second term is an added Drude function used to describe the response of the mobile carriers. Here Ω_P denotes the plasma wave number. γ_0 and γ_P corresponds to the plasmon damping calculated at $\omega = 0$ and $\omega = \Omega_P$. Finally, ϵ_∞ is the high wave number dielectric constant that includes the electronic contributions.

Once $\tilde{\varepsilon}(\omega)$ is known, the complex refractive index, $\tilde{n}(\omega)$, is easily computed since:

$$\tilde{n}(\omega)=\sqrt{\tilde{\varepsilon}(\omega)}=n(\omega)+ik(\omega) \tag{3}$$

with $n(\omega)$ the index of refraction and $k(\omega)$ the index of absorption. This latter optical function is straightforwardly connected to the absorption coefficient of the bulk, $K(\omega)$, since :

$$K(\omega)=4\pi k(\omega)\omega \tag{4}$$

The knowledge of $K(\omega)$ allows the determination of the critical thickness for which a dense slab can be considered as optically thick i.e. its transmittance is less than 1% on a given spectral range. This information is important, for example, for estimating the thickness of coating which can impose its optical response to the one of the supporting substrate. Let us recall that for an optically thick slab with optically polished faces, the emittance, $E(\omega)$, is given by :

$$E(\omega)=\frac{4n(\omega)}{(n(\omega)+1)^2+k(\omega)^2} \tag{5}$$

It means that, once the opacity regime (transmittance less than 1%) is reached, $E(\omega)$ is defined by Eq. 5 whatever the slab thickness is.

Influence of the texture of open-cell foams: towards the use of a Ray Tracing Code

How can we compute the effective emittance, $\tilde{E}(\omega)$, of an open-cell foam with optically thick struts from the only knowledge of the optical properties of its constituting struts and of its textural features (porosity, nominal diameter, volumetric surface)? This question cannot be easily solved and requires to put certain statements. Guevelou et *al.* have recently assessed this problem for α-SiC open-cell foams with the help of a Ray Tracing code (iMorphRad, C++). [20] Thus, for the case of foams, for which both the mean pores size and the mean struts size are higher than the incident wavelength, $\tilde{E}(\omega)$ can be computed by applying the geometrical optic approximation at each interfaces (air/matter) encountered. This numerical scheme needs to have a complete description of the 3D geometry of the foam. It can be done with an image which can be easily obtained by X-ray μ-tomography [21] or by a suitable foam generator. [22] Briefly, to compute $E(\omega)$, the Kirchhoff laws is applied since :

$$\tilde{E}(\omega) = 1 - \hat{R}(\omega) - \hat{T}(\omega) \quad (6)$$

where $\hat{R}(\omega)$ and $\hat{T}(\omega)$ are respectively the normal hemispherical reflectance and transmittance. To obtain $\hat{R}(\omega)$ and $\hat{T}(\omega)$, the method consists in launching a huge number of collimated rays towards one irradiated face of the 3D image and then to follow step by step, the displacement of rays within the foam. At each interaction with a strut, the code records the amount of absorbed intensity (initially the intensity is settled to one) and defines a new direction for the reflected rays. If the rays exhaust from the foam with a carried intensity higher than 1% and with a direction pointing toward the irradiated face, it is considered as reflected ; otherwise it is transmitted. By the way a counter records the number of reflected rays and the number of transmitted rays. The originality of the code relies in the fact that stories of each ray are registered i.e. the number of interaction with the matter, the cumulative distances as both the spatial position and the direction when rays escape from the foam. It allows a statistical study of the travels performed by the rays which can be used as an interesting guideline for further optimization of the radiative properties of the open-cell foams.

RESULTS AND DISCUSSION

Optical properties of ZrC, TaC, ZrB_2 single crystals at T = 300 K

Reflectivity spectra of zirconium carbide[12] (ZrC), tantalum carbide[13] (TaC) and zirconium diboride[14] (ZrB_2) are extracted from previous published papers, dedicated to the understanding of their respective optical mechanisms. Studied samples were optically polished single crystals with crystallographic structures which are face-centered cubic for TaC and ZrC and hexagonal for ZrB_2. It may be noted that densified ceramics based on these three compounds have been recently studied (TaC and ZrC in [11], ZrB_2 in [18]) but some precautions must be taken to directly applied Equation 1 and Equation 2. Indeed ceramics are well-known for exhibiting polycrystallinity and residual porosity both of which may affect the optical properties of the corresponding single crystals.[23, 24] The good agreements between the reported reflect spectra and the modelled ones are reported in Figure 1. The parameters injected in Eq. 2 to reproduce the optical responses are given in Table 1.

Table 1 : Parameters used for fitting the reflectivity spectra of ZrC, TaC and ZrB_2

	Ω_P (cm^{-1})	γ_0 (cm^{-1})	γ_P (cm^{-1})	Ω_{TO} (cm^{-1})	Ω_{LO} (cm^{-1})	γ_{jTO} (cm^{-1})	γ_{jLO} (cm^{-1})	ϵ_∞
ZrC	5000	1000	25000	×	×	×	×	10
TaC	22350	2480	12150	×	×	×	×	4.07
ZrB_2	10890	0	5700	16350	21000	20600	20600	4.54

To perform each modelling, the FOCUS software was freely applied. [25] It is important to note that even if ZrB_2 has an hexagonal crystalline structure, effects of anisotropy on the optical properties are neglected. Furthermore, for ZrC a large polaronic band with symmetric dampings[19] had been added in its dielectric function. O nce the complex dielectric functions are known it is straightforward to compute the complex refractive indices. In Figure 2 are plotted the critical thicknesses for which three slabs composed of the UHTCs studied in this paper can be considered as optically thick.

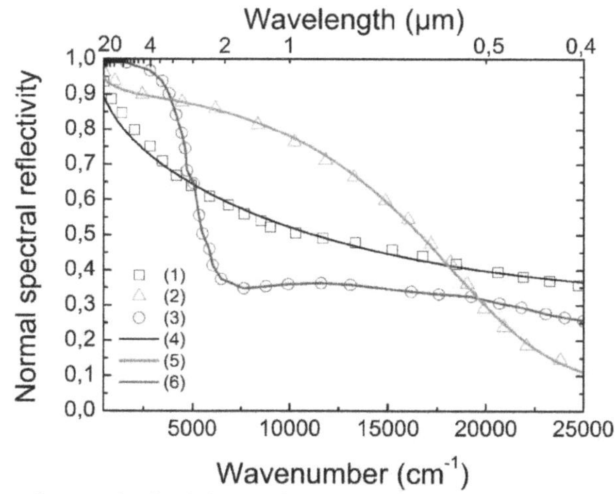

Figure 1 : Normal spectral reflectivity acquired at room temperature for (1) ZrC, (2) TaC and (3) ZrB_2. Modelled spectra with the help of Equation 1 and Equation 2 for (4) ZrC, (5) TaC and (6) ZrB_2

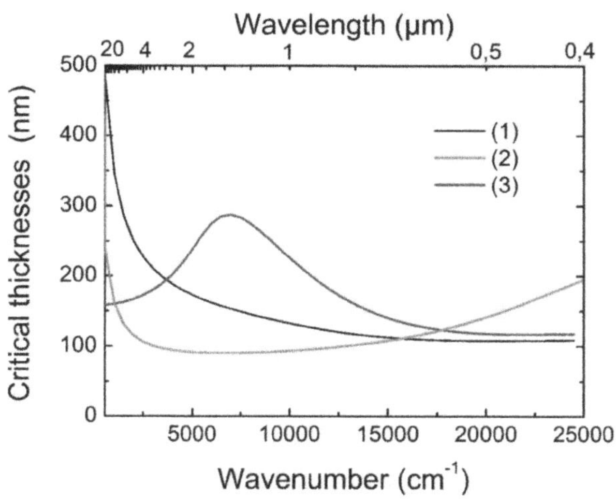

Figure 2 : Critical thicknesses beside which a dense and homogenous slab can be considered as optically thick (1) ZrC, (2) TaC and (3) ZrB_2.

For the three chemical compositions, dense and homogenous slabs with a thickness higher than 500 nm are therefore opaque. In this case, the normal emittances are simply given by Equation 5 since complex refractive indices are now known. It is worth mentioning that emittances are rather high in the visible spectral range and low in the infrared range (for wavenumbers below 12500 cm-

^{1}). This optical feature appears here very interesting to develop solar receivers with a good spectral selectivity, which is theoretically defined by a thermal emittance equal to zero up to a cutoff wavenumber and a solar absorptance equal to one for wavenumbers just above. The cutoff wavenumber is determined when is established the thermal balance of the studied surfacic receiver at is operating temperature and on w hich are focused concentrated rays. [26] For a factor of concentration of 1000 and a temperature of air of 1300 K, the cutoff wavenumber is found to be 5600 cm^{-1} (≅ 1.8 μm). The discussion concerning the spectral selectivity of UHTCs will be discussed later.

Radiative properties of open-cell foams coated with UHTCs

Let us start this part, by presenting the 3D image of α-Silicon Carbide open-cell foam. The present image (see Figure 3a) has been obtained by investigating the porous network of the real by X-ray μ-tomography. [7] The foam had been beforehand tested for concentrated solar applications at PROMES laboratory.[8] The sample volume is 1 cm^3 and the porosity is found to be 68%. The specific surface is 2800 m^{-1}. Details concerning the use of the iMorph software [27] for the construction of the 3D image are reported in [22].

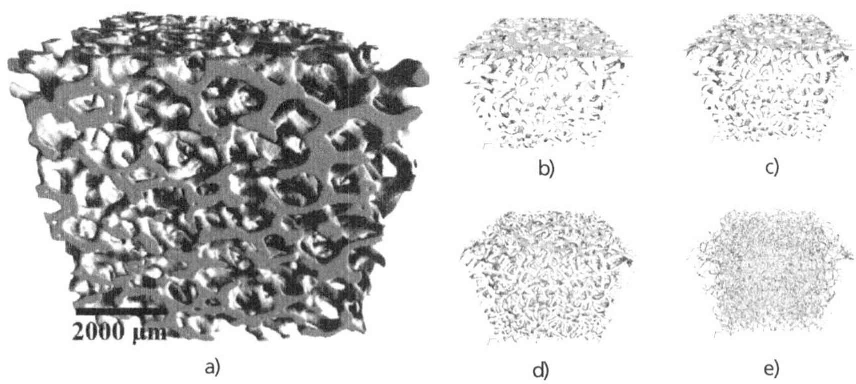

Figure 3 : (a) 3D image of α-SiC open-cell foam (1×1×1 cm^3) obtained with the iMorph software. Reconstructed α-SiC open-cell foam (1×1×1 cm^3) with (b) 24 PPI, $p = 49$ %, (c) 24 PPI, $p = 61$ %, (d) 24 PPI, $p = 81$ % (e) 24 PPI, $p = 96$ %

In particular, the air/matter surface is totally meshed with regular triangles that makes easy, afterwards, the management of the interactions between rays and the foam. For the reconstructed samples, a specific foam's generator allowing the design of realistic structure has been used. [22] Briefly, the foam generation was initialized by a seeding corresponding to the centers of the foam cells. A regular tetrahedral seeding is first performed to comply with the mean cell connectivity that one can measure on real foams. Then following the seeding, a fast-marching technique[28] permits the growth of domains corresponding to the foam's cells around each center. After a segmentation process, a step aiming to perform the construction of the 3D thin skeleton of struts is done. Finally, a final phase enables accurate control over the porosity of the virtual foam while preserving the previously shaped textural features. The efficiency of the generator, in terms of quality of reconstruction, has been validated previously.[22] As it can be viewed in Figure 3, the generator allows the design of 4 synthetic open-cell foams with same number of Pore Per Inches (PPI) but with different porosities (49 %, 61 %, 81% and 96 %).

Then, two cases of simulation can be proposed for the UHTCs foams, once their respective critical thicknesses criterion for opacity are known. Indeed, one can model the optical responses of open-cell foams which are totally composed of UHTCs or those of open-cells foams in which coatings composed of UHTCs are deposited on the whole solid network. For the latter case, the coating must have just a minimal thickness of 500 nm. For these two cases, which implies different routes of elaboration, the results will be finally the same.

To reproduce the emittance of the UHTCs foams, the code iMorphRad (C++) is therefore used. The textural investigation of the foam indicates that the approximation of the geometrical optics is valid. Indeed the mean pore size is centered at 1250 μm and the mean struts size at 250 μm for a spectral range of interest going from 0.4 up to 20 μm. For each simulation, the diameter of the virtual incident beam is fixed at 1 cm and 10^6 rays are normally launched toward the illuminated face. The complex refractive indices computed by combining Equation (1), Equation (2) and Equation (3) are then injected in the code. The Figure 4 shows the role played by the porous network on the radiative behavior of the coated foams when comparison with corresponding homogenous slabs are done.

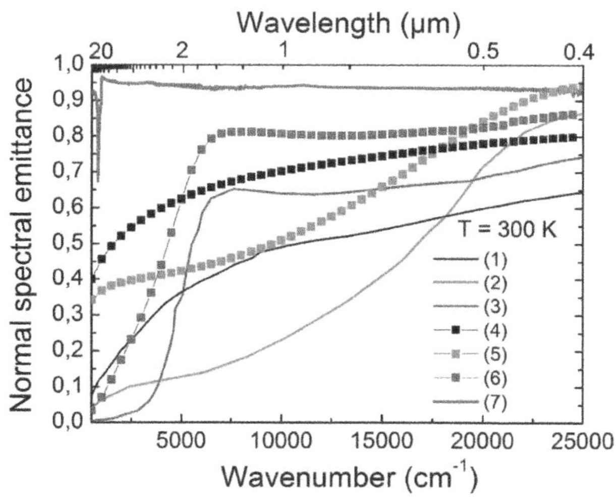

Figure 4 : Normal spectral emittance of fictitious slabs of (1) ZrC, (2) TaC and (3) ZrB_2 and modelled emittance of an α-SiC open-cell foams internally coated with (4) ZrC, (5) TaC and (6) ZrB_2. The curve 7 depicts the normal spectral emittance of the α-SiC open-cell foam without coating.

Globally, the geometry of the foam induces an increase of the emittances of the corresponding slabs for the full spectral range of this work (0.4-20 μm). However the growths of the emittances are not the same according to specific spectral range. To go one step further in the investigation, let us now discuss of the optical effects due to the three UHTCs coatings on the original radiative behavior of the α-SiC open-cell foam. We have first to define the thermal emittance, ε_T, and then the solar absorptance, α_S, which are both computed according to the cutoff wavenumber defined just before.

$$\varepsilon_T(T) = \frac{\int_{\omega_{min}}^{\omega_{cutoff}} E(\omega,T)B(\omega,T)d\omega}{\int_{\omega_{min}}^{\omega_{cutoff}} B(\omega,T)d\omega} \tag{7}$$

and,

$$\alpha_S(T) = \frac{\int_{\omega_{cutoff}}^{\omega_{max}} E(\omega,T)S(\omega)d\omega}{\int_{\omega_{cutoff}}^{\omega_{max}} S(\omega)d\omega} \tag{8}$$

where $S(\omega)$ is the direct solar spectral irradiance[29] and $B(\omega, T)$ is the well-known Planck's law[30] calculated at a given temperature T. The Table 2 lists the values of ε_T and α_S for the three slabs and three foams respectively composed of ZrC, TaC, ZrB_2 when the receiver operates at T = 1300 K. The comparison of the results is also extended to the case of the native α-SiC open-cell foam.

Table2: Thermal emittances and solar absorptances of the slabs and the foams based on UHTCs

	Slab		Foam	
	ε_T	α_S	ε_T	α_S
ZrC	0.26	0.53	0.56	0.73
TaC	0.10	0.41	0.40	0.64
ZrB_2	0.10	0.66	0.31	0.81
α-SiC	×	×	0.95	0.94

Let us precise that each numerical simulation realized on the foams have been performed for some volumes of foam largely higher than the Representative Elementary Volumes mandatory to compute normal spectral emittances.[20] Moving from the geometry of a slab to that of a foam is expressed by the multiplication by 4 of ε_T for ZrC and by 1.6 of α_S for TaC. The best result for the spectral selectivity is achieved for the foam coated with ZrB_2 the ratio α_S/ε_T being equal to 2.6 against nearly 1 for α-SiC. However, the later ratio is still high when is calculated in the case of a slab. The transition from a 2D to a 3D structure tends to decrease the spectral selectivity behavior. For ZrB_2, even if the result is promising for the simulation conducted at T= 300 K, more computations must be done by combining ZrB_2 with α-SiC or $MoSi_2$, these compounds being important to prevent prejudicial degradation of ZrB_2 when it is heated under air up to high temperature.[18]

Impact of the textural features of the foams on the radiative properties.

The Figure 5 shows the number of impacts undergone by rays when they travel through ZrB_2-based foams with growing porosity (from 49 to 91 %) but with same PPI (24). At each interaction, between the ray and the solid phase, the code activates a counter until the ray can move within the foam. For each foam, at ω = 10000 cm^{-1} (i.e. a wavenumber belonging to spectral range where α_S is determined), most of the interactions occurs at the first impact (i.e. 70 % for p = 49 %, 64% for p = 61 %, 57 % for p = 70 %, 46 % for p= 81 % and 30 % for p = 96 %). If the local absorptivity of the solid phase is high, it means that for the foams with the lowest porosities, the absorption of the incident light mainly occurs on the first cell row. In other words, this first observation gives some interesting elements as far as the minimal volume of foam required to maximize the absorption of the light. For a foam with ten cells split over a length of 1 cm (~ 24 PPI), the first cell rows are crucial. What happens when the porosity grows up for same value of

PPI? It can be viewed in Figure 5 that the proportion of rays absorbed at the first impact decreases whereas the one absorbed in two or more impacts raises up. It means that the penetration of the light is promoted and that the possibility to be absorbed by the solid phase is enhanced. Then for the whole set of porosities, the number of absorbed rays does not exceed 7 meaning that a cubic foam of 24 PPI, 1 cm of length extinguishes completely a radiation beam which is normally collimated on its irradiated face.

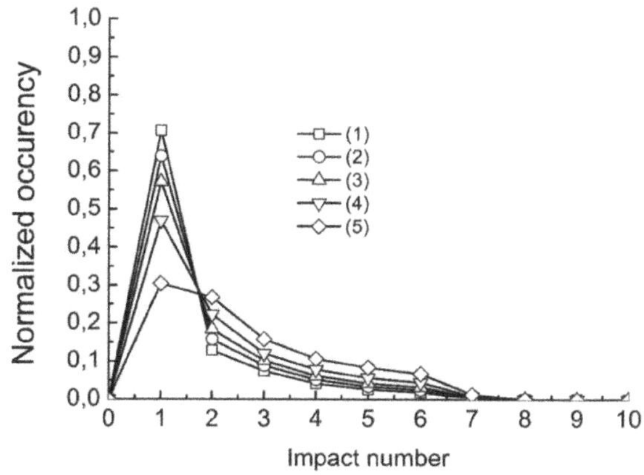

Figure 5: Normalized distribution of the impacts undergone par rays during their travels within ZrB_2 open-cell foams (1×1×1 cm^3, 24 PPI) with prescribed porosities: (1) $p = 49$ %, (2) $p = 61$ %, (3) $p = 70$ %, (4) p= 81 % and (5) p = 96 %. The wavenumber is 10000 cm^{-1}

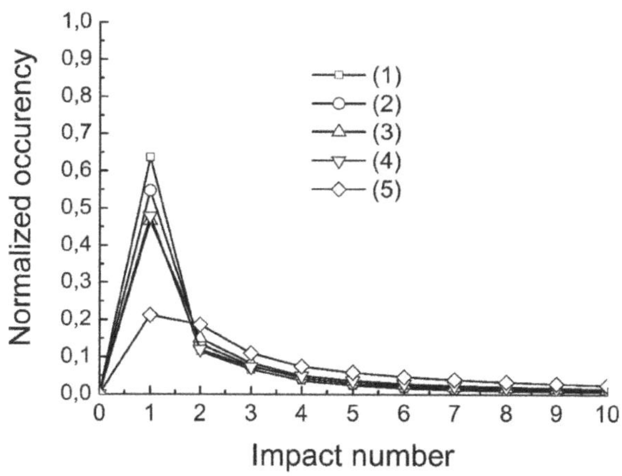

Figure 6: Normalized distribution of the impacts undergone par rays during their travels within ZrB_2 open-cell foams (1×1×1 cm^3, 24 PPI) with prescribed porosities: (1) $p = 49$ %, (2) $p = 61$ %, (3) $p = 70$ %, (4) p= 81 % and (5) p = 96 %. The wavenumber is 4000 cm^{-1}.

Therefore $\hat{T}(\omega)$ is equal to 0 at $\omega = 10000\ \mathrm{cm^{-1}}$ for each foam whatever their porosities are. It shows that the emittance grows up for an open-cell foam all the more so as its porosity is high. Furthermore, it explains why the emittance of each slabs composed of UHTCs is globally lower than the one of the homologous coated foams. Open-cell foams act in reality as efficient light traps when they have a volume higher than the Representative Elementary Volume (REV) needed for safely characterizing emittance.[20]

Similar exploitation of the ray tracing code can be also done at $\omega = 4000\ \mathrm{c\,m^{-1}}$ (a wavenumber useful for computing ε_T). The results are plotted in Figure 6. Similar observations can be proposed for the first impacts but for the number of next impacts is less important. It means that the volume investigated by rays before being absorbed is higher. Indeed, since struts are more reflective, the probability for being scattered from struts to struts is higher.

Finally, in Figure 7, the effect of the porosity is investigated. The foams previously shown in Figure 3 are selected and injected as input data in iMorphRad. Two typical wavenumbers are also choosen ($\omega = 2000\ \mathrm{cm^{-1}}$ and $\omega = 20000\ \mathrm{cm^{-1}}$) as well as the corresponding complex refractive indexes for ZrC, TaC , ZrB_2 and α-SiC. [20] For each case, the porosity increases the values of emittances but with different behavior. It indicates that, for a fixed thickness, the porosity globally rises ε_T and α_S. As shown by Guevelou et *al.*,[20] a linear relationship, independent from PPI 's number, can link $\tilde{E}(\omega)$ and p, the slope being defined by a weighted reflectivity coefficient. In this expression, the weighted reflectivity coefficient is straightforwardly computed with $\tilde{\varepsilon}(\omega)$. It can be expected that for a critical volume of light scattering, open-cell foams exhibit a black-body behavior whatever the optical properties of the struts (coated or not) are.

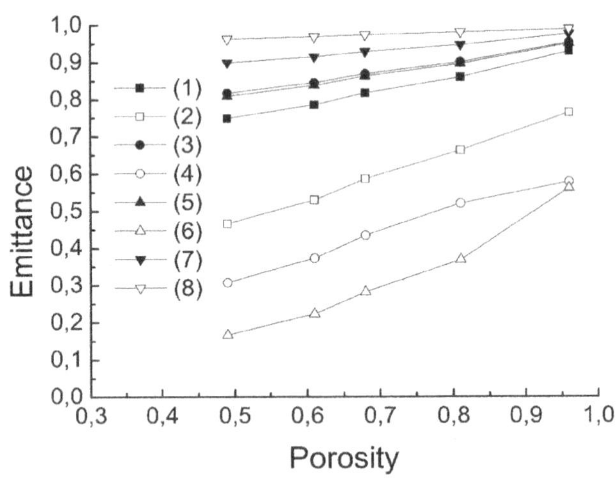

Figure 7 : Calculated effective emittances at T = 300 K of coated foams with growing porosities (from 49 % up to 96 %, 24 PPI, 1×1×1 $\mathrm{cm^3}$) for $\omega = 2000\ \mathrm{cm^{-1}}$ (hollow symbols) and for $\omega = 20000\ \mathrm{cm^{-1}}$ (filled symbols). The curves (1) and (2) corresponds to a coating composed of ZrC, (3) and (4) corresponds to a coating composed of TaC, (5) and (6) corresponds to a coating composed of ZrB_2. The curves (7) and (8) corresponds to α-SiC open-cell foams

CONCLUSIONS

In this work, the relevance of using open-cell foams in which UHTCs coatings are deposited has been numerically demonstrated. A multi-length scale approach has been therefore developed to simulate the optical response of open-cell foams where both the chemical and the textural parameters can be finely tuned. Once the porosity and the thickness of a given foam are known as well as its Representative Elementary Volume, a suitable UHTC coating increases its spectral selectivity behavior. Particularly ZrB_2 appears to present promising features. Future works will consist in dealing with high temperature optical properties of UHTC-based coatings, which will be furthermore able to be used under air in the real operating conditions imposed in concentrated solar tower plants. Collaborations with teams specialized in the elaboration of such foams are also planned in order to check all the predicted data.

ACKNOWLEDGEMENTS

This work was supported by the OPTISOL project funded the ANR (French National Research Agency, program SEED 2011, project ref. ANR-11-SEED-0009)

REFERENCES

1. K. Lovegrove and W. Stein, "Concentrating solar power technology: Principles, developments and applications." Elsevier, (2012).
2. A. L. Ávila-Marín, "Volumetric receivers in Solar Thermal Power Plants with Central Receiver System technology: A review," *Solar energy,* 85[5] 891-910 (2011).
3. A. Kribus, R. Zaibel, D. Carey, A. Segal, and J. Karni, "A solar-driven combined cycle power plant," *Solar energy,* 62[2] 121-29 (1998).
4. C. K. Ho and B. D. Iverson, "Review of high-temperature central receiver designs for concentrating solar power," *Renewable and Sustainable Energy Reviews,* 29[0] 835-46 (2014).
5. A. Kribus, Y. Gray, M. Grijnevich, G. Mittelman, S. Mey-Cloutier, and C. Caliot, "The promise and challenge of solar volumetric absorbers," *Solar Energy,* 110 463-81 (2014).
6. T. Fend, R. Pitz-Paal, O. Reutter, J. Bauer, and B. Hoffschmidt, "Two novel high-porosity materials as volumetric receivers for concentrated solar radiation," *Solar Energy Materials and Solar Cells,* 84[1] 291-304 (2004).
7. B. Rousseau, S. Guevelou, G. Domingues, J. Vicente, C. Caliot, and G. Flamant, "Prediction of the radiative properties of reconstructed alpha-SiC foams used for concentrated solar applications," in MRS Spring Meeting.
8. Z. Wu, C. Caliot, G. Flamant, and Z. Wang, "Coupled radiation and flow modeling in ceramic foam volumetric solar air receivers," *Solar energy,* 85[9] 2374-85 (2011).
9. A. R. Studart, U. T. Gonzenbach, E. Tervoort, and L. J. Gauckler, "Processing routes to macroporous ceramics: a review," *Journal of the American Ceramic Society,* 89[6] 1771-89 (2006).
10. C. P. Cagran, L. M. Hanssen, M. Noorma, A. V. Gura, and S. N. Mekhontsev, "Temperature-Resolved Infrared Spectral Emissivity of SiC and Pt–10Rh for Temperatures up to 900° C," *International Journal of Thermophysics,* 28[2] 581-97 (2007).
11. E. Sani, L. Mercatelli, P. Sansoni, L. Silvestroni, and D. Sciti, "Spectrally selective ultra-high temperature ceramic absorbers for high-temperature solar plants," *Journal of Renewable and Sustainable Energy,* 4[3] 033104 (2012).
12. F. Modine, T. Haywood, and C. Allison, "Optical and electrical properties of single-crystalline zirconium carbide," *Physical Review B,* 32[12] 7743 (1985).

13. F. A. Modine, R. W. Major, T. W. Haywood, G. R. Gruzalski, and D. Y. Smith, "Optical properties of tantalum carbide from the infrared to the near ultraviolet," *Physical Review B,* 29[2] 836-41 (1984).
14. S. M. Sichkar, V. N. Antonov, and V. P. Antropov, "Comparative study of the electronic structure, phonon spectra, and electron-phonon interaction of ZrB2> and TiB2," *Physical Review B,* 87[6] 064305 (2013).
15. B. Rousseau, D. De Sousa Meneses, A. Blin, M. Chabin, P. Echegut, P. Odier, and F. Gervais, "High-temperature behavior of infrared conductivity of a Pr_{2}NiO_{4+ delta } single crystal," *Physical Review B,* 72[10] 104114 (2005).
16. B. Rousseau, J. Y. Rolland, P. Echegut, E. Brun, and J. Vicente, "Numerical prediction of the radiative behavior of metallic foams from the microscopic to macroscopic scale," *Journal of Physics: Conference Series,* 369[1] 012003 (2012).
17. B. Pierrat, M. Balat-Pichelin, L. Silvestroni, and D. Sciti, "High temperature oxidation of ZrC–20% MoSi 2 in air for future solar receivers," *Solar Energy Materials and Solar Cells,* 95[8] 2228-37 (2011).
18. D. Sciti, L. Silvestroni, L. Mercatelli, J.-L. Sans, and E. Sani, "Suitability of ultra-refractory diboride ceramics as absorbers for solar energy applications," *Solar Energy Materials and Solar Cells,* 109 8-16 (2013).
19. F. Gervais, "Optical conductivity of oxides," *Materials Science and Engineering: R: Reports,* 39[2-3] 29-92 (2002).
20. S. Guevelou, B. Rousseau, G. Domingues, J. Vicente, and C. Caliot, "Representative elementary volumes required to characterize the normal spectral emittance of silicon carbide foams used as volumetric solar absorbers," *International Journal of Heat and Mass Transfer* (submitted).
21. J. Baruchel, "X-ray tomography in material science ". Lavoisier, (2000).
22. S. Guévelou, B. Rousseau, G. Domingues, J. Vicente, C. Caliot, and G. Flamant, "Identification of the radiative properties of α-SiC foams realitically designed with a numerical generator " *Journal of Porous Media* (in press).
23. T. G. Mayerhöfer and J. Popp, "Modelling IR spectra of polycrystalline materials in the large crystallites limit-quantitative determination of orientation," *Journal of Optics A: Pure and Applied Optics,* 8[8] 657 (2006).
24. B. Rousseau, H. Gomart, D. De Sousa Meneses, P. Echegut, M. Rieu, R. Dugas, P. Lenormand, and F. Ansart, "Modelling of the radiative properties of an opaque porous ceramic layer," *Journal of Electroceramics,* 27[2] 89-92 (2011).
25. D. De Sousa Meneses, "FOCUS, Software Utility for the Creation of Optical Function : http://www.cemhti.cnrs-orleans.fr/pot/software/focus.html." in.
26. J. Spitz, A. Aubert, J. Behaghel, S. Berthier, J. Lafait, and J. Rivory, "Matériaux sélectifs pour la conversion photothermique de l'énergie solaire," *Revue de Physique Appliquée,* 14[1] 67-80 (1979).
27. J. Vicente, "iMorph a visualisation and a morphological characterisation tool, www.imorph.fr." in.
28. E. Konukoglu, M. Sermesant, O. Clatz, J.-M. Peyrat, H. Delingette, and N. Ayache, "A recursive anisotropic fast marching approach to reaction diffusion equation: Application to tumor growth modeling," pp. 687-99 in Information processing in medical imaging.
29. NREL, "http://rredc.nrel.gov/solar/spectra/am1.5/." in.
30. R. Siegel and J. R. Howell, "Thermal Radiation Heat Transfer " 4 edition ed. Taylor & Francis, (2001).

THERMO-MECHANICAL ANALYSIS OF A SILICON CARBIDE HONEYCOMB COMPONENT APPLIED AS AN ABSORBER FOR CONCENTRATED SOLAR RADIATION

Thomas Fend[1], Peter Schwarzboezl[1], Olena Smirnova[1], Martin Schmuecker,[2] Ferdinand Flucht[2], Sven Dathe[1]
[1]German Aerospace Center, Institute of Solar Research, DLR, 51147 Koeln, Linder Hoehe, Germany
[2]German Aerospace Center, Institute of Materials Research, DLR, 51147 Koeln, Linder Hoehe, Germany

ABSTRACT

The study focuses on a numerical model describing the thermo-mechanical processes in an open volumetric solar absorber. In this application thermal loads of up to 850 kW/m² induce mechanical stresses inside the volume of the component. The objective of the study is to identify critical thermal load cases which may especially occur during start-up or shut down cases as well as during the transition of clouds. The study can be subdivided into three major parts. In a first step, the mechanical strength of the investigated materials has been determined in double-ring, three or four-point bending tests. In a subsequent numerical stress analysis the maximum stresses in the walls of the honeycomb structure at fracture load have been determined. The results of this combined experimental/numerical approach have been compared with the strength properties of dense materials. In a second step various transient and stationary load cases were considered and the corresponding temperature distributions were calculated. Thirdly, the thermo-mechanical model was used to determine the stress distributions which have been induced. Special attention was paid on the influence of the macro geometry and cell-density. The results of the study allow important conclusions for the operation of the receiver.

INTRODUCTION

Thanks to its high application temperature of more than 1300°C, Silicon Carbide (SiC) is used in solar energy, electro-technical, chemical and automobile industries. However, since this ceramic material is not able to undergo plastic deformations, it is very sensible to thermo-shocks. The present study has been carried out to investigate, whether typical thermal load cases, as they occur during the application of SiC honeycomb structures as volumetric receivers in solar tower technology, lead to thermal shocks with critical stress distributions.

The volumetric solar absorber is the key component of solar tower technology, which uses concentrated solar radiation to generate thermal energy which operates a thermal engine.

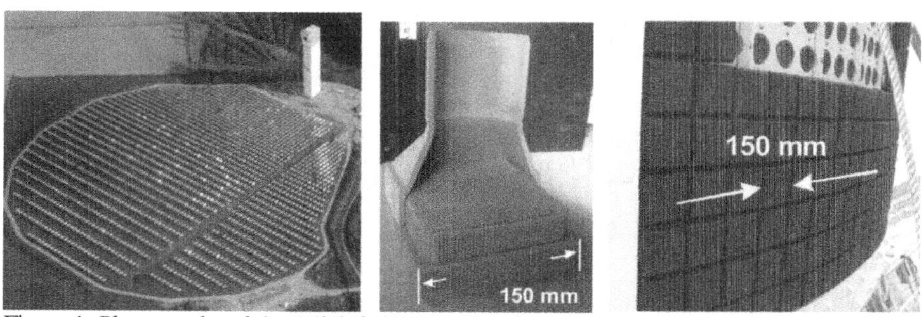

Figure 1. Photographs of the Jülich Solar Tower (left), a part of a single absorber module and the solar receiver during installation

The concentrated radiation is generated by a large field of individually moving mirrors (heliostats). The focus is located on the top of a tower (Fig. 1, left). The volumetric solar absorber, located in the focus, generally consists of a porous material, which absorbs the radiation and transfers it to air. After that, the air is used to heat up steam in a boiler. The heat transfer in the solar absorber has been investigated already since 1985 and since then, numerous studies have been published [1-4]. A comparative review is given in [5]. The investigated material with the honeycomb structure based on the state-of-the-art-technology shown in Fig. 1 (middle and right), which has also been applied in the solar tower in Jülich [6]. It has been manufactured by St. Gobain IndustrieKeramik Rödenthal GmbH.

The operation experience of the solar volumetric absorber in solar tower technology has shown a good reliability of the honeycomb structure. But the influence of various spatial and temporal temperature gradients at the different load cases on the structure stability is not known. Therefore the aim of the presented thermo-mechanical analysis can be divided into three parts: 1.The determination of the mechanical fracture stress for different geometries of the solar absorber; 2.The determination of the maximum mechanical stress and its localization at the studied load cases and 3.The influence of structure changes on the thermo-mechanical stability for various load cases.

Two macro geometries and two microstructures with different cell-densities of the volumetric solar absorber have been investigated in the current study. The first one is simple rectangular block geometry (simple geometry) and the second one is a similar one with an additional truncated pyramid in the air outlet zone (real geometry). In the model, the simple geometry was directly drawn in ANSYS 15; the second geometry was made in Inventor 14 and imported in ANSYS afterwards. Both geometries are shown in Fig. 2.
The simple geometry was investigated with a cell-density of 80cpsi[1]*. The real geometry was investigated with a cell-density of 80cpsi (Fig. 3 "standard geometry") and with a cell-density of 200cpsi (Fig. 3 "advanced geometry").

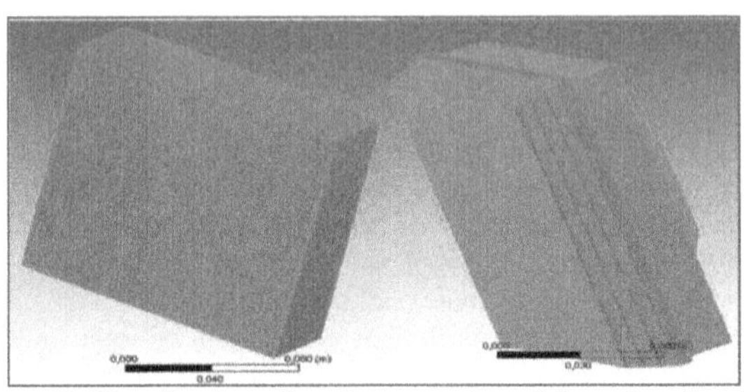

Figure 2. The simple and real geometries of the solar absorber

The various geometries shown above were numerically investigated for different thermal load cases, which are deduced from the practical operation of the component in real service conditions. The numerical results of the thermo-mechanical analysis were compared with the mechanical fracture stress, which was found from bending tests. The values of the fracture force, which were measured experimentally, were then used in the numerical models as the boundary condition for the determination of the fracture stress.

[1] channels per square inch

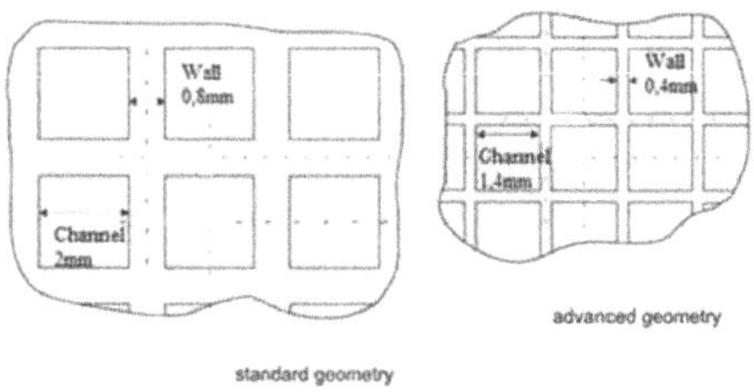

Figure 3. Cross-sectional view of the absorber with cell density 80cpsi (standard geometry) and 200cpsi (advanced geometry)

The numerical work was carried out in two steps using the two different ANSYS modules: thermal and mechanical for each of the studied thermal load cases. The obtained results allowed estimating both the critical temperature differences and the localization of the critical and overcritical mechanical stresses for each of the investigated structures.

The estimation of the maximum thermo-mechanical stress in the porous structure during the critical thermal loads was one of the main aims of the represented investigations. Therefore the following load cases commonly encountered during operation in real service conditions have been investigated:

- Stationary homogenous radiation of app. 850kW/m²
- Stationary homogenous radiation of app. 850 kW/m² with an additional non-homogeneous temperature distribution in diagonal or axial directions
- Stationary non-homogeneous radiation
- Transient cooling and heating at homogenous time-dependent radiation (cloud-effect)

METHODOLOGY

Experimental and numerical estimation of the fracture stress

The fracture stress is used as the main criterion for the assessment of the solar absorber stability. Three kinds of standard bending tests have been carried out to estimate this quantity: double ring and 3- and 4- points bending tests [7]. The test set-ups and theoretical equations for the determination of the maximum stress in the dense material are shown in Fig.4.

Double ring bending test | 3-point bending test | 4-point bending test

$$\sigma = 1.04 \cdot \frac{F_{max}}{h^2} [MPa] \qquad \sigma = 1.5 \cdot \frac{F_{max} \cdot L_1}{b \cdot h^2} [MPa] \qquad \sigma = 1.5 \cdot \frac{F_{max} \cdot (L_1 - L_2)}{b \cdot h^2} [MPa] \qquad (1\text{-}3)$$

Figure 4. Schematic draws of the bending tests and theoretical equations for the calculation of the maximum stress in a dense material.

The tests were made for the two honeycomb structures with cell densities 80cpsi and 200cpsi. The numeric models simulated both the dense body and the honeycomb structures for the estimation of the maximum stress in the wall of the cellular structure. The validation of the numeric models was realised by the comparison between the simulated and calculated data for the dense body.

Numerical Models

Each numerical model can be divided into two parts. The first is the numerical simulation of the bending test with the mechanical module and the second one is the thermo–mechanical model, which consists of two modules: thermal and mechanical. The models are 3-D models. The numerical model was calculated for the four load cases and the three geometries described in the introduction. The bending test was simulated for the dense body and for the honeycomb structure. The sample geometry as well as the dimensions of the test equipment and the fracture force was taken from the experiments.

The thermo-mechanical model was studied in two modules. The temperature distribution was calculated in a thermal module and then it was imported into the mechanical module, where the resulting stress distribution was obtained. The transient tasks were considered in three steps:

- The stationary thermal module, which calculated the initial temperature distribution for the transient task
- The transient thermal module for the time dependent temperature distribution
- The stationary mechanical module for the stress distribution at selected points of time

Boundary and domain conditions of the models for the investigated load cases

The boundary conditions of the models can be divided into two groups. The main conditions, which have used in all load cases, belong to the first group. The specific conditions, each of which used only in some of the load cases, belong to the second group. The main boundary conditions of the thermal modules are: the inlet solar radiation and forced convective heat transfer in the channels of the solar absorber. The solar radiation was described as a homogeneous or non-homogeneous heat flow density on the front surface. The homogeneous heat flow density was q_0=650 kW/m^2 (normal incidence). A non-normal incident angle (φ) of the radiation was taken into account for the sloped front surface (in the real structure):

$$q_{Sl.} = q_0 \cdot \sin(\varphi) \tag{4}$$

The non-homogeneous heat flow density was described with Gaussian function (5) or according to the measured temperature distribution as a temperature boundary condition.

$$q = q_0 \cdot e^{-k\cdot\left(\left[x-\frac{x_{\max}}{N}\cdot L_x\right]^2+\left[y-\frac{y_{\max}}{N}\cdot L_y\right]^2\right)} \tag{5}$$

Here, N is the number of cells on the front surface, k is a measure for the non-homogeneity, $x_{\max}, y_{\max}$ are the coordinates in x and in y directions, $L_x = L_y = 0{,}14$ [m] is the dimension of the front surface.

The convective boundary conditions in the channel were determined with a previous COMSOL model [8] in dependence of the channel size. The corresponding diagrams and equations for the simple geometry with cell density 80cpsi are shown in Fig. 5 and in equations (6) and (7).

$$\alpha = 35 - 2{,}95\cdot 10^6 \cdot z^3 - 2{,}2\cdot 10^5 \cdot z^2 - 5{,}2\cdot 10^3 \cdot z \tag{6}$$

$$T_{air} = 332.15 - 3.050728\cdot 10^7 \cdot z^3 - 2.60648\cdot 10^6 \cdot z^2 - 7.3595\cdot 10^4 \cdot z \tag{7}$$

The convective heat transfer depends on the absorber geometry, therefore each of the studied structures has own equations.

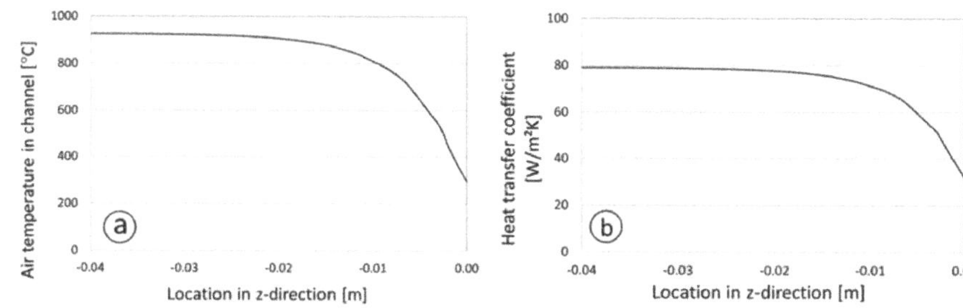

Figure 5. The air temperature in the channel (a) and the heat transfer coefficient (b) from one channel model COMSOL 3.5 for the standard channel (2:0,8mm)

The air temperatures in the channels for the load cases of the cloud-effect have been estimated as 15°C lower compared to the ones described with the equation above.

To describe the transient intensity of the solar radiation during a cloud transition, a transient heat flow was used as a specific boundary condition:

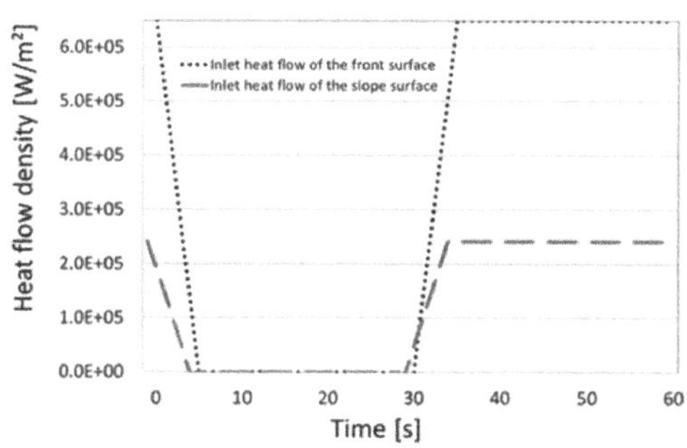

Figure 6. The inlet heat flow density for the cloud-effect model

The time–dependent decrease and increase of the heat flow was assumed to last 5 seconds (Fig. 6). This short time has been considered to be the maximum possible thermal load, which represents a very fast transition of a cloud. The following equations describe the time-dependent heat flow:

$$q_0(t) = q_0 - 130000 \cdot t \quad (8)$$

$$q_{Sl.} = q_0 \cdot \sin(\varphi) - 48488 \cdot t \quad (9)$$

A homogeneous heat flow was assumed as the first boundary condition to obtain the initial temperature distribution for all studied structures.

Two additional temperature boundary conditions have been used in two specific load cases: a temperature gradient in x direction and a "diagonal" boundary temperature with two different temperatures at the 4 lateral boundaries of the receiver module with different temperatures at opposite boundaries:

$$T=873+1428.5 \cdot x \text{ [K]}, \quad T=1073-1428.5 \cdot x \text{ [K]} \quad (10)$$

$$T{=}873\ [\mathrm{K}], \qquad T{=}\ 1073\ [\mathrm{K}] \qquad (11)$$

RESULTS AND DISCUSSIONS

The results of the bending test simulation were used for the estimation of the maximum mechanical stresses in the real micro geometry of the material, which do not come out of the bending theory. This allows us to estimate whether the thermal loads are below or above critical values. The study of the influence of the geometry on the maximum stress allows an estimation of the reliability of the investigated structures.

The results of the bending test simulation

The simulation of the bending test enables a comparison between the fractures stress in the real geometry of the material and the fracture stress in a virtual "dense" material, which is delivered by the fracture force of the experiments and bending theory (equations in Fig. 4). Additionally and for a validation of the model, the fracture stress of a corresponding dense material was modeled. The fracture stress results for the dense body are summarized in Table 1. The comparison gives a good correspondence between the theoretical and simulated data for the dense body with the difference ≈ 7-8%.

Table 1. The theoretical and simulated fracture stress for the dense body

	Double ring flexural test			3-Points flex. test			4-Points flex. test	
	1.Test	2.Test	3.Test	1.Test	2.Test	3.Test	1.Test	2.Test
Theoretic fracture stress [MPa]	24	21	22	31	46	25	22	9
Simulated fracture stress [MPa]	22	19	20	28	42	23	20	9

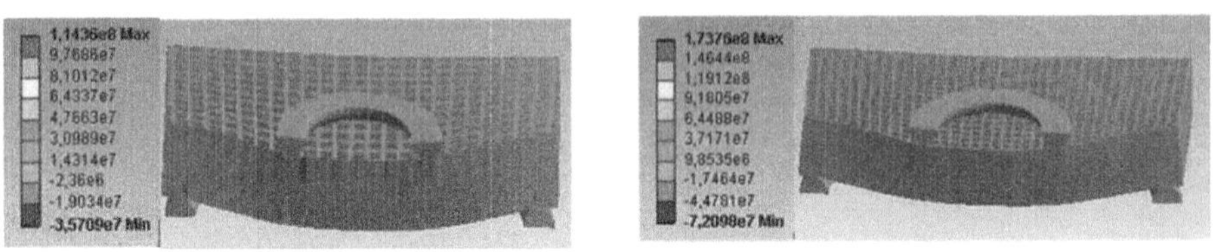

(a) (b)

Figure 7. Maximum stress in the principal coordinate system for the double ring bend test for the 80cpsi (a) and for the 200cpsi sample (b)

The simulation results of the double ring bending tests with the fracture stress for two studied honeycomb structures are shown in Fig.7. The comparison of the fracture parameters can be found in Table 2. The comparison of these results for the honeycomb structure shows, that the stress in the real geometry is about a factor of 5 (80 cpsi) and 11 (200 cpsi) higher than for the corresponding dense material.

Table 2. Comparison of the fracture parameters for the investigated structures

Porosity [cpsi]		Experimental data		Numerical data	
		thickness h [mm]	break force F_{max} [N]	calculated break stress σ [MPa]	simulated break stress σ [MPa]
80	solid body	9.36	1839	22	20
	honeycomb structure	9.36	1839	-	108
200	solid body	8.67	1026	14	16
	honeycomb structure	8.67	1026	-	174

The results of the thermo-mechanical models

The thermo-mechanical analysis of the studied solar absorber was made on the basis of the comparison between the different structures at an equal load case.

Load case: Stationary homogeneous radiation

This load case was considered for the simple structure with cell density 80cpsi and for the real structure with both cell densities 80cpsi and 200cpsi. The temperature distributions at stationary homogeneous radiation are given in Fig.8.

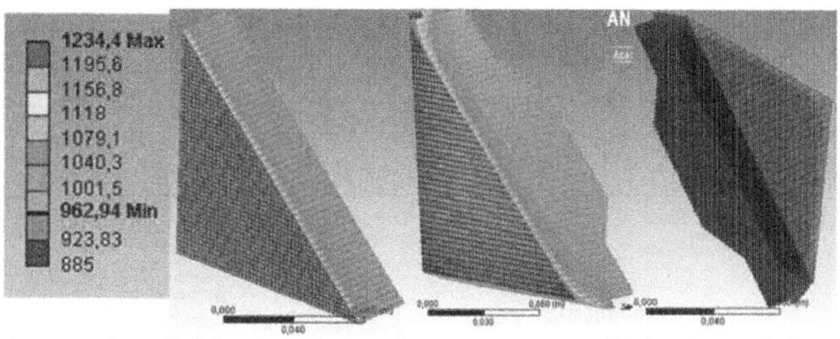

Figure8. The temperature fields at stationary homogeneous radiation from left to right: simple geometry, real geometry 80cpsi and real geometry 200cpsi.

Values up to 1234 K for the real geometry (80cpsi) and up to 1043 K for the real geometry (200cpsi) are reached. The lower temperature with the 200 cpsi structure can be explained by the larger heat transfer surface.

As expected, the temperature distributions have a layer character for all structures. This character is homogeneous for the simple structure and non-homogeneous for the real structure. The inhomogeneous temperature distribution can be explained by the influence of the truncated pyramid. The front temperature increases at the corners and decreases in the center of the front surface. This effect is more pronounced for the 200 cpsi absorber. The temperature field in the absorber with the porosity 200cpsi shows an unexpected decline in the depth ≈ 13-16mm. The temperature increase behind this minimum may be explained by a conductive heat flow in the solid from the outer regions.

Figure 9. Maximal stresses in the absorber (from left to right) with simple geometry (80 cpsi) and with real geometry (80cpsi and 200cpsi).

The maximum stress distributions for the investigated load case are summarized in Fig. 9. The maximum stress in the absorber with simple geometry is located in 16-18mm distance from the irradiated surface and achieves 51MPa. The maximum stress for both real geometries is locates in 10-12mm distance from the irradiated surface and achieves 104MPa for the coarse structure and 66MPa for the finer structure. The largest stress in the 80cpsi real structure is directly due to the largest temperature difference in this sample.

Results at non-homogeneous radiation/non homogeneous temperature boundary conditions

Fig. 10 shows the temperature and the stress distributions for the load case stationary non-homogeneous radiation. The density of the incoming inlet heat flow due to concentrated solar radiation ranges from 800kW/m^2 to 1.57MW/m^2. The corresponding peak temperature measured temperature reaches 1309K. The simulated maximum stress achieves 97MPa at a temperature difference of 650K.

The distributions of temperature and mechanical stress for the load case with stationary homogeneous radiation and an additional temperature difference at the boundaries are shown in Fig.11. Both of these extreme load cases show overcritical mechanical stresses of up to 142.2MPa in the diagonal direction and up to 208.4MPa in the axial direction.

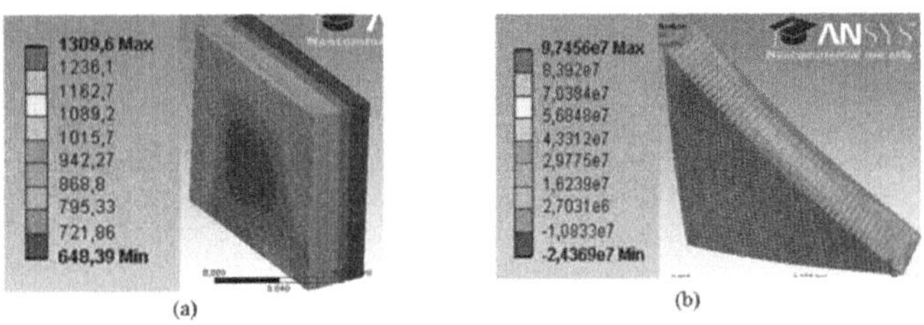

Figure 10. Temperature (a) and stress distribution (b) at the non-homogeneous radiation

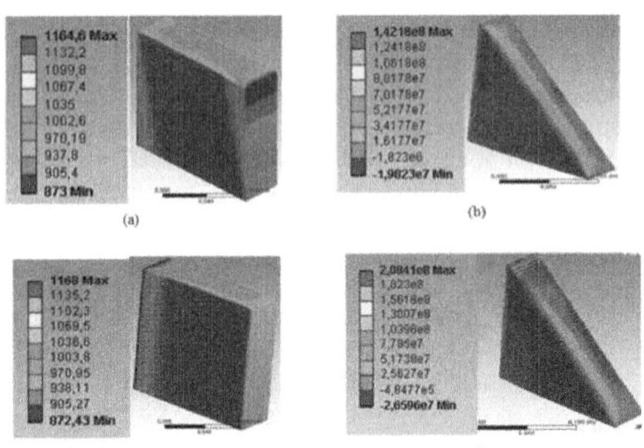

Figure 11. Temperature distributions at stationary homogeneous radiation with additional temperature gradients in the diagonal (a) and in the axial direction (c). (b) and (d) show corresponding stresses.

The results of the "cloud effect" case

The cloud-effect is modeled as a time dependent cooling and heating processes. It uses a time dependent homogeneous heat flux shown in Fig.6. The time dependent temperature distributions of the maximum and minimal temperatures for both studied real structures are shown in Fig.12.

The comparison yields larger values for the coarser structure. The minimal temperature difference is 5 seconds and the maximum one 30 seconds after cooling start. The values are for the finer structure $\Delta T_{max} = 195°C$ and $\Delta T_{min} = 116°C$, for the coarser structure they are $\Delta T_{max} = 273°C$; $\Delta T_{min} = 65°C$.

The values of the mechanical stress have been taken at the starting time and at the end of the process, and in four intermediate points during the cooling and the heating. Figure 13 shows the change of the temperature fields at these points.

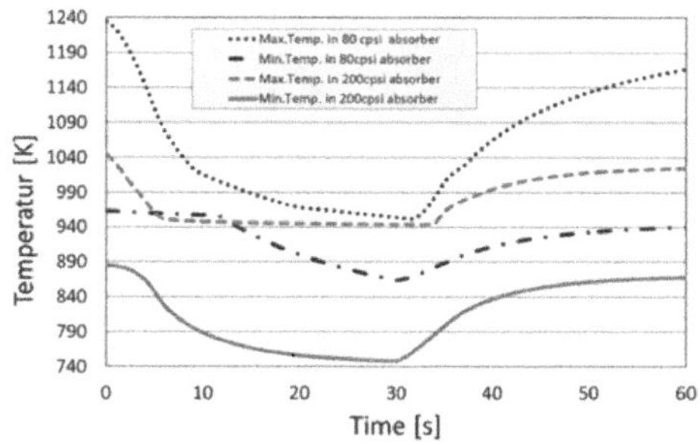

Figure12. The comparison of the maximum and minimal temperatures for both real structures

Figure 13. The temperature fields in the following time points from left to right: at start, 3s after the cooling start, 15s after the cooling start, 3s after the heating start, 15s after the heating start and at the end (30s after the heating start). The figures of the upper series belong to the simple geometry, the figures of the middle series to the real geometry with 80cpsi and the lower series to the real geometry with 200cpsi.

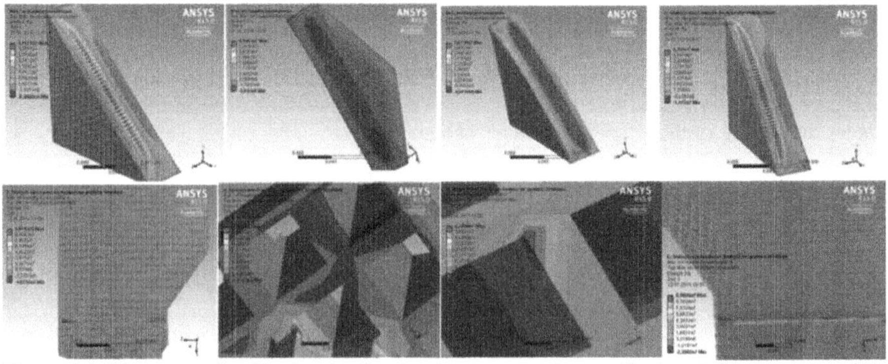

Figure 14. The stress distribution in the simple structure (upper line) and the increase of the maximum mechanical stress locations in the real geometry 80cpsi of the solar absorber at the 4 intermediate time points mentioned above (from left to right).

The distribution of the corresponding stresses is shown in Fig. 14. Together with the analysis of Fig. 15 the results of the cloud effect model can be summarized as follows:

- During the cooling period the point of maximum stress moves from a distance of 16-18mm from the irradiated surface to the air outlet surface. During the heating period it moves back for the load case with the simple structure and with the cell-density 80cpsi (the upper pictures in Fig.14)
- In the case of the pyramid geometry the point of maximum stress is located at channel walls in 15mm distance from the front surface at the edges of the truncated pyramid. During the cooling process it moves to the air outlet surface of these channels and during the heating it moves back. This is true for the load case with the real structure (80cpsi, the bottom pictures in Fig.14)
- The 200cpsi real structure behaves similar as the 80 cpsi one, only with slightly larger stress values

Finally, the fracture stress of the double ring bending test was compared with the maximum mechanical stress of the cloud-effect. The result is shown in Fig.15 (left). It becomes clear, that the maximum mechanical stress of the sample with the simple structure is lower than fracture stress. The maximum mechanical stress of the sample with the standard real structure (cell density 80cpsi) achieves the critical value at the start and at the end of the process. The maximum stress is higher than the fracture stress for the finer real structure in the middle of the process. The dependence between of the maximum mechanical stress and maximum temperature difference is especially important for the thermo-mechanical analysis. It is shown in Fig.15 (right). As expected, the dependence is almost linear for each of the studied structures. It may be also stated, that the cooling process is more critical than the heating process for all studied structures. The critical and over critical values of the mechanical stress correspond to temperature differences of app. 230°C for the standard structure and 130°C for the finer structure.

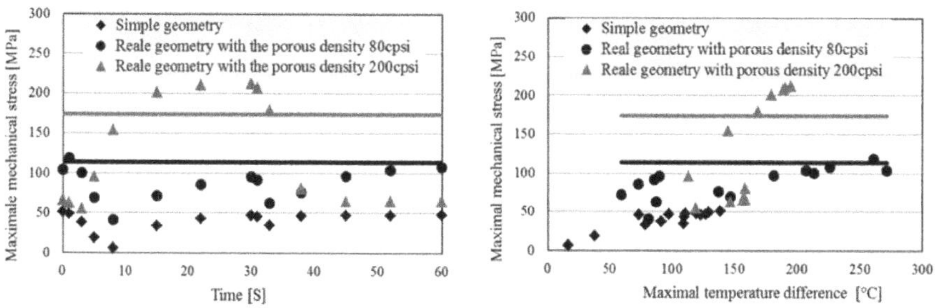

Figure15. Comparison between the fracture stress and the maximum mechanical stress during the cloud-effect simulation (left) and the maximum mechanical stress as a function of the maximum temperature difference (right).

CONCLUSIONS

The following conclusions can be made according to the represented thermo-mechanical analysis:

1. The thermo-mechanical models presented allow an investigation of the influence of representative stationary and transient load cases on the stresses in a solar receiver.
2. The fracture stress determined from the bending test simulation can be used to assess the amount of mechanical stress occurring during representative load cases
3. During a cloud transition, the point of maximum mechanical stress moves from a localization in the middle of the receiver to the outlet surface (cooling phase) and back during the heating phase.
4. The comparison of the two macro geometries shows that for the truncated pyramid case the temperature difference and maximum stress is significantly larger.
5. The maximum mechanical stress in the sample with the finer structure is always larger than in the one with the coarse structure at the equal thermal load.
6. Both the transient thermal load and the inhomogeneous boundary conditions increase the maximum mechanical stress.
7. Although the finer structure of the solar absorber shows a higher thermal effectiveness, the coarser one promises a safer operation and a larger reliability.

REFERENCES:

[1] H. Fricker, Bull. SEV/VSE, 76, pp. 10-16
[2] M. Romero, R. Buck, J. E. Pacheco, An Update on Solar Central Receiver Systems, Projects, and Technologies. Journal of Solar Energy Engineering124, 98-108 (2002)
[3] A. Kribus, R. Zaibel, D. Carrey, A. Segal, J. Karni, A solar – driven combined cycle power plant, Solar Energy Vol. 62, No 2, 1998, pp.121–129
[4] M. Becker, Th. Fend, B. Hoffschmidt, R. Pitz–Paal, Theoretical and numerical investigation of flow stability in porous materials applied as volumetric solar receivers, 1 Solar Energy 80 (2006) p.1241 - 1248
[5] Th. Fend, B. Hoffschmidt, R. Pitz–Paal, O. Reutter, Chapter: "Cellular ceramics use in solar radiation conversion", from the book M. Scheffler and P. Colombo: "Cellular ceramics: structure, manufacturing and applications", Willey – VCH GmbH & Co. KgA, Weinheim, 2005
[6] Th. Fend, B. Hoffschmidt, R. Pitz–Paal, O. Reutter, P. Rietbrock, Porous materials as open volumetric solar receivers experimental determination of thermophysical and heat transfer properties, Energy 29 (2004), (5-6), pp.823 - 833

[7] Metallic materials - Bend test (ISO 7438:2005); German version EN ISO 7438:2005
[8] Th. Fend, P. Schwarzbözl, O. Smirnova, D. Schöllgen, Ch. Jakob (2013) Numerical Investigation of Flow and Heat Transfer in a Volumetric Solar Receiver. Renewable Energy, 60, Seiten 655-661. ISSN 0960-1481

High-Temperature Superconductors: Materials, Technologies, and Systems

ANOMALOUS PROXIMITY EFFECT AND MORE THAN ONE MAJORANA FERMION

S. Ikegaya and Y. Asano
Department of Applied Physics, Hokkaido University
Sapporo 060-8628, Japan

ABSTRACT

We discuss the highly degenerated Majorana bound states appearing in a dirty normal metal attached to a spin-triplet p_x-wave superconductor. A p_x-wave superconductor intrinsically preserves chiral symmetry and traps more than one Majorana fermion at its edges. All the Majorana bound states at an edge belong to the same chirality due to the p_x-wave pairing symmetry. The pure chiral nature of the wave function enables the zero-energy states to penetrate into a dirty normal metal while retaining their high degree of degeneracy. We prove the perfect Andreev reflection into the dirty normal metal at the Fermi level.

INTRODUCTION

When a normal metal is attached to a superconductor, Cooper pairs penetrate into the normal metal and modify physical properties there. This phenomenon is called superconducting proximity effect. In a superconductor, energy spectra are gapped near the Fermi level (zero energy). Thus the penetrating Cooper pairs usually form the gapped quasiparticle density of states (DOS) in the normal metal. This is well known in spin-singlet s-wave superconducting junctions. The normal metal shows similar electromagnetic properties to the superconductor as a result of the proximity effect. In spin-triplet p-wave superconductor junctions, however, the penetrating Cooper pairs form a peak structure in the DOS at the zero energy and induce various anomalous electromagnetic properties in the normal metal[1-6]. This effect is called anomalous proximity effect. For instance, a perfect Andreev reflection from a p_x-wave superconductor into a dirty normal metal causes anomalous low energy transport in the x-direction such as a zero-bias conductance quantization in normal-metal/superconductor (NS) junctions[4] and a fractional current-phase relationship in superconductor/normal-metal/superconductor (SNS) junctions[3]. In this case, penetrating Cooper pairs in the normal metal belong to unusual symmetry class so called odd-frequency symmetry[7]. The zero-energy peak in DOS implies the penetration of a quasiparticle into the normal metal. Such a quasiparticle state is originally formed at a surface of a p_x-wave superconductor as a result of topologically nontrivial character of the p_x-wave superconducting state[8]. By attaching a normal metal, such a surface quasiparticle also penetrates into the normal metal accompanied by odd-frequency Cooper pairs.

Recently the anomalous proximity has been growing its importance in the Majorana physics[9]. A quasiparticle localized at the surface of isolating p_x-wave superconductor is turned out to be a Majorana fermion, which is equivalent to its own anti-particle. Detecting a Majorana fermion and controlling of the Majorana bound states (MBSs) has been a desirable subject to realize the fault-tolerant topological computation[10-11]. According to a definition of topological number characterizing p_x-wave superconducting state, a p_x-wave superconductor traps more than one Majorana fermion as its surface at zero energy[8]. The anomalous proximity effect of the p_x-wave superconductor is explained by the penetration of Majorana fermions into the dirty normal metal while retaining their high degree of degeneracy[1,3,5]. In other words, the anomalous proximity effect is nothing other than the physics of Majorana fermion. Thus experimental demonstration of the anomalous proximity effect proves directly the existence of Majorana fermions.

Quantum mechanics tells us that degeneracy of quantum states is a direct consequence of symmetry. To characterize a p_x-wave superconductor topologically, we need to assume translational symmetry in the superconductor[8]. Therefore the translational symmetry seems to protect highly degenerate Majorana bound states at its surface. In an attached normal metal, the translational symmetry is broken by impurity scatterings. Therefore, it has been unclear what symmetry protects the high degeneracy of MBSs in the dirty normal metal and why the perfect Andreev reflection persists at zero energy. Recent theoretical studies have shown that the chiral symmetry is responsible for the stability of more than one Majorana fermion[12-14]. On the basis of the novel insight, we provide a microscopic understanding of the anomalous proximity effect[15].

In this paper, we analyze the chiral property of MBSs both at the edge of the superconductor and at the normal metal attached to it. The results show that the MBSs localized at a same place belong to a same chirality due to a p_x-wave pairing symmetry. The pure chiral nature of the wave function is responsible for the robustness of highly degenerate MBSs even in the presence of potential disorder. We will also discuss the perfect Andreev reflection from a p_x -wave superconductor into a dirty normal metal at zero energy.

MORE THAN ONE MAJORANA FERMION

At first, we analyze the chiral property of the MBSs in an isolating spin-less p_x-wave superconductor. The Hamiltonian of the p_x-wave superconductor in continuous space is represented by

$$H_{BdG}(\boldsymbol{r}) = \begin{bmatrix} -\frac{\hbar^2}{2m}\nabla^2 - \mu & -i\frac{\Delta_0}{k_F}\partial_x \\ -i\frac{\Delta_0}{k_F}\partial_x & \frac{\hbar^2}{2m}\nabla^2 + \mu \end{bmatrix}, \tag{1}$$

where the diagonal and the off-diagonal component denote the kinetic energy and the pair potential, respectively. The energy is measured from the Fermi energy μ, k_F denotes the Fermi wave length and Δ_0 is an amplitude of pair potential at zero temperature. The Bogoliubov-de

Gennes (BdG) Hamiltonian preserves a chiral symmetry which is defined by

$$\Gamma H_{BdG}(\boldsymbol{r})\Gamma^{-1} = -H_{BdG}(\boldsymbol{r}), \qquad \Gamma = \begin{bmatrix} 0 & -i \\ i & 0 \end{bmatrix}. \tag{2}$$

Symmetry of Hamiltonian is usually represented by the commutation relation between an operator and a Hamiltonian. In Eq. (2), however, Γ anti-commutes with the Hamiltonian in Eq. (1). It is easy to confirm that Γ commutes with $[H_{BdG}]^2$. Therefore an electronic state with eigenvalue E and that with –E are correlated with each other. Here we summarize two important features of the eigenstates of H_{BdG} proved in Ref. 8. (i) An eigenstate of H_{BdG} at zero energy is simultaneously an eigenstate of the chiral symmetry operator Γ with eigenvalue $\lambda = +1$ or -1. (ii) In contrast to the zero-energy states, the nonzero-energy states are not the eigenstates of Γ. They are described by the linear combination of two states: one belongs to $\lambda = +1$ and the other belongs to $\lambda = -1$. In what follows, we prove robustness of the highly degenerated MBSs against potential disorder by taking these features into account.

We consider an isolating p_x-wave superconductor with the length 2L (M) in the x (y) direction as shown in Fig. 1. The left and right edge in the x direction locate at x = -L and x = L, respectively. We assume L is much longer than the superconducting coherence length:

$$L \gg \xi_0 = \frac{\hbar v_F}{\Delta_0}. \tag{3}$$

We apply the hard-wall boundary condition in both the x and y directions. By solving the BdG equation under this boundary condition, we can find the zero-energy edge states for each transport channel as

$$\varphi_n^L(\boldsymbol{r}) = C_n \begin{bmatrix} e^{i\pi/4} \\ e^{-i\pi/4} \end{bmatrix} \sin\left[\sqrt{k_n^2 - \xi_0^{-2}}(x+L)\right] e^{-x/\xi_0} \sin\left(\frac{n\pi}{M}y\right), \tag{4}$$

$$\varphi_n^R(\boldsymbol{r}) = C_n \begin{bmatrix} e^{-i\pi/4} \\ e^{i\pi/4} \end{bmatrix} \sin\left[\sqrt{k_n^2 - \xi_0^{-2}}(x-L)\right] e^{x/\xi_0} \sin\left(\frac{n\pi}{M}y\right), \tag{5}$$

$$k_n = \frac{\sqrt{2m\mu_n}}{\hbar}, \quad \mu_n = \mu - \frac{\hbar^2}{2m}\left(\frac{n\pi}{M}\right)^2, \tag{6}$$

where n=1, 2, 3... indicate the transport channel and C_n denotes a normalization coefficient. Such solutions exist when the condition $\mu_n > 0$ is satisfied therefore the number of the zero-energy states at each edge is equivalent to the number of the propagating channels N_c. Note that the zero-energy states satisfy the condition

$$\Xi\varphi_n^{L(R)}(\boldsymbol{r}) = \varphi_n^{L(R)}(\boldsymbol{r}), \tag{7}$$

where Ξ denote the charge conjugation operator defined by

$$\Xi H_{BdG}(\boldsymbol{r})\Xi^{-1} = -H_{BdG}(\boldsymbol{r}), \qquad \Xi = \begin{bmatrix} 0 & K \\ K & 0 \end{bmatrix}, \tag{8}$$

with the operator K representing the complex conjugation. Therefore, the all of the zero-energy states are Majorana fermions. It is important to note that all the zero-energy states at the left (right) edge have λ = -1 (+1), which is shown schematically in Fig. 1. Generally speaking, such highly degenerated states are fragile in the presence of a random impurity potential near the edge. However, at the left edge for example, all of the zero-energy states have λ =-1. According to chiral property (ii), therefore, such highly degenerated states are robust against potential disorder because the random potential preserves chiral symmetry and the zero-energy states with λ =+1 are absent. We conclude that the highly degeneracy of MBSs are protected by a chiral symmetry.

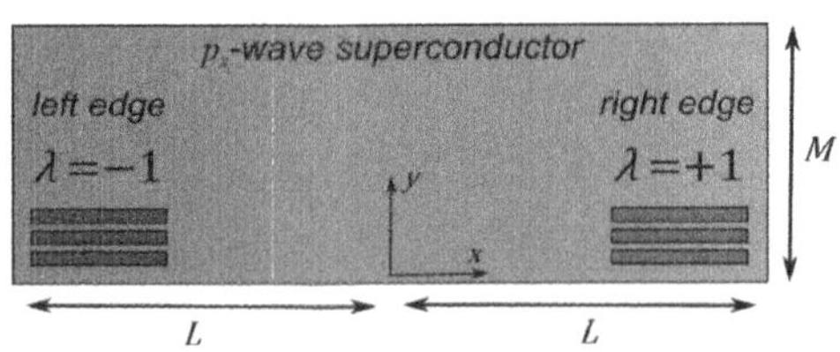

Figure 1. Schematic image of the zero energy states at two edges of an isolating p_x-wave superconductor. The number of the MBSs at either edge is equal to N_C. All of the MBSs at the left (right) edge have λ =-1 (λ =+1)

PERFECT ANDREEV REFLECTION AND CHIRAL NATURE

Here we prove the stability of the highly degenerated zero-energy edge states in a dirty normal metal. To analyze the conductance of a normal-metal/superconductor (NS) junction, we attach a normal metal to the left side of the superconductor as shown in Fig. 2. The Hamiltonian of normal metal is given by H_{BdG} in Eq. (1) with Δ_0 is 0. In the absence of impurity potentials, the wave function in the normal segment at E=0 is described by

$$\varphi_N(\boldsymbol{r}) = \sum_n C_n' \left[\begin{bmatrix} 1 \\ r_n^{he} \end{bmatrix} e^{ik_n x} + \begin{bmatrix} r_n^{ee} \\ 0 \end{bmatrix} e^{-ik_n x} \right] \sin\left(\frac{n\pi}{M} y\right), \tag{9}$$

where r^{ee} and r^{he} denote the normal and the Andreev reflection coefficients, respectively. The index n labels the transmission channel. From boundary conditions at the NS interface, the reflection coefficients are calculated to be

$$r_n^{ee} = 0, \quad r_n^{he} = -i, \tag{10}$$

for all channel. The results are valid even if we introduce a potential barrier at the NS interface. The resonant transmission through the MBSs results in the perfect Andreev reflection. The wave function φ_N satisfies $\Gamma\varphi_N = -\varphi_N$, therefore, all of the zero-energy states in the normal metal have $\lambda = -1$, as well as the MBSs at the left edge of the superconductor. According to the chiral property (ii), they cannot form the nonzero-energy states. Thus, the MBSs can penetrate into the normal metal while retaining a high degree of degeneracy. The conclusion is also valid even in the presence of potential disorder because the impurity potential preserves chiral symmetry and does not damage the pure chiral feature of the MBSs. This fact is unique to the p_x-wave pairing symmetry. In the ballistic limit, the perfect conductance quantization at zero bias is a common property of unconventional superconductors that have zero-energy edge states with a flat dispersion. For instance, a spin-singlet d-wave superconductor preserves chiral symmetry and host the N_c degenerated zero-energy states. However, the highly degenerated zero-energy states are fragile under the potential disorder because the zero-energy states belonging to two different eigenvalues of the chiral symmetry operator coexist at the same edge. Therefore the presence of the chiral symmetry is not a sufficient condition for the anomalous proximity effect but a necessary one.

In a p_x-wave superconductor junction, all the zero-energy states in a normal metal have the same chirality of $\lambda = -1$ in the same way as the MBSs at the left edge of the superconductor. The pure chiral feature of the zero-energy states enables us to explain the perfect Andreev reflection into the dirty normal segment. According to property (i), the zero-energy states must be the eigenstates of the chiral symmetry operator Γ. We emphasize that the wave function in Eq. (9) can be the eigenstates of Γ with $\lambda = -1$ only when Eq. (10) is satisfied. Although the channel index n is no longer a good quantum number under the potential disorder, all the wave functions in the normal segment have the same vector structure reflecting their pure chiral nature. This is a mathematical requirement arising from the chiral symmetry. The physical consequence of the vector structure is the perfect Andreev reflection into the dirty metal at E=0. This explains the perfect quantization of the zero-bias conductance at $2e^2N_c/h$.

NUMERICAL SIMULATION

Finally, we briefly demonstrate the perfect conductance quantization at zero bias in the differential conductance of a NS junction as illustrated in Fig. 2. We assume random impurity potentials in the normal segment. We describe the junction on the two dimensional tight-binding model and numerically calculate the differential conductance G_{NS} of the junction based on the Blonder-Tinkham-Klapwijk (BTK) formula[16]. The transmission coefficients appearing in the

BTK formula are calculated by using the lattice Green's function method[17-19]. In Fig. 3, we show the differential conductance G_{NS} of the junction as a function of the bias voltage for several length of the dirty normal metal segment L. The results are normalized to $2e^2N_c/h$, where N_c denotes the number of propagating channels in the junction. The differential conductance at finite bias voltages decreases by increasing L. However, as an important fact, the differential conductance at zero-bias voltage takes the value of $2e^2N_c/h$ irrespective of L. Such conductance quantization occurs only when the Andreev reflection probability is unity and suggests the existence of the perfect transmission channels in the dirty normal metal. The perfect quantization of the zero-bias conductance in the disordered junction is an aspect of the anomalous proximity effect.

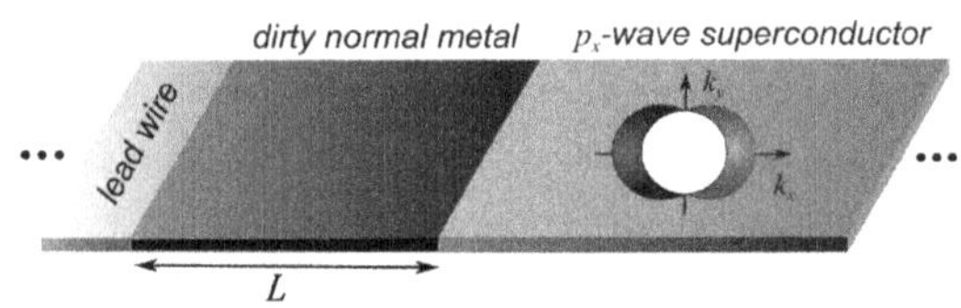

Figure 2. Schematic picture of the NS junction of a p_x-wave superconductor

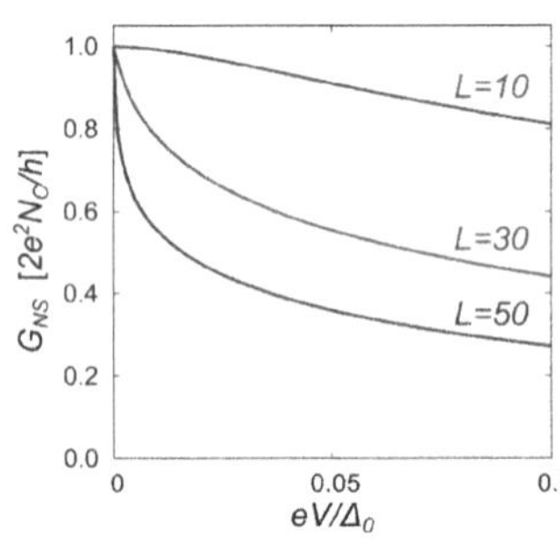

Figure 3. Differential conductance is plotted as a function of the bias voltage for several length of disordered segment L.

CONCLUSION

In conclusion, we have discussed the stability of highly degenerated zero-energy states appearing in the dirty normal-metal/ p_x-wave superconductor junctions. The px-wave superconductor hosts more than one zero-energy Majorana fermion bound state at its edges. When we make a NS junction, such highly degenerated MBSs penetrate into the dirty normal segment and form the resonant transmission channels there. An analysis of the wave function in the normal segment shows the all the MBSs have the same chirality due to the p_x-wave pairing symmetry. The perfect Andreev reflection into the dirty normal metal is a direct consequence of the pure chiral feature of the MBSs. This physical description provides a microscopic understanding the anomalous proximity effect of a spin-triplet superconductor[15].

ACKNOWLEDGEMENT

The authors are grateful to J. D. Sau for useful discussions. This work was supported by "Topological Quantum Phenomena" (No. 22103002) and KAKENHI (Nos. 26287069, 15H03525) from the Ministry of Education, Culture, Sports, Science and Technology (MEXT) of Japan and by the Ministry of Education and Science of the Russian Federation (Grant No. 14Y.26.31.0007).

REFERENCE

[1] Tanaka, Y., Asano, Y., Golubov, A. A., & Kashiwaya, S. (2005). Anomalous features of the proximity effect in triplet superconductors. Physical Review B, volume 72, 140503(R).

[2] Tanaka, Y., & Golubov, A. A. (2007). Theory of the Proximity Effect in Junctions with Unconventional Superconductors. Physical Review Letters, volume 98, 037003.

[3] Asano, Y., Tanaka, Y., & Kashiwaya, S. (2006). Anomalous Josephson Effect in p-Wave Dirty Junctions. Physical Review Letters, volume 96, 097007.

[4] Tanaka, Y., & Kashiwaya, S. (2004). Anomalous charge transport in triplet superconductor junctions. Physical Review B, volume 70, 012507.

[5] Asano, Y., Tanaka, Y., Golubov, A. A., & Kashiwaya, S. (2007). Conductance Spectroscopy of Spin-Triplet Superconductors. Physical Review Letters, volume 99, 067005.

[6] Asano, Y., Golubov, A. A., Fominov, Ya. V., & Tanaka, Y. (2011). Unconventional Surface Impedance of a Normal-Metal Film Covering a Spin-Triplet Superconductor Due to Odd-Frequency Cooper Pairs. Physical Review Letters, volume 107, 087001.

[7] Tanaka, Y., Tanuma Y., & Golubov, A. A. (2007). Odd-frequency pairing in normal-metal/superconductor junctions. Physical Review B, volume 76, 054522.

[8] Sato, M., Tanaka, Y., Yada, K., & Yokoyama, T. (2011). Topology of Andreev bound states with flat dispersion. Physical Review B, volume 83, 224511.

[9] Wiczek, F. (2009). Majorana returns. Nature Physics. volume 5, 614.

[10] Ivanov, D. A. (2001). Non-Abelian Statistics of Half-Quantum Vortices in p-Wave Superconductors. Physical Review Letters, volume 86, 268.

[11] Sau, J. D., Clarke, D. J., & Tewari, S. (2011). Controlling non-Abelian statistics of Majorana fermions in semiconductor nanowires. Physical Review B, volume 84, 094505.

[12] Tewari S., & Sau, J. D. (2012). Topological Invariants for Spin-Orbit Coupled Superconductor Nanowires. Physical Review Letters, volume 109, 150408.

[13] Niu, Y., Chung, S.-B., Hsu, C.-H., Mandel, I., Raghu, S., & Chakraverty, S. (2012). Majorana zero modes in a quantum Ising chain with longer-ranged interactions. Physical Review B, volume 85, 035110.

[14] Diez, M., Dahlhaus, J. P., Wimmer, M., & Beenakker, C. W. J. (2012). Andreev reflection from a topological superconductor with chiral symmetry. Physical Review B, volume 86, 094501.
[15] Ikegaya, S., Asano, Y., & Tanaka, Y. (2015). Anomalous proximity effect and theoretical design for its realization. Physical Review B, volume 91, 174511.
[16] Blonder, G. E., Tinkham, M., & Klapwijk, T. M. (1982). Transition from metallic to tunneling regimes in superconducting microconstrictions: Excess current, charge imbalance, and supercurrent conversion. Physical Review B, volume 25, 4515.
[17] Lee, P. A., & Fisher, D. S. (1981). Anderson Localization in Two Dimensions. Physical Review Letters, volume 47, 882.
[18] Ando, T. (1991). Quantum point contacts in magnetic fields. Physical Review B, volume 44, 8017.
[19] Fisher, D. S., & Lee, P. A. (1981). Relation between conductivity and transmission matrix. Physical Review B, volume 23, 6851.

ATOMIC-SCALE STUDY OF THE SUPERCONDUCTING PROXIMITY EFFECT IN MANGANITE/CUPRATE THIN-FILM HETEROSTRUCTURES

Hao Zhang[1], Igor Fridman[1], Nicolas Gauquelin[2], Gianluigi Botton[2], John Y.T. Wei[1, 3]

[1] *Department of Physics, University of Toronto, Toronto ON, M5S1A7, Canada*
[2] *Canadian Centre for Electron Microscopy, McMaster University, Hamilton ON, L8S4M1 Canada*
[3] *Canadian Institute for Advanced Research, Toronto ON, M5G1Z8, Canada*

ABSTRACT

Superconducting proximity effects that are extremely long-ranged have recently been reported in manganite/cuprate thin-film heterostructures, and attributed to spin-triplet correlations involving odd-frequency pairing. To probe this novel scenario microscopically, we studied *c*-axis multilayer $La_{2/3}Ca_{1/3}MnO_3/YBa_2Cu_3O_{7-\delta}$ (LCMO/YBCO) thin films using scanning tunneling spectroscopy (STS), scanning transmission electron microscopy (STEM), x-ray diffraction (XRD) and electrical transport. STS measurements of bilayers at 4.2 K showed no proximity-induced pairing gap on the ferromagnetic LCMO overlayer down to 5 nm thickness. Cross-sectional STEM revealed double CuO-chain intergrowths in the YBCO layer that form regions with the so-called 247 lattice structure. These nanoscale 247 regions do not show up in XRD, but can physically account for the reduction of superconducting critical temperature (T_c) vs. YBCO thickness in LCMO/YBCO multilayers. As further corroboration, we observed similar T_c reduction in $LaNiO_3$/YBCO multilayers, where $LaNiO_3$ is not ferromagnetic. These results suggest that microstructural defects, rather than the ferromagnetism, are responsible for the long-ranged attenuation of superconductivity reported in *c*-axis LCMO/YBCO heterostructures.

INTRODUCTION

In recent years, there have been numerous studies of the interplay between ferromagnetism and superconductivity in heteroepitaxial thin films comprising the ferromagnetic manganites and superconducting cuprates.[1-7] An observation of particular interest is the dependence of the superconducting critical temperature (T_c) on the *c*-axis layer thicknesses in $La_{2/3}Ca_{1/3}MnO_3/YBa_2Cu_3O_{7-\delta}$ (LCMO/YBCO) multilayers.[8,9] The length scale of this dependence

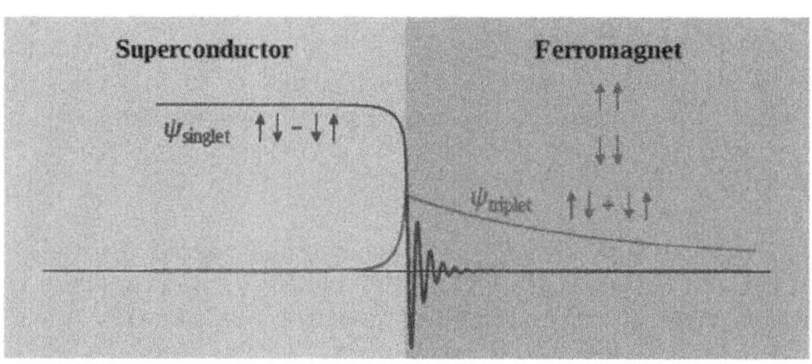

Figure 1: Schematic illustration of long-ranged proximity effect across a superconductor/ ferromagnet junction. Odd-frequency spin-triplet pairs with even-parity orbital symmetry are induced near the interface and penetrate into the ferromagnet far deeper than spin-singlet pairs, since spin-triplet pairs are not easily broken by the exchange field in the ferromagnet.

has been attributed to extremely long-ranged superconducting proximity effect associated with spin-triplet pair formation.[6,10-25] A schematic illustration of this effect is shown in Figure 1. Unlike spin-singlet pairs, spin-triplet pairs are not easily broken by an exchange field, and can thus penetrate deep into the ferromagnet. Theoretical models based on odd-frequency pairing have been proposed, to reconcile the triplet pairing with *d*-wave orbital symmetry, which is known to predominate in YBCO.[26,27] However, other interfacial mechanisms such as charge transfer,[28-30] orbital reconstruction,[31] spin diffusion[9] and induced magnetic modulation[32,33] are also believed to affect the superconductivity in LCMO/YBCO heterostructures.

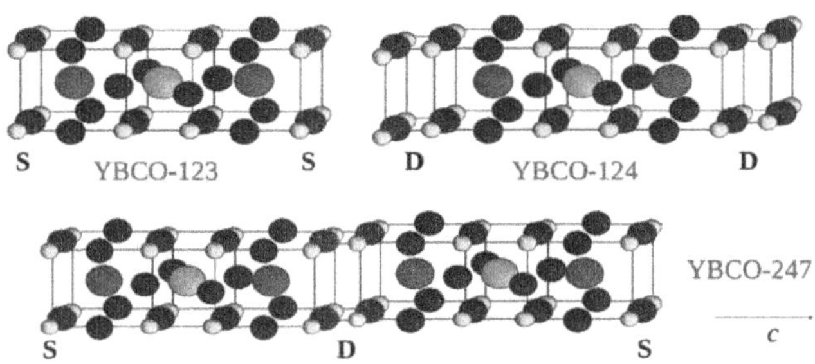

Figure 2: The lattice structures of YBCO-123, YBCO-124 and YBCO-247, with the Cu, Y, and Ba atoms color-labeled as yellow, green, and red, respectively. The CuO chain sites are labelled as S and D for single-chains and double-chains respectively.

Despite these prior studies, direct microscopic evidence for long-ranged proximity effect in *c*-axis manganite/cuprate heterostructures has been lacking. In this paper, we report on an atomic-scale study of the phenomenon using scanning tunneling spectroscopy (STS) and scanning transmission electron microscopy (STEM), and correlate the results with macroscopic measurements including x-ray diffraction (XRD) and electrical transport. First, STS can be used to directly probe proximity effects in ferromagnet-on-superconductor bilayers, by measuring the quasiparticle density of states (DOS) on the top layer.[34-36] A gap in the DOS spectrum would be a distinct signature of proximity-induced pair potential, whose dependence on top-layer thickness could give a measure of the proximity length scale.[37] Second, a crucial aspect of LCMO/YBCO heterostructures that has not been well studied is the microscopic stoichiometry of the YBCO layer. The Y-Ba-Cu-O compounds are exceptional among the cuprates in having CuO chains, the number of which per unit cell allows the cation stoichiometry to vary between the so-called YBCO-123, YBCO-124 and YBCO-247 phases (see Figure 2).[38,39] Since these phases differ in their optimal T_c,[40-45] microscopic variations in the YBCO stoichiometry of LCMO/YBCO heterostructures could affect the resistively-measured T_c and thus the physical interpretation.

EXPERIMENTAL

The *c*-axis LCMO/YBCO multilayer and YBCO unilayer films used in our study were epitaxially grown on (001)-oriented (La,Sr)(Al,Ta)O_3 (LSAT) and $SrTiO_3$ (STO) substrates using pulsed laser-ablated deposition (PLD). Multilayer films comprising $LaNiO_3$ (LNO) and YBCO, where LNO is paramagnetic, were also grown for comparison. Our PLD system is equipped with a 248 nm KrF excimer laser, and the deposition was done with a repetition rate of 2-5 Hz and fluence of

~2 J/cm^2 at 700°-800°C in 200-500 mTorr of O_2. Following deposition, the films were annealed in-situ by slow cooling from the growth temperature to 300°C in 45 minutes in 760 Torr of O_2 to fully oxygenate the YBCO layer. In our films, the LCMO layer ranged from 5 to 25 nm, the YBCO layer from 10 to 50 nm, while the LNO layer was 25 nm in thickness.[46]

The STS measurements were made with a home-built scanning tunneling microscope operating at 4.2 K in ~1 mTorr of ^{4}He exchange gas. Pt-Ir tips were used, and the typical junction impedance was ~10 GΩ. The conductance *dI/dV* spectra were obtained from numerically differentiating the average of 50 *I-V* curves at each tip location. The spectra were measured on several samples for each layer thickness, and at multiple locations on each sample to ensure reproducibility.[47]

The cross-sectional STEM images were taken using a FEI Titan 80–300 microscope fitted with a high-brightness field emission gun and CEOS aberration correctors for both condenser and objective lens aberrations. The microscope operated at 200 keV in scanning mode using the high-angle annular dark-field (HAADF) imaging method, which is sensitive to the atomic number contrast. The elemental identification made by the HAADF imaging was corroborated by atomic-scale electron energy loss spectroscopy (EELS).[48]

The XRD patterns were taken using θ - 2θ method with Cu K_α radiation from a Philips PW2273/20 X-ray tube. Finally, electrical resistance of the films was measured using standard AC lock-in technique in the four-contact configuration, in which silver contacts were sputtered onto the surface of the films in parallel strips. Temperature dependence of the resistance was recorded using a variable-temperature dipper probe inside a liquid helium dewar, and the sample temperature was monitored with a Cernox sensor to 0.1 K accuracy.

RESULTS AND DISCUSSION

The STS data taken on bilayer LCMO/YBCO films grown on STO substrates are shown in Figure 3. Figure 3(a) shows the *dI/dV* spectra for three films, ranging from 10 nm to 20 nm of LCMO

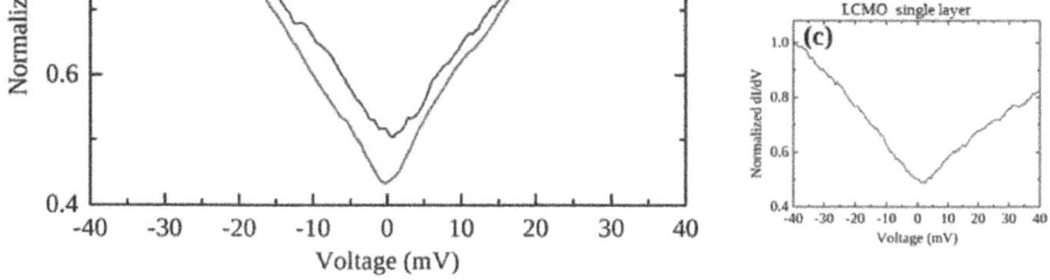

Figure 3: Scanning tunneling conductance spectra measured with a Pt-Ir tip on thin-film samples at 4.2 K. Panel (a) shows spectra taken on bilayer LCMO/YBCO films, for 10-nm (blue curve) and 20-nm (red curve) LCMO thickness. These spectra are distinct from data taken on a unilayer YBCO film (b), but similar to data taken on a unilayer LCMO film (c).

deposited over 20 nm of YBCO. The plotted spectra have been normalized by their values at 50 mV for comparison. These spectra have a characteristic V shape with some asymmetry, as well as substantial conductance at zero bias indicating finite DOS at the Fermi level. No clear gap structure is seen, in contrast to data taken on a unilayer YBCO film as shown in Figure 3(b). The bilayer spectra data are also qualitatively similar to data taken on a unilayer LCMO film as shown in Figure 3(c). Similar V-shaped spectra have been seen in other transition-metal oxides, and appear to be characteristic of strongly-correlated oxide materials.[3,49] When the LCMO layer is reduced to 5 nm, the *dI/dV* spectra (not shown) are also V-shaped similar to unilayer LCMO films. Thus, down to 5 nm of LCMO thickness, there is no clear spectral signature of superconducting proximity effect in *c*-axis LCMO/YBCO bilayers. These observations are consistent with conventional short-ranged proximity effect associated with strong suppression of spin-singlet pairing in the nearly half-metallic LCMO layer.[44]

Figure 4 shows a high-resolution HAADF-STEM image taken over the cross section of a 25-nm/50-nm bilayer LCMO/YBCO film grown on LSAT substrate. The color labels indicate different elements as identified from the HAADF-STEM images, with the Cu, Y, and Ba atoms color-labeled as yellow, green, and red, respectively. To the right of the LCMO/YBCO interface (dotted blue line), there are three unit cells of YBCO-123 characterized by single CuO chains between Ba atoms, followed by a unit cell of YBCO-124 characterized by double CuO chains between Ba atoms. This alternation of single and double CuO chains forms a unit cell of YBCO-247, which is essentially a superlattice of YBCO-123 and YBCO-124 blocks. Further on the right

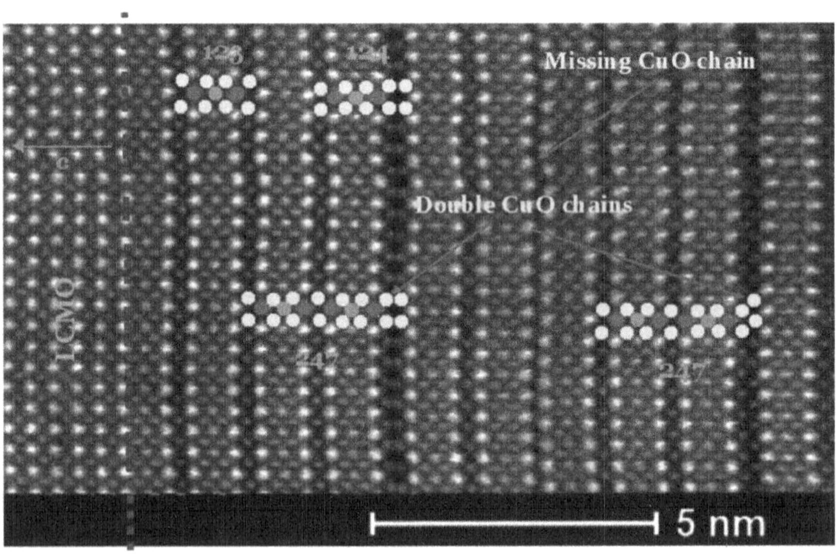

Figure 4: High-resolution HAADF-STEM image of a 25-nm/50-nm bilayer LCMO/YBCO film grown on (001)-oriented LSAT substrate. Blue dotted line indicates the LCMO/YBCO interface, and the Cu, Y, and Ba atoms are respectively color-labeled as yellow, green and red. While the YBCO layer nominally has 123 stoichiometry, the intergrowths of double-CuO chains effectively form nanoscale YBCO-247 regions.

of Figure 4 is another YBCO-247 unit cell, also containing both a single and a double CuO chain. Between the two YBCO-247 unit cells, a unit cell of YBCO with missing CuO chain is seen, in which two adjacent Ba layers join together with no CuO chain in between. Such chain-less region can be explained in terms of compensation for the double CuO chains in the YBCO-247 regions, as the YBCO target used for PLD had 123 stoichiometry.

To examine the pervasiveness of the nanoscale YBCO-247 intergrowths, we compare the STEM images of a bilayer LCMO/YBCO and of a unilayer YBCO. In the bilayer LCMO/YBCO film shown in Figure 5(a), YBCO-247 regions appear throughout the YBCO layer, with the double-CuO chains indicated by the arrows. In contrast, these double-CuO intergrowths are mostly absent in unilayer YBCO films grown under similar conditions, as shown in Figure 5(b). Interestingly, the nanoscale YBCO-247 intergrowths that appear in the high-resolution STEM image of a bilayer LCMO/YBCO film do not appear in its XRD pattern. In fact, the locations of the diffraction peaks for the bilayer LCMO/YBCO film (Figure 5c) and for a unilayer YBCO film (Figure 5d) are indistinguishable. All the major peaks are identified in terms of either YBCO-123, LCMO or LSAT, although the peaks for LCMO and LSAT cannot be distinguished because of their close lattice parameters. For the bilayer case, by relating the YBCO (005)- and (007)-peaks

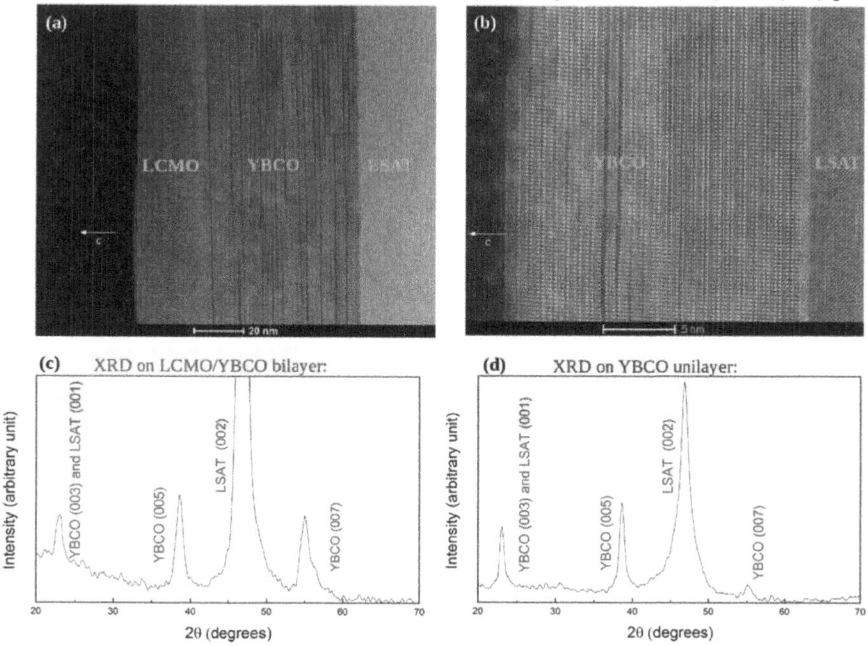

Figure 5: Comparison between bilayer LCMO/YBCO and unilayer YBCO films. Panel (a) shows the STEM image of a bilayer film: nanoscale YBCO-247 regions appear throughout the YBCO layer, with the double-CuO chains indicated by arrows. These double-CuO intergrowths are largely absent in a unilayer YBCO film, as shown in panel (b). The peak locations in the XRD patterns for the bilayer film (panel c) and for the unilayer film (panel d) are indistinguishable. Note that the nanoscale YBCO-247 regions appearing in the bilayer film do not appear in the XRD data within the resolution of our instrument.

with $2\theta = 38.55°$ and $2\theta = 55.06°$ respectively, we find the *c*-axis lattice parameter in the YBCO layer to be 11.68 Å, in agreement with fully oxygenated YBCO-123.[50] It should be emphasized that no peaks associated with YBCO-247 are visible in the XRD data for either the bilayer or unilayer film, within the resolution of our instrument.

The resistively-measured T_c in our bilayer LCMO/YBCO films is evidently affected by the presence of nanoscale YBCO-247 regions. Figure 6 plots the resistance R vs. temperature T data taken on various films. Each R-vs.-T curve is normalized to its R value at room temperature to facilitate comparison. The unilayer YBCO films grown on LSAT or STO substrates show sharp superconducting transitions with T_c near 90 K, consistent with fully-oxygenated YBCO-123. These results indicate that the resistive T_c of the YBCO layer is largely insensitive to its lattice mismatch with the substrate material, at least down to 25-nm YBCO thickness. However, the 25-nm/25-nm bilayer LCMO/YBCO film shows a much lower and broader T_c, near 60 K, than any of the unilayer YBCO films, indicating that the addition of an epitaxial LCMO overlayer significantly reduces the resistive T_c in the YBCO layer. We can plausibly attribute this T_c reduction to the nanoscale YBCO-247 regions seen in our high-resolution STEM images, since YBCO-247 has generally shown lower T_c than either YBCO-124 or fully-oxygenated YBCO-123 according to bulk studies.[51,52] We note that pure YBCO-123 samples can also show reduced T_c when under-oxygenated, with $T_c \sim 60$ K corresponding to $\delta \sim 0.5$ or $YBa_2Cu_3O_{6.5}$.[53] However, this alternative scenario of under-oxygenation is not likely to explain our R-vs.-T data, since the LCMO overlayer was grown in-situ at an even higher oxygen pressure than the YBCO layer, and the entire bilayer was given ample time to fully oxygenate during in-situ post-annealing in 760 Torr of oxygen.

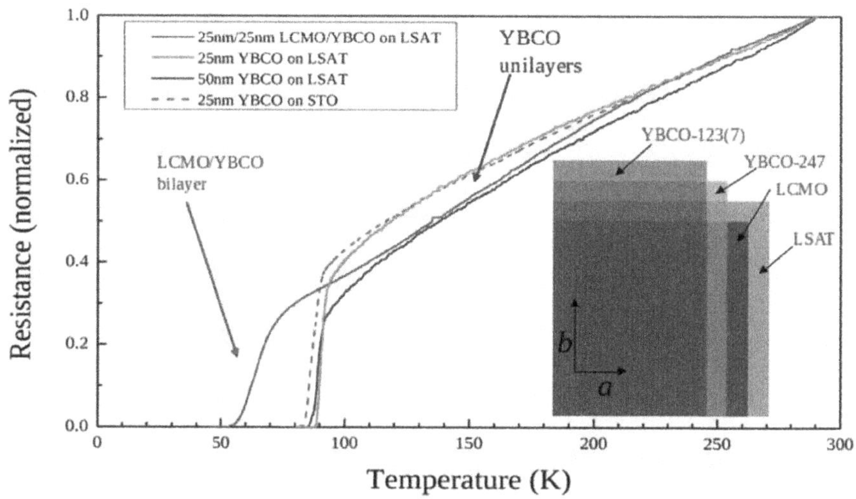

Figure 6: Plot of normalized resistance vs. temperature for unilayer YBCO and bilayer LCMO/YBCO films. All the unilayer films show sharp superconducting transitions with T_c near 90 K. The 25-nm/25-nm bilayer LCMO/YBCO film shows a broadened and reduced T_c near 60 K. The inset shows a schematic diagram illustrating the differences in both lattice symmetry and lattice parameters of the *ab*-plane lattice structure between YBCO-123, YBCO-247, LCMO, and LSAT. In the diagram, the relative length scales between the *a*- and *b*-axes for each material, and between the materials, are exaggerated for clarity.

To explain the formation of the nanoscale YBCO-247 regions in our LCMO/YBCO films and the absence of these intergrowths in unilayer YBCO films, we examine the lattice structures of the materials involved in the heteroepitaxy. It is well known that all the superconducting phases of Y-Ba-Cu-O are orthorhombic due to the CuO chain that runs along the *b*-axis. YBCO-247 is less orthorhombic than YBCO-123, because of its shorter *b*-axis due to the inter-chain attraction within the double-CuO chains. As shown in the schematic diagram of the *ab*-plane lattice structures of YBCO-123, YBCO-247, LCMO, and LSAT in the inset of Figure 6, since both the LCMO overlayer and the LSAT substrate have cubic lattices, their combined mismatch in lattice symmetry with the YBCO layer favors the formation of the less orthorhombic YBCO-247 phase. In addition to this lattice-symmetry mismatch, the *a*- and *b*-axes lattice parameters of both LCMO and LSAT are closer to YBCO-247 than to YBCO-123. Thus the lattice-parameter mismatch, from both sides of the YBCO layer, also favors the formation of YBCO-247. In the YBCO layer, the conversion to YBCO-247 via the intergrowth of double CuO chains provides an effective mechanism for relieving the heteroepitaxial strain imposed by both the LCMO overlayer and LSAT substrate.

To examine the role played by ferromagnetism in the T_c reduction of LCMO/YBCO heterostructures and to further corroborate our microstructural interpretation of T_c reduction, we also grew LNO/YBCO heterostructures and measured their R vs. T for comparison. LNO is a paramagnetic perovskites with similar lattice parameters (within ~ 0.26%) as LCMO. Thus the YBCO layer in LNO/YBCO heterostructures is subject to similar heteroepitaxial strain as but

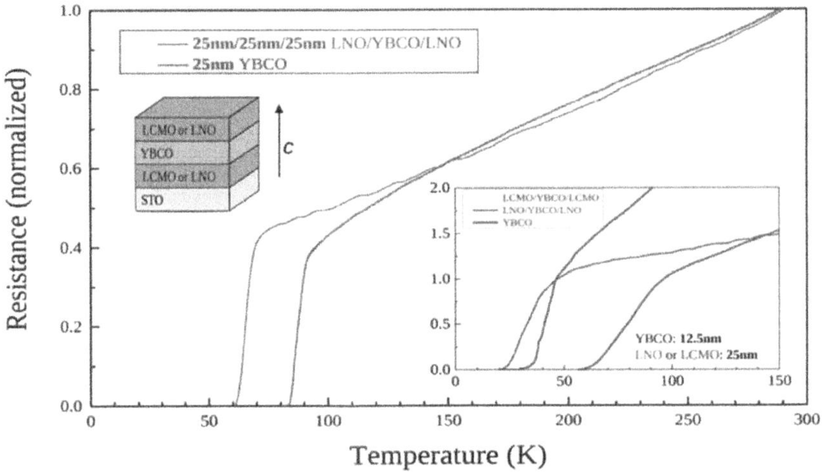

Figure 7: Plot of normalized resistance vs. temperature for trilayer LNO/YBCO/LNO and unilayer YBCO films grown on (001)-oriented STO substrates. In the trilayer, YBCO is sandwiched between two LNO layers, each 25 nm thick, to symmetrize the heteroepitaxial strain. At 25 nm YBCO thickness, the trilayer film has a $T_c \sim 60$ K, which is much lower than that of the unilayer YBCO film. The inset shows similar behavior at 12.5 nm YBCO thickness, with data from trilayer LCMO/YBCO/LCMO included for comparison: superconductivity onsets at similarly low temperatures (~45 K) for both LNO/YBCO/LNO and LCMO/YBCO/LCMO, as compared with unilayer YBCO.

without the ferromagnetism in LCMO/YBCO heterostructures. Figure 7 compares the R-vs.-T plots between trilayer LNO/YBCO/LNO and unilayer YBCO films. All the samples were grown on (001)-oriented STO substrates. For the trilayer films, the YBCO layer was sandwiched between two LNO layers, each 25 nm thick, to symmetrize the strain from both the top and the bottom sides. When the YBCO layer is 25 nm thick, the LNO/YBCO/LNO film has a $T_c \sim 60$ K, which is much lower than a unilayer YBCO film but close in value to a LCMO/YBCO/LCMO film with the same YBCO thickness. The inset of Figure 7 shows a comparison of the R-vs.-T data for LNO/YBCO/LNO, LCMO/YBCO/LCMO and unilayer YBCO, with the YBCO layer thickness fixed at 12.5 nm. Here the trilayer films also show much lower T_c than the unilayer film. In particular, the onset of superconductivity for LNO/YBCO/LNO occurs at a similarly low temperature (~ 45 K) as LCMO/YBCO/LCMO. Since LNO is not ferromagnetic, the similar dependence of T_c on YBCO layer thickness, between the LNO/YBCO and LCMO/YBCO heterostructures, indicates that the ferromagnetism plays a negligible role in the T_c reduction seen in the latter.

CONCLUSION

In summary, we have carried out an atomic-scale study of the superconducting proximity effect in *c*-axis LCMO/YBCO heterostructures. STS, STEM, XRD and *R*-vs.-*T* measurements were performed on bilayer LCMO/YBCO and unilayer YBCO thin films, grown epitaxially on either LSAT or STO substrates by PLD. The STS measurements at 4.2 K on bilayer LCMO/YBCO films observed no proximity-induced pairing gaps down to 5-nm LCMO thickness. The HAADF-STEM images taken on the bilayer films revealed nanoscale YBCO-247 regions characterized by double CuO-chain intergrowths, which are largely absent in the unilayer YBCO films and can be attributed to heteroepitaxial lattice mismatch of the YBCO layer with both the LCMO overlayer and LSAT substrate. These nanoscale 247 regions do not appear in XRD, but can physically explain the lower resistive T_c measured in LCMO/YBCO multilayers. As further corroboration, we observed similar T_c reduction in LNO/YBCO multilayers, where LNO is not ferromagnetic. These results suggest that the anomalously long-ranged attenuation of superconductivity reported in *c*-axis LCMO/YBCO thin-film heterostructures is more likely due to strain-induced microstructural defects than to the ferromagnetism.

ACKNOWLEDGEMENTS

Work supported by NSERC, CFI-OIT and the Canadian Institute for Advanced Research.

REFERENCES

[1] V. A. Vas'ko, V. A. Larkin, P. A. Kraus, K. R. Nikolaev, D. E. Grupp, C. A. Nordman, and A. M. Goldman, Phys. Rev. Lett. 78, 1134 (1997).
[2] P. Kraus, A. Bhattacharya, and A. M. Goldman, Phys. Rev. B 64, 220505(R) (2001).
[3] Z. Y. Chen, A. Biswas, I. Zutić, T. Wu, S. B. Ogale, R. L. Greene, and T. Venkatesan, Phys. Rev. B 63, 212508 (2001).
[4] Z. Sefrioui, D. Arias, V. Pena, J. E. Villegas, M. Varela, P. Prieto, C. Leon, J. L. Martinez, and J. Santamaria, Phys. Rev. B 67, 214511(2003).
[5] A. I. Buzdin, Rev. Mod. Phys. 77, 935 (2005).
[6] K. Dybko, K. Werner-Malento, P. Aleshkevych, M. Wojcik, M. Sawicki, and P. Przyslupski, Phys. Rev. B 80, 144504 (2009).
[7] For recent studies, see C. Visani, Z. Sefrioui, J. Tornos, C. Leon, J. Briatico, M. Bibes, A. Barthelemy, J. Santamaria, and J. E. Villegas, Nat. Phys. 8, 539 (2012), and references therein.
[8] H.-U. Habermeier, G. Cristiani, R. K. Kremer, O. Lebedev, and G. van Tendeloo, Physica C 364, 298 (2001).
[9] V. Pena, C. Visani, J. Garcia-Barriocanal, D. Arias, Z. Sefrioui, C. Leon, and J. Santamaria, Phys. Rev. B 73, 104513 (2006).
[10] F. S. Bergeret, A. F. Volkov, and K. B. Efetov, Phys. Rev. Lett. 86, 4096 (2001).
[11] F. S. Bergeret, A. F. Volkov, and K. B. Efetov, Rev. Mod. Phys. 77, 1321 (2005).
[12] R. S. Keizer, S. T. B. Goennenwein, T. M. Klapwijk, G. Miao, G. Xiao and A. Gupta, Nature 439, 825 (2006).
[13] Shiro Kawabata, Yasuhiro Asano, Yukio Tanaka, and Alexander A. Golubov, J. Phys. Soc. Jpn. 82, 124702 (2013).
[14] Z. Ping Niu and D. Y. Xing, Phys. Rev. Lett. 98, 057005 (2007).
[15] Y. Kalcheim, T. Kirzhner, G. Koren, and O. Millo, Phys. Rev. B 83, 064510 (2011).
[16] A. S. Mel'nikov, A. V. Samokhvalov, S. M. Kuznetsova, and A. I. Buzdin, Phys. Rev. Lett. 109, 237006 (2012).

[17] F. S. Bergeret and I. V. Tokatly, Phys. Rev. Lett. 110, 117003 (2013).
[18] Roland Grein, Tomas Löfwander, and Matthias Eschrig, Phys. Rev. B 88, 054502 (2013).
[19] V. I. Zdravkov, J. Kehrle, G. Obermeier, D. Lenk, H.-A. Krug von Nidda, C. Müller, M. Yu. Kupriyanov, A. S. Sidorenko, S. Horn, R. Tidecks, and L. R. Tagirov Phys. Rev. B 87, 144507 (2013).
[20] T. Golod, A. Rydh, V. M. Krasnov, I. Marozau, M. A. Uribe-Laverde, D. K. Satapathy, Th. Wagner, and C. Bernhard, Phys. Rev. B 87, 134520 (2013).
[21] M. Egilmez, J. W. A. Robinson, Judith L. MacManus-Driscoll, L. Chen, H. Wang and M. G. Blamire, Europhys. Lett. 106, 37003 (2014).
[22] Yoav Kalcheim, Israel Felner, Oded Millo, Tal Kirzhner, Gad Koren, Angelo Di Bernardo, Mehmet Egilmez, Mark G. Blamire, and J. W. A. Robinson, Phys.Rev. B 89, 180506(R) (2014).
[23] A. Singh, S. Voltan, K. Lahabi, and J. Aarts, Phys. Rev. X 5, 021019 (2015).
[24] C. Visani, F. Cuellar, A. Pérez-Muñoz, Z. Sefrioui, C. León, J. Santamaría, and Javier E. Villegas, Phys. Rev. B 92, 014519 (2015).
[25] Ya. V. Fominov, Y. Tanaka, Y. Asano, and M. Eschrig, Phys. Rev. B 91, 144514 (2015).
[26] C. C. Tsuei and J. R. Kirtley, Rev. Mod. Phys. 72, 969 (2000).
[27] J. Y. T. Wei, N.-C. Yeh, D. F. Garrigus, and M. Strasik, Phys. Rev. Lett. 81, 2542 (1998).
[28] T. Holden, H.-U. Habermeier, G. Cristiani, A. Golnik, A. Boris, A. Pimenov, J. Humlicek, O. I. Lebedev, G. Van Tendeloo, B. Keimer, and C. Bernhard, Phys. Rev. B 69, 064505 (2004).
[29] M. Varela, A. R. Lupini, V. Pena, Z. Sefrioui, I. Arslan, N. D. Browning, J. Santamaria, and S. J. Pennycook, e-print arXiv:cond-mat/0508564 (2006).
[30] V. Pena, T. Gredig, J. Santamaria, and I. K. Schuller, Phys. Rev. Lett. 97, 177005 (2006).
[31] J. Chakhalian, J. W. Freeland, H.-U. Habermeier, G. Cristiani, G. Khaliullin, M. van Veenendaal, and B. Keimer, Science 318, 1114 (2007).
[32] J. Chakhalian, J. W. Freeland, G. Srajer, J. Strempfer, G. Khaliullin, J. C. Cezar, T. Charlton, R. Dalgliesh, C. Bernhard, G. Cristiani, H.-U. Habermeier, and B. Keimer, Nat. Phys. 2, 244 (2006).
[33] J. Stahn, J. Chakhalian, C. Niedermayer, J. Hoppler, T. Gutberlet, J. Voigt, F. Treubel, H.-U. Habermeier, G. Cristiani, B. Keimer, and C. Bernhard, Phys. Rev. B 71, 140509(R) (2005).
[34] L. Crétinon, A. K. Gupta, H. Sellier, F. Lefloch, M. Faure, A. Buzdin, and H. Courtois, Phys. Rev. B 72, 024511 (2005).
[35] I. Asulin, O. Yuli, G. Koren, and O. Millo, Phys. Rev. B 74, 092501 (2006).
[36] I. Asulin, O. Yuli, I. Felner, G. Koren, and O. Millo, Phys. Rev. B 76, 064507 (2007).
[37] P. C. van Son, H. van Kempen, and P. Wyder, Phys. Rev. Lett. 59, 2226 (1987).
[38] J. Karpinski, S. Rusiecki, B. Bucher, E. Kaldis, and E. Jilek, Physica C 161, 618 (1989).
[39] Donald E. Morris, Andrea G. Markelz, Boris Fayn, Janice H. Nickel, Physica C 168, 153 (1990).
[40] R. J. Cava, B. Batlogg, R. B. van Dover, D. W. Murphy, S. Sunshine, T. Siegrist, J. P. Remeika, E. A. Rietman, S. Zahurak, and G. P. Espinosa, Phys. Rev. Lett. 58, 1676 (1987).
[41] P. Bordet, C. Chaillout, J. Chenavas, J. L. Hodeau, M. Marezio, J. Karpinski, and E. Kaldis, Nature 334, 596 (1988).
[42] D. E. Morris, J. H. Nickel, J. Y. T. Wei, N. G. Asmar, J. S. Scott, U. M. Scheven, C. T. Hultgren, A. G. Markelz, J. E. Post, P. J. Heaney, D. R. Veblen, and R. M. Hazen, Phys. Rev. B 39, 7347 (1989).
[43] D. E. Morris, N. G. Asmar, J. Y. T. Wei, J. H. Nickel, R. L. Sid, J. S. Scott, and J. E. Post, Phys. Rev. B 40, 11406 (1989).
[44] J. Karpinski, E. Kaldis, S. Rusiecki, E. Jilek, P. Fischer, P. Bordet, C. Chaillout, J. Chenavas, J. L. Hodeau, and M. Marezio, Physica C 160, 449 (1989).

[45] J. L. Tallon, D. M. Pooke, R. G. Buckley, M. R. Presland, and F. J. Blunt, Phys. Rev. B 41, 7220 (1990).
[46] H. Zhang, N. Gauquelin, G. A. Botton, and J. Y. T. Wei, Appl. Phys. Lett. 103, 052606 (2013).
[47] I. Fridman, L. Gunawan, G. A. Botton, and J. Y. T. Wei, Phys. Rev. B 84, 104522 (2011).
[48] N. Gauquelin, G.-z. Zhu, H. Zhang, J. Y. T. Wei and G. A. Botton, manuscript in preparation.
[49] A. K. Raychaudhuri, K. P. Rajeev, H. Srikanth, and N. Gayathri, Phys. Rev. B 51, 7421 (1995).
[50] J. J. Capponi, C. Chaillout, A. W. Hewat, P. Lejay, M. Marezio, N. Nguyen, B. Raveau, J. L. Soubeyroux, J. L. Tholence, and R. Tournier, Europhys. Lett. 3, 1301 (1987).
[51] M. Kato, M. Nakanishi, T. Miyano, T. Shimizu, M. Kakihana, K. Yoshimura, and K. Kosuge, J. Solid State Chem. 139, 266 (1998).
[52] A. Irizawa, T. Ohmura, T. Shibata, M. Kato, K. Yoshimura, K. Kosuge, Y. Ito, H. Michor, and G. Hilscher, J. Phys. Soc. Jpn. 71, 574 (2002).
[53] Ruixing Liang, D. A. Bonn, and W. N. Hardy, Phys. Rev. B 73, 180505(R) (2006).

TUNNELING AND PHOTOEMISSION SPECTRA IN CUPRATE SUPERCONDUCTORS: EVIDENCE FOR STRONG MULTIPLE-PHONON COUPLING AND POLARONIC EFFECTS

Guo-meng Zhao

Department of Physics and Astronomy, California State University, Los Angeles, CA 90032, USA

ABSTRACT

We have analyzed scanning tunneling spectra of two electron-doped cuprates $Pr_{0.88}LaCe_{0.12}CuO_{4-y}$ (T_c = 21 K and 24 K) and compared them with tunneling spectrum of hole-doped $La_{1.84}Sr_{0.16}CuO_4$. We have also analyzed tunneling and angle-resolved photoemission spectra for hole-doped $Bi_2Sr_2CaCu_2O_{8+y}$. The tunneling spectra show no evidence for strong coupling to magnetic resonance modes in both electron- and hole-doped cuprates. Furthermore, the peak-dip-hump features observed in tunneling and angle-resolved photoemission spectra of hole-doped $Bi_2Sr_2CaCu_2O_{8+y}$ can be well explained in terms of lattice polaronic effects.

INTRODUCTION

Developing the microscopic theory for high-T_c superconductivity (HTS) is one of the most challenging problems in condensed matter physics. Twenty-nine years after the discovery of HTS by Bednörz and Müller,[1] no consensus on the microscopic pairing mechanism has been reached despite tremendous experimental and theoretical efforts. There are essentially two opposite views about the pairing mechanism. Many researchers believe that antiferromagnetic (AF) fluctuations predominantly mediate the electron pairing.[2] In contrast, other researchers believe that strong electron-phonon coupling is mainly responsible for HTS in cuprates. The polaron-bipolaron theory of superconductivity,[3] which is based on strong electron-electron correlation and significant electron-phonon interaction (EPI), has gained strong support from various experimental results. In particular, extensive studies of unconventional oxygen-isotope effects (OIEs) in hole-doped cuprates have clearly shown strong EPI and the existence of polarons/bipolarons in both normal and superconducting states.[4-13] Neutron scattering,[14-16] angle-resolved photoemission (ARPES),[17,18] pump probe,[19] and optical specroscopies[20,23] have also demonstrated strong EPI. Further, ARPES data[24] and tunneling spectra[25-29] have consistently provided direct evidence for strong coupling to multiple-phonon modes in hole-doped cuprates.

On the other hand, many researchers still maintain the *d*-wave magnetic pairing mechanism, allegedly supported by some highly publicized experimental papers.[30,32] In one of the papers,[31] the authors used an unrealistic parameter, which overestimated the magnetic coupling constant by two orders of magnitude (for detailed arguments, see Ref.[33] and references therein). Other two papers[30,32] reported the combined neutron and tunneling data for two electron-doped $Pr_{0.88}LaCe_{0.12}CuO_{4-y}$ (PLCCO) crystals with different superconducting transition temperatures (21 and 24 K). These data seemingly suggest that the energies of the magnetic resonance modes are the same as those of the bosonic modes revealed in the second derivative (d^2I/dV^2) of electron tunneling current (I) with respective to bias voltage (V). They thus concluded that the magnetic resonance modes rather than phonons mediated electron pairing in electron-doped cuprates.[30,32] However, Zhao[34] pointed out a basic mistake in the data analyses of Ref.[30] and showed that the combined neutron and tunneling data had actually disproved this magnetic pairing mechanism. Despite citing Ref.[34], the same basic mistake was repeated in Ref.[32]. Based on the repeated incorrect analyses, these authors concluded that their data supported d-wave magnetic pairing mechanism.

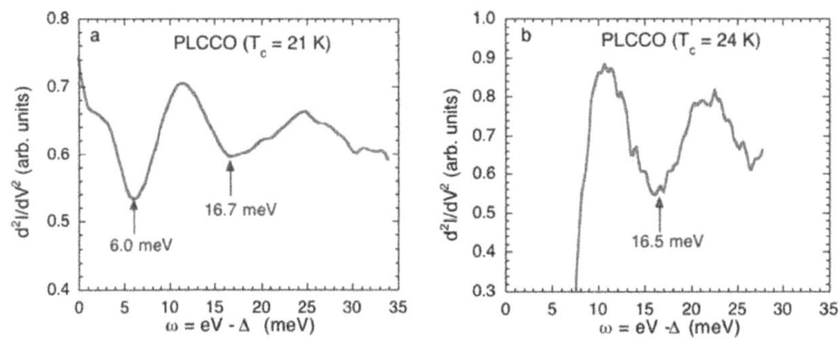

Figure 1. a) d^2I/dV^2 spectrum for an electron-doped PLCCO sample with T_c = 21 K, which is reproduced from Ref.[32] b) d^2I/dV^2 spectrum for an electron-doped PLCCO sample with T_c = 24 K, which is reproduced from Ref.[30] The energy positions of the d^2I/dV^2 spectra are measured from their superconducting gaps.

Here we re-analyze scanning tunneling spectra of two electron-doped cuprates $Pr_{0.88}LaCe_{0.12}CuO_{4-y}$ (T_c = 21 K and 24 K) and compare them with tunneling spectrum of hole-doped $La_{1.84}Sr_{0.16}CuO_4$. The tunneling data show no evidence for strong coupling to the magnetic resonance modes. We have also analyzed tunneling spectra and ARPES for hole-doped $Bi_2Sr_2CaCu_2O_{8+y}$ (Bi-2212 or BSCCO). The data do not show evidence for strong coupling to the magnetic resonance mode either. Instead, the peak-dip-hump features observed in tunneling spectra and ARPES of hole-doped Bi-2212 can be well explained in terms of lattice polaronic effects.

It has been well established[26,28,29,35,36] that the energies of the dip positions (rather than the peak positions) in d^2I/dV^2 correspond to the energies of the modes strongly coupled to electrons. In Fig. 1, we adopt this well-established protocol to identify the mode energies from the d^2I/dV^2 spectra of two electron-doped samples. For the electron-doped sample with T_c = 21 K, the mode energies identified from the d^2I/dV^2 spectrum below 35 meV are 6.0 meV and 16.7 meV (indicated by the arrows). For the electron-doped sample with T_c = 24 K, the mode energy identified from the d^2I/dV^2 spectrum between 7 and 28 meV is 16.5 meV. It is apparent that the mode energies revealed in the d^2I/dV^2 spectra of the two electron-doped cuprates are nearly the same (16.7 and 16.5 meV) and significantly different from the magnetic resonance energies (9.0 or 10.5 meV) revealed by inelastic neutron scattering experiments.[30,32] The mode energy of about 16.5 meV, which is independent of doping and T_c, agrees with the energies of the two transverse optical (TO) phonon modes (15.6 meV for the E_u mode and 17.0 meV for the A_{2u}).[37] This implies that the phonon modes rather than the magnetic-resonance modes have strong coupling to electrons in the electron-doped cuprates. Therefore, their combined neutron and tunneling data[32] actually disprove rather than support their claim that magnetic resonance modes mediate *d*-wave pairing in electron-doped cuprates.

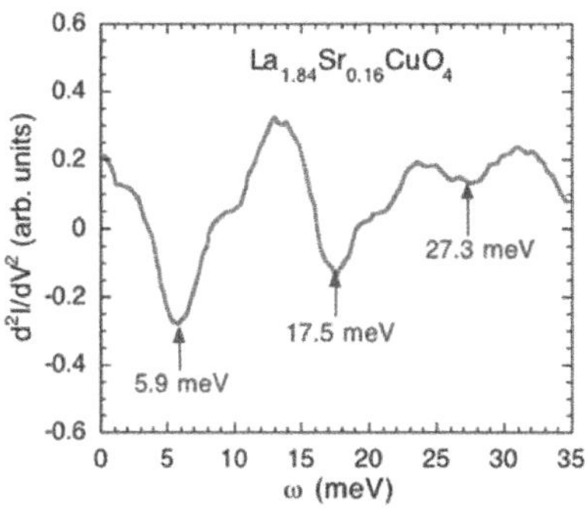

Figure 2. d^2I/dV^2 spectrum at 7.2 K for a hole-doped $La_{1.84}Sr_{0.16}CuO_4$ with T_c = 24 K, which is reproduced from Fig. 4 of Ref. [29]. The energy position of the d^2I/dV^2 spectrum is measured from the superconducting gap.

It is interesting that in the structurally similar single-layer hole-doped $La_{1.84}Sr_{0.16}CuO_4$ (LSCO), there is a similar mode with energy of 17.5 meV (see Fig. 2). Since the magnetic-resonance mode energy was found[41] to be proportional to T_c, insensitivity of the mode energies to T_c further argues against the magnetic origin of the modes. The similar mode energies across different doping levels (Fig. 1) and across different single-layer structures (PLCCO and LSCO) are consistent with the phonon mode assignment and contradict the magnetic resonance mode assignment.

For bi-layer hole-doped BSCCO, strong coupling features are also seen in the d^2I/dV^2 spectrum (see Fig. 3a). It is striking that the energies of all the dip features in the d^2I/dV^2 spectrum precisely match the energies of the peak features in the phonon density of states obtained from high-resolution inelastic neutron scattering.[43] This suggests strong coupling to multiple phonon modes in this bi-layer hole-doped cuprate. Therefore, strong coupling to multiple phonon modes is universally seen in single-layered electron- and hole-doped cuprates and in bi-layered hole-doped cuprates. It is the multiple phonon coupling that leads to a large electron-phonon coupling constant and to HTS.[26] In contrast, the Cu-O buckling mode and the oxygen half-breathing mode, which have been considered to the most important phonon modes in contributing to a *d*-wave pairing, cannot explain high-temperature superconductivity.

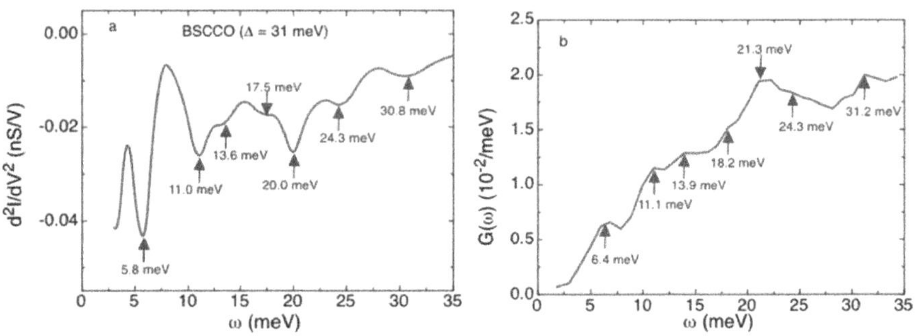

Figure 3. a) d^2I/dV^2 spectrum of bi-layer hole-doped $Bi_2Sr_2CaCu_2O_{8+y}$.The d^2I/dV^2 spectrum is obtained by numerically taking the derivative of the dI/dV spectrum of Ref.[39] after the spectrum is smoothened by cubic spline. The energy position of the d^2I/dV^2 spectrum is measured from the superconducting gap. b) Phonon density of states for slightly overdoped $Bi_2Sr_2CaCu_2O_{8+y}$. The curve is reproduced from Ref.[40]

Within the *d*-wave channel, the phonon contribution to the pairing is too small, so magnetic contribution must be dominant in order to explain HTS. But within an *s*-wave channel, all the phonon modes contribute to the pairing, so the total electron-phonon coupling constant could be large although the coupling constant of single phonon mode could be small. The high-quality tunneling spectrum[26] indicates that the total electron-phonon coupling constant in Bi-2212 is larger than 3. In particular, low energy phonon modes below 40 meV make dominant contributions[26] and the HTS appears to be well explained in terms of the strong coupling phonon-mediated mechanism.[26] One of the major problems within the phonon-mediated mechanism is the *d*-wave pairing symmetry, which has been accepted by the most scientists in the field. In fact, based on the quantitative analyses of many bulk-sensitive experiments in optimally doped and overdoped cuprates, Zhao has concluded that the intrinsic pairing gap symmetry in the bulk of cuprates is not *d*-wave, but extended *s*-wave (having eight line nodes) in hole-doped cuprates[33,41] and nodeless *s*-wave in electron-doped cuprates.[41-43] The phase-sensitive experiments and ARPES data that appear to point towards *d*-wave pairing symmetry are discussed in APPENDIX III below. The second major problem is the calculated electron-phonon coupling constant from local density approximation (LDA), which is about one order of magnitude smaller than that inferred from the tunneling data. It is important to notice that LDA could significantly underestimate electron-phonon coupling due to a large overestimate of electron screening in quasi-two-dimensional electronic systems. In fact, theoretical studies[44,45] show that the optical phonons are strongly coupled to electrons due to the unscreened long-range interaction along the *c*-axis. Strong electron-electron correlation[46] and nonadiabatic effects[45] further enhance electron-phonon coupling. Moreover, strong long-range coupling to high-energy phonon modes can lead to Fröhlich polarons, which can significantly enhance the density of states and the effective coupling constant. If one considers all these enhancement effects together, the effective electron-phonon coupling constant is likely to be enhanced by an order of magnitude compared with that calculated from LDA.

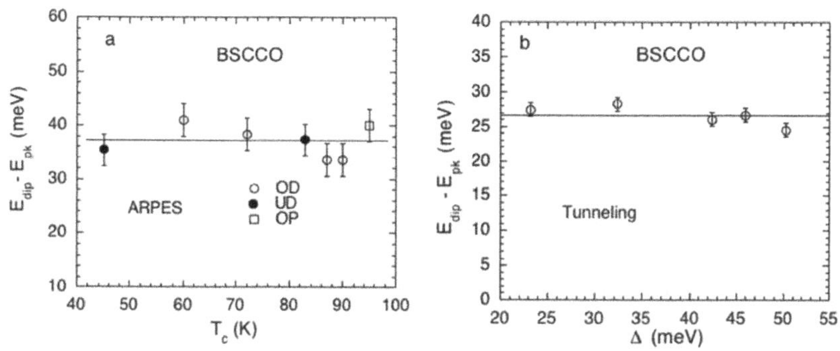

Figure 4. a) Dependence of E_{dip}-E_{pk} on T_c, which is reproduced from Ref.[10] The E_{dip}-E_{pk} values were determined from ARPES data. b) Dependence of E_{dip}-E_{pk} on the superconducting gap. The E_{dip}-E_{pk} values are obtained from Supplementary Figure 2 of Ref.[49] The different E_{dip}-E_{pk} values extracted from ARPES and tunneling data may arise from different energy resolutions that affect peak width.

The above analyses indicate that the d^2I/dV^2 tunneling spectra do not reveal strong coupling features of magnetic resonance modes in both electron- and hole-doped cuprates. Now a question arises as to whether strong coupling to magnetic resonance modes can be seen in the dI/dV tunneling spectra and angle-resolved photoemission spectra (energy distribution curves). The phenomenological spin-fermion model of Abanov and Chubukov[47] showed that strong coupling to the magnetic resonance mode yields a peak-dip-hump (PDH) structure in ARPES spectra along the antinodal direction.[47] This theory is based on the one-loop correction to the *t-t-J* mean-field theory. This PDH structure was also predicted to be present in dI/dV tunneling spectra based on the conventional strong coupling theory.[48] Within both approaches, the energy separation between the dip and peak features is exactly equal to the energy E_r of the magnetic resonance mode,[47,48] that is, E_{dip}-E_{pk} = E_r. Since E_r was found[38] to be proportional to T_c, E_{dip}-E_{pk} should also be proportional to T_c. This is in sharp contrast to the experimental results shown in Fig. 4. The E_{dip}-E_{pk} values obtained from ARPES (Fig. 4a) and tunneling spectra (Fig. 4b) are nearly independent of doping or T_c, in contradiction with the *d*-wave magnetic pairing mechanisms based on the spin-fermion model[47] and on the conventional approach.[48]

More quantitative approach to the *t-t-J* model is the slave-boson mean-field theory developed by Brinckmann and Lee.[50] The pairing interaction is the AF exchange energy J. This mean-field theory also predicts the PDH structure in ARPES spectra along the antinodal direction.[51] The predicted E_{dip}-E_{pk} is nearly independent of doping and in the range of 0.32-0.46J depending on the energy resolution.[51] With J = 130 meV, one yields E_{dip}-E_{pk} = 41.6-59.8 mV, which is in reasonable agreement with the ARPES data shown in Fig. 4a. On the other hand, the t-t-J model predicts that E_r = 0.52-0.54J = 67.6-70.2 meV in the overdoped range.[52] This prediction is in contradiction with neutron data,[53] which show that E_r is equal to 5.4$k_B T_c$ in the overdoped range. For overdoped $Y_{0.85}Ca_{0.15}Ba_2Cu_3O_7$ with T_c = 75 K, E_r was found to 34 meV (Ref. [54]), which is only half the predicted value of 70 meV. Therefore, the neutron data do not support the magnetic pairing mechanism based on the t-t-J model. In order to reconcile the slave-boson mean-field theory with the neutron data, one must assume that the parameter J is adjustable.

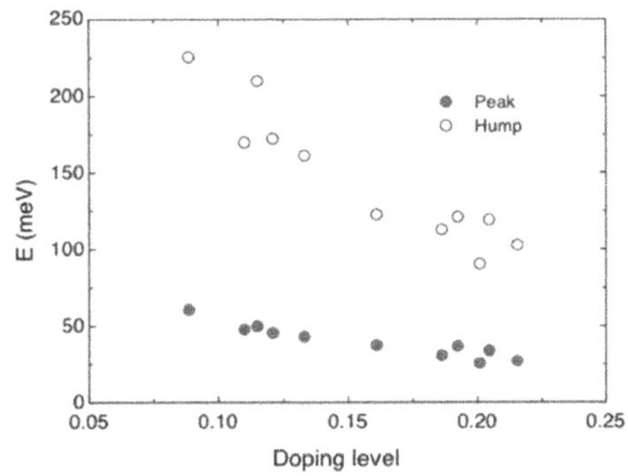

Figure 5. Doping dependence of the energy position of the peak and hump features of the angle-resolved photoemission spectra along the antinodal direction. The data were reproduced from Ref.[55]

The second important prediction of the t-t-J model is the existence of the pronounced hump feature at energy close to J (about 130 meV) at all the doping levels.[51] This is also in contradiction with the ARPES data shown in Fig. 5. The energy position of the hump feature increases rapidly with the decrease of doping and reaches a value of about 230 meV at a doping level of 0.089. The hump energy of 230 meV is too high to be explained by the t-t-J model. Since the t-t-J model is characteristic of the CuO_2 and universal to all cuprate superconductors, the predicted PDH structure from the t-t-J should be seen in all the cuprate superconductors. The absence of the PDH structure in the antinodal EDCs of an overdoped single-layer $Tl_2Ba_2CuO_6$ (Ref.[56]) further argues against magnetic origin of the structure. This fact actually provides strong evidence for weak coupling to the magnetic resonance mode. Since the PDH structures are observed in double- and triple-layer cuprates, the PDH structures have been explained in terms of strong coupling to the B_{1g} Cu-O buckling vibration mode in double- and triple-layer cuprates and weak coupling in single-layer cuprates due to the difference in the environment surrounding the CuO_2 planes.[56] Within this picture, E_{dip}-E_{pk} is predicted to be the energy of the B_{1g} mode, which is 36 meV in $Bi_2Sr_2CaCu_2O_{8+y}$. The result shown in Fig. 4a appears to be in quantitative agreement with this prediction. However, the result in Fig. 4b contradicts the prediction. The mean value of E_{dip}-E_{pk} is about 26 meV, 10 meV lower than the energy of the B_{1g} mode.

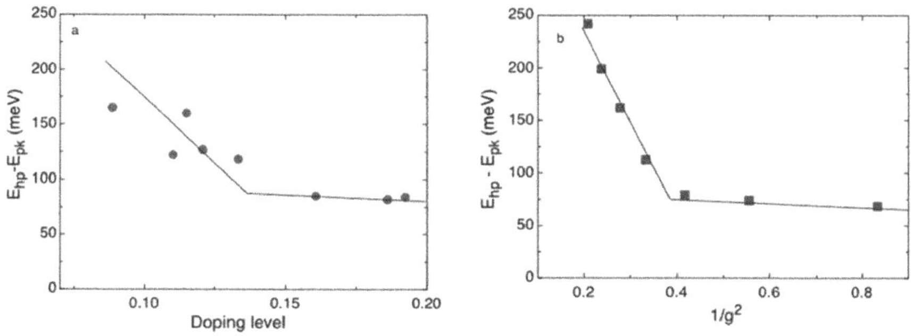

Figure 6. a) The peak-hump energy separations in the ARPES of BSCCO, which are evaluated from the data in Fig. 5. b) Dependence of the peak-hump energy separation on $1/g^2$ in SIN tunneling spectra (similar to ARPES), which is predicted from a lattice polaronic model where 72-meV phonon modes are involved in polaron formation.

Alternatively, the peak-dip-hump features consistently observed in scanning tunneling, break-junction tunneling, and angle-resolved photoemission spectra of $Bi_2Sr_2CaCu_2O_{8+y}$ can be well explained by the polaron-bipolaron theory of superconductivity, where the hump is the incoherent broad feature that reflects the local hopping of electrons in various frozen lattice configurations.[57] The weak PDH structure in the overdoped single-layer $Tl_2Ba_2CuO_6$ could be explained as due to a much weaker polaronic effect and a much broader coherent peak. The much lower T_c in the bi-layer $Bi_2Sr_2CaCu_2O_{8+y}$ than in the bi-layer $Tl_2Ba_2CaCu_2O_8$ is consistent with the fact that the polaronic effect in the former is much stronger than in the latter. A too strong polaronic effect leads to too heavy effective masses of the supercarriers and to a lower T_c.

The experimental peak-hump energy separations in the ARPES of the BSCCO are shown in Figure 6a. The experimental result can be naturally explained by the polaron-bipolaron theory of superconductivity. The theory predicts[57] that the peak-hump energy separation E_{hp}-E_{pk} in superconductor-insulator-superconductor (SIS) tunneling spectra is close to the energy Ω of the phonon modes in polaronic cloud when the electron-phonon coupling strength g^2 is not too large. It is also true for the tunneling spectrum of a superconductor-insulator-normal (SIN) junction, as demonstrated in APPENDIX II below. Figure 6b shows the theoretical prediction of E_{hp}-E_{pk} with varying g^2 and $\Omega = 72$ meV. The numerically calculated SIN tunneling spectra are shown in Figure A3 below. It is apparent that E_{hp}-E_{pk} is close to Ω when g^2 is below 2.5 and increases to $>2\Omega$ when $g^2 >3.6$. The theoretical prediction is in quantitative agreement with the data shown in Figure 6a. The 72-meV phonon mode should be mainly associated with oxygen vibration so that the whole broad hump feature should shift down by about $72[1-(18/16)^{1/2}]$ meV =

4.3 meV upon replacing ^{16}O by ^{18}O. Indeed, the energy of the peak feature at 52 meV in d^2I/dV^2 spectra was found to shift shown by 3.7±0.8 meV upon replacing ^{16}O by ^{18}O (Ref.[49]). Since the peak position in d^2I/dV^2 simply corresponds to the steepest point of the hump feature in dI/dV, the oxygen isotope shift of the steepest point at 52 meV implies that the hump feature is also shifted down by about 3.7±0.8 meV upon replacing ^{16}O by ^{18}O. Therefore, the observed oxygen-isotope shift of the steepest point of the hump feature in dI/dV provides essential proof that peak-dip-hump features observed in tunneling spectra and ARPES arise from the lattice polaronic effect.

CONCLUSIONS

In summary, our analyses of tunneling and ARPES spectra in both electron- and hole-doped cuprates unambiguously rule out the magnetic pairing mechanism and support polaronic/bipolaronic superconductivity in hole-doped $Bi_2Sr_2CaCu_2O_{8+y}$. An intermediate polaronic effect can significantly increase T_c through the enhancement in the density of states. A too strong polaronic effect tends to suppress T_c due to a too large enhancement in the effective mass of supercarriers. For detailed discussion about the polaron-bipolaron theory of superconductivity and the intrinsic pairing symmetry in cuprate superconductors, see APPENDIX below.

APPENDIX

APPENDIX I—Polaron-bipolaron theory of superconductivity

Despite the fact that many experiments have supported polaronic/bipolaronic superconductivity, the theory has not been generally accepted. There are several reasons for this. First, much attention has been paid to bipolaronic theories based on strong short-range Holstein-type EPI. These theories fail to predict a high T_c because the bipolarons are too heavy. Secondly, there is strong experimental evidence for the Cooper pairing rather than the real-space pairing in overdoped cuprates.[10] Thirdly, ARPES, optical, and transport experiments have consistently pointed towards a large Fermi surface in overdoped cuprates.[10] These results appear to argue against the bipolaronic theory of superconductivity. However, an analytical multi-polaron model of HTS in highly polarizable ionic lattices has recently been proposed with generic (bare) Coulomb and Fröhlich EPI avoiding any ad-hoc assumptions on their range and relative magnitude.[58] The generic Hamiltonian comprising any-range Coulomb repulsion and the Fröhlich EPI can be reduced to a short-range t-J_p model at a very large static dielectric constant for the moderate and strong EPI. In this limit the bare static Coulomb repulsion and EPI negate each other giving rise to a novel physics described by the polaronic t-J_p model with a short-range polaronic spin exchange J_p of phononic origin.[58] Moreover, if one considers realistic finite dielectric constant, the cancelation of the bare Coulomb repulsion by the Fröhlich EPI is not complete so that a residual on-site repulsion $\hat{U}$ of polarons could be substantial if the size of the Wannier orbitals is small enough.[59] The residual Hubbard $\hat{U}$ drives the system to undergo a crossover from Bose-Einstein condensation (BEC) to the Cooper pairing.[59] This naturally reconciles the polaron-bipolaron theory of superconductivity with the Cooper pairing and the large Fermi surface observed in overdoped cuprate superconductors.[10] The strong coupling of polarons with low-energy (<40 meV) multiple phonon modes is the key to the understanding of high-T_c

mechanism in optimally doped cuprates.[11] Interestingly, similar polaronic mechanism has been applied to a nonmagnetic high-T_c (30 K) superconductor $Ba_{1-x}K_xBiO_3$ (Ref.[60]).

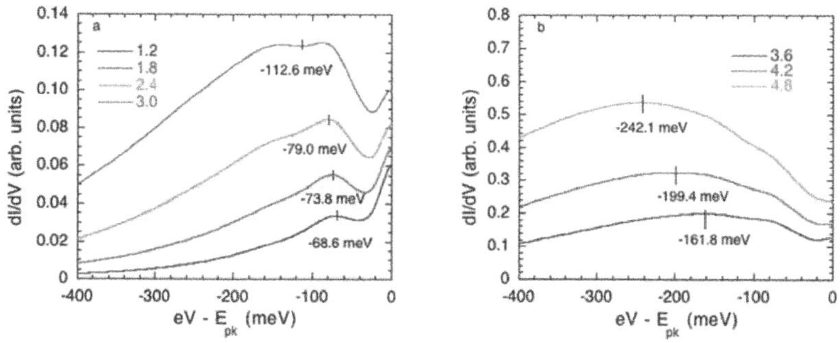

Figure A1. The calculated tunneling conductance dI/dV for a superconductor-insulator-normal (SIN) junction based on the polaron-bipolaron theory of superconductivity. The numbers in the figure are the g^2 values.

Furthermore, recent analytical[61] and numerical (Monte-Carlo)[62,63] studies also cast serious doubts on the possibility of HTS from repulsive interactions only. Optical experiments[22] showed that the electron-boson spectral function was independent of magnetic field, in contradiction with its strong field dependence expected from the magnetic pairing mechanism.

APPENDIX II— Tunneling spectra calculated based on the polaron-bipolaron theory

The tunneling conductance dI/dV of an SIS junction was derived based on the polaron-bipolaron theory of superconductivity.[57] By replacing the parameters of a superconductor by those of a normal-metal, we can readily deduce dI/dV for an SIN junction as

$$dI/dV \propto \sum_{n=0}^{\infty} \frac{g^{2n}(\Gamma + n\delta\Omega)}{n![(\Gamma + n\delta\Omega)^2 + (e|V| - \Delta - n\Omega)^2]}, \quad (1)$$

where Ω is the characteristic energy of phonon modes in the polaron cloud, $\delta\Omega$ is the phonon dispersion, g^2 is proportional to the electron-phonon coupling strength, Γ is the life-time broadening parameter, $\Delta = [(\Delta_p)^2 + (\Delta_c)^2]^{1/2}$, Δ_p is the single-particle gap (normal-state gap), and Δ_c is the superconducting gap that closes at T_c.

Figure A1 shows the numerically calculated SIN tunneling spectra with a varying g^2 (labeled in the figure) and fixed parameters of $\Omega = 72$ meV, $\delta\Omega = 30$ meV, $\Delta = 35$ meV, and $\Gamma = 20$ meV. The energy of the coherence peak position (E_{pk}) is slightly higher than Δ due to a finite Γ value. As g^2 increases, the spectral weight of the coherence peak is reduced while the weight of the incoherent broad hump increases. We have used a large Γ value (20 meV) to simulate all the ARPES spectra.[55] The hump feature is dominated at $eV \approx E_{pk}-\Omega$ for $g^2 < 3.0$, while it is dominated at $eV < E_{pk}-2\Omega$ for $g^2 > 3.6$.

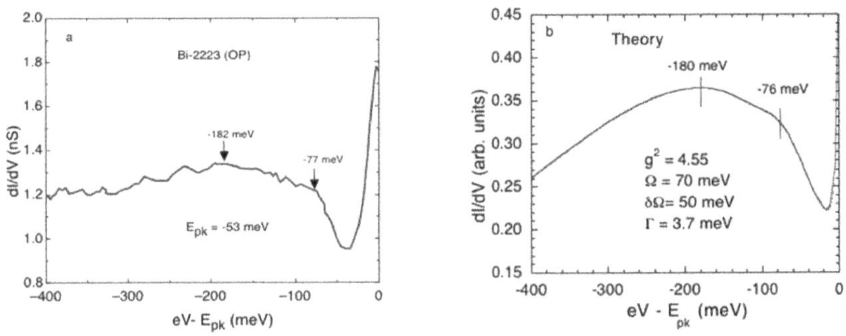

Figure A2. a) Scanning tunneling spectrum of optimally doped Bi-2223. b) The calculated scanning tunneling spectrum based on polaronic superconductivity. The parameters for the calculation are listed in the figure. The measured SIN spectrum was reproduced from Ref. [66]

In Figure A2a, we reproduce scanning tunneling spectrum in the negative bias for optimally doped $Bi_2Sr_2Ca_2Cu_3O_{10+y}$ (Bi-2223).[66] It is apparent that the spectrum shows double hump features, as indicated by the arrows. In Figure A2b, we present the calculated SIN tunneling spectrum from Eq. 1 that is deduced based on the polaronic superconductivity. It is remarkable that the energy positions of the double hump features in the measured tunneling spectrum are in quantitative agreement with the theoretical predictions. We do not believe that any other theoretical model can quantitatively and consistently explain the double hump features observed in the tunneling spectrum.

APPENDIX III—Intrinsic pairing gap symmetry in the bulk

Another important issue is the intrinsic pairing gap symmetry in the bulk of superconducting cuprates, which is directly related to the pairing interaction. The symmetry of the superconducting condensate (gap), which can be directly probed by phase-sensitive experiments such as Josephson-junction experiments, is not necessarily the same as the pairing gap symmetry. Within the Bardeen-Cooper-Schrieffer (BCS) picture, both symmetries happen to be the same. But in the strong-coupling limit, real-space pairing becomes possible and the symmetry of the superconducting condensate can be completely different from the pairing gap symmetry.[67,68] Without considering the above possibility, one should not have taken the *d*-wave symmetry of the superconducting condensate (probed by the in-plane phase-sensitive experiments for both electron- and hole-doped cuprates[69]) to be as indisputable evidence for the magnetic pairing mechanism. In contrast, from quantitative analyses of many bulk-sensitive experiments in optimally doped and overdoped cuprates, Zhao has concluded that the intrinsic pairing gap symmetry in the bulk of cuprates is not *d*-wave, but extended *s*-wave (having eight line nodes) in hole-doped cuprates[33,41] and nodeless *s*-wave in electron-doped cuprates.[41,43] In particular, a bulk and phase-sensitive experiment in an overdoped BSCCO along *c*-axis supports *s*-wave gap symmetry.[70] In fact, this phase-sensitive experiment is in quantitative agreement with an extended *s*-wave gap.[41] Bulk-sensitive Raman spectra[71] in the B_{1g} and B_{2g} symmetries for a heavily overdoped BSCCO with T_c = 55 K provides indisputable evidence for an extended *s*-wave gap: $\Delta = 15\cos 4\theta$ meV (where θ is the angle measured from the Cu-O bonding direction). Earlier angle-resolved photoemission spectra for a heavily overdoped BSCCO with T_c = 60 K also suggested a large gap Δ_D (about 10 meV) along the diagonal direction.[72] On the other hand, ARPES data for a slightly overdoped BSCCO (Ref.[73]) are consistent with either *d*-wave or an extended *s*-wave gap with $\Delta_D = 4.0\pm2.5$ meV. More recent high-resolution ARPES data for slightly overdoped BSCCO and Bi-2223 confirm finite gap sizes along the diagonal direction (see Figure A3). The symmetrized EDCs shown in Figure A3 clearly show dip features at zero energy, which proves finite gaps along the diagonal direction in both compounds. Quantitative analyses[33] of the EDCs show that Δ_D is 6.0 meV for the slightly overdoped BSCCO, 14.0 meV for the heavily overdoped BSCCO with T_c = 60 K, and 7.0 meV for the slightly overdoped Bi-2223. Therefore, the pairing gap symmetry in overdoped bi- and tri-layer cuprates is extended *s*-wave with eight line nodes. Since the gap sizes around $\theta = 45°$ can be readily suppressed to zero by disorder and/or contamination on the cleaved surface, the gap function probed by the surface-sensitive experiments may vary from run to run and is different from that of the bulk.

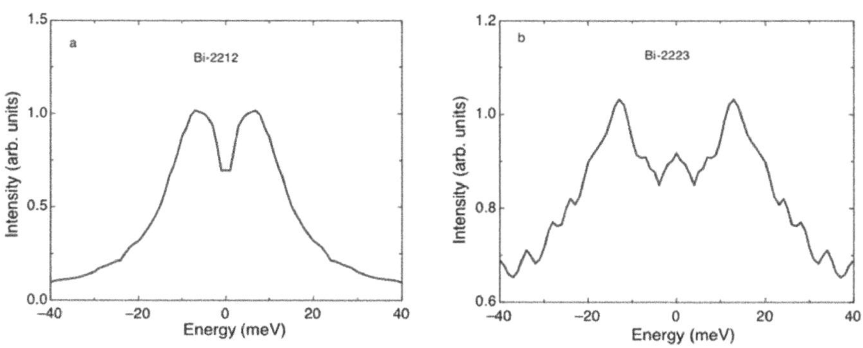

Figure A3. The symmetrized energy distribution curves (EDCs) along the diagonal direction for slightly overdoped Bi-2212 and Bi-2223. The data are taken from Refs. [76,77] The dip feature at zero energy proves a finite gap.

Interestingly, EDCs for underdoped bi-layer BSCCO show normal-state pseudogaps in the anti-nodal region and zero gap in an extended angle-range centered at $\theta = 45°$ (Ref.[74]). In the superconducting state, zero gap is seen at $\theta = 45°$, consistent with a *d*-wave superconducting gap symmetry.[75] It is interesting that the pseudogap state even exists in a slightly overdoped single-layer $(Bi,Pb)_2(Sr,La)_2CuO_{6+y}$ (Bi-2201) with T_c =

29 K (Ref.[78]). These ARPES data can be well explained in terms of a local-pair superconductivity model, where inter-site pairs (or inter-site bipolarons) are condensed into superfuid through the BEC. Within this model, the superconducting gap Δ_c has a *d*-wave symmetry, which is controlled by the anomalous kinetic process rather than by the pairing interaction[68]. Because Δ_c in the anti-nodal region is much smaller than the pseudogap Δ_p,[68] the total gap Δ is slightly larger than the normal-state pseudogap in this angle region, in agreement with the data.[74,78] In contrast, Δ_p is zero in an extended angle-range centered at $\theta = 45^o$ (possibly due to the pair-breaking effect in the presence of impurities and disorder), so the total gap Δ is simply equal to Δ_c in this angle region. The fact that $\Delta \geq \Delta_p$ in the antinodal region of underdoped cuprates[74,78] provides strong evidence that the normal-state gap is associated with the superconducting pairing rather than with charge-density wave.

The local-pair superconductivity model for underdoped cuprates can also naturally explain the *d*-wave superconducting gap symmetry probed by surface- and phase-sensitive experiments based on planar Josephson tunneling. Because the surface- and phase-sensitive experiments are probing the superconducting state at surfaces and interfaces which were found to be intrinsically underdoped,[79,80] they naturally see the *d*-wave superconducting gap symmetry in underdoped cuprates. Since the superconducting gap symmetry in underdoped cuprates is controlled by the anomalous kinetic process rather than by the pairing interaction,[68] these phase-sensitive experiments do not determine the pairing symmetry associated with the pairing interaction. Only for overdoped cuprates (BCS-like superconductors), the superconducting gap symmetry is associated with the pairing symmetry, so the *s*-wave pairing gap symmetry identified for overdoped cuprates[33,41,43] places an essential constraint on the pairing interaction and rules out *d*-wave magnetic pairing mechanism.

APPENDIX IV—Pairing interactions

Now we address a basic question concerning the key pairing interactions in cuprates. Some density functional (DFT) calculations[81,82] found small EPI, which is too small to explain high T_c's while some other first-principles studies found large EPI in cuprates.[45] It is common that DFT underestimates the role of the Coulomb interactions. The inclusion of a short-range repulsion (Hubbard U) via the LDA+U algorithm[46] and/or nonadiabatic effects[45] significantly enhances the EPI strength due to a poor screening of some particular phonons. Furthermore, due to layered structures, electron-phonon coupling to *c*-axis phonons becomes significant due to the poorly screened long-range Madelung potential. The detailed discussion on the strong electron-phonon coupling to *c*-axis phonons can be found in a review article.[83] What is more important is the polaronic effect, which significantly enhances the density of states at the Fermi level and the effective electron-phonon coupling constant for low-energy phonon modes. That is the key to the understanding of HTS. It is interesting to note that the experimental observations of low energy excitations in pairing appear to be well explained in terms of strong nonlinear coupling to anharmonic phonons with *d*-wave pairing symmetry.[84,85] It is unclear whether this pairing in an *s*-wave channel can lead to a higher T_c.

REFERENCES

[1] J. G. Bednörz and K. A. Müller, Z. Phys. B: Condens.Matter 64, 189 (1986).
[2] For a recent review see N. Plakida, High temperature cuprate superconductors: experiment, theory and applications (Springer, Heidelberg 2010).
[3] For a recent review see A. S. Alexandrov and J. T. Devreese, Advances in Polaron Physics (Springer, Berlin 2009).
[4] G. M. Zhao, K. K. Singh, and D. E. Morris, Phys. Rev. B 50, 4112 (1994).
[5] G. M. Zhao and D. E. Morris, Phys. Rev. B 51, 16487(R)(1995).
[6] G. M. Zhao, K. K. Singh, A. P. B. Sinha, and D. E. Morris, Phys. Rev. B 52, 6840 (1995).
[7] G. M. Zhao, M. B. Hunt, H. Keller, and K. A. Müller, Nature 385, 236 (1997).
[8] G. M. Zhao, K. Conder, H. Keller, and K. A. Müller, J. Phys.: Condens. Matter 10, 9055 (1998).
[9] G. M. Zhao, H. Keller, and K. Conder, J. Phys.: Condens. Matter 13, R569 (2001).
[10] G. M. Zhao, Phil. Mag. B 81, 1335 (2001).
[11] G. M. Zhao, V. Kirtikar, and D. E. Morris, Phys. Rev. B 63, 220506(R) (2001).
[12] R. Khasanov, D. G. Eshchenko, H. Luetkens, E. Morenzoni, T. Prokscha, A. Suter, N. Garifianov, M. Mali, J. Roos, K. Conder, and H. Keller, Phys. Rev. Lett. 92, 057602 (2004).
[13] A. S. Alexandrov and G. M. Zhao, New J. Phys. 14, 013046 (2012).
[14] R. J. McQueeney, Y. Petrov, T. Egami, M. Yethiraj, G. Shirane, and Y. Endoh, Phys. Rev. Lett. 82, 628 (1999).
[15] T. R. Sendyka, W. Dmowski, T. Egami, N. Seiji, H. Yamauchi, and S. Tanaka, Phys. Rev. B 51, 6747 (1995).
[16] D. Reznik, L. Pintschovius, M. Ito, S. Iikubo, M. Sato, H. Goka, M. Fujita, K. Yamada, G. D. Gu, and J. M. Tranquada, Nature 440, 1170 (2006).
[17] A. Lanzara et al., Nature 412, 510 (2001).
[18] O. Rösch, O. Gunnarsson, X. J. Zhou, T. Yoshida, T. Sasagawa, A. Fujimori, Z. Hussain, Z.-X. Shen, and S. Uchida, Phys. Rev. Lett. 95 227002 (2005).
[19] C. Gadermaier, A. S. Alexandrov, V. V. Kabanov, P. Kusar, T. Mertelj, X. Yao, C. Manzoni, D. Brida, G. Cerullo, and D. Mihailovic, Phys. Rev. Lett. 105, 257001 (2010).
[20] D. Mihailovic, C. M. Foster, K. Voss, and A. J. Heeger, Phys. Rev. B 42, 7989 (1990).
[21] O. V. Misochko, E. Ya. Sherman, N. Umesaki, K. Sakai, S. Nakashima, Phys. Rev. B 59, 11495 (1999).
[22] Y. S. Lee, K. Segawa, Z. Q. Li, W. J. Padilla, M. Dumm, S. V. Dordevic, C. C. Homes, Yoichi Ando, and D. N. Basov, Phys. Rev. B 72, 054529 (2005).
[23] A. S. Mishchenko, N. Nagaosa, Z.-X. Shen, G. De Filippis, V. Cataudella, T. P. Devereaux, C. Bernhard, K.W. Kim, and J. Zaanen, Phys. Rev. Lett. 100, 166401 (2008).
[24] X. J. Zhou et al., Phys. Rev. Lett. 95, 117001 (2005).
[25] S. I. Vedeneev, P. Samuely, S. V. Meshkov, G. M. Eliashberg, A. G. M. Jansen, and P. Wyder, Physica C 198, 47 (1992).
[26] D. Shimada, Y. Shiina, A. Mottate, Y. Ohyagi, and N. Tsuda, Phys. Rev. B 51, 16495 (1995).

[27]R. S. Gonnelli, G. A. Ummarino, and V. A. Stepanov, Physica C 275, 162 (1997).
[28]G. M. Zhao, Phys. Rev. B 75, 214507 (2007).
[29]H. Shim, P. Chaudhari, G. Logvenov and I. Bozovic, Phys. Rev. Lett. 101, 247004 (2008).
[30]F. C. Niestemski, S. Kunwar, S. Zhou, S. L. Li, H. Ding, Z. Q. Wang, P. C. Dai, and V. Madhavan, Nature 450, 1058 (2007).
[31]T. Dahm, V. Hinkov, S. V. Borisenko, A. A. Kordyuk, V. B. Zabolotnyy, J. Fink, B. Buchner, D. J. Scalapino,W. Hanke, and B. Keimer, Nature Phys. 5, 780 (2009).
[32]J. Zhao, F. C. Niestemski, S. Kunwar, S.-L. Li, P. Steffens, A. Hiess, H. J. Kang, S. D. Wilson, Z.-Q. Wang, P. C. Dai, and V. Madhavan, Nature Phys. 7, 719 (2011).
[33]G. M. Zhao, Phys. Scr. 83, 038304 (2011).
[34]G. M. Zhao, Phys. Rev. Lett. 103, 236403 (2009).
[35]W. L. McMillan and J. M. Rowell, Phys. Rev. Lett. 14, 108 (1965).
[36]J. P. Carbotte, Rev. Mod. Phys. 62, 1027 (1990); W. L. McMillan and J. M. Rowell, in Superconductivity, edited by R. D. Parks, (Marcel Dekker, New York, 1969), Vol. 1, p. 561; E. L. Wolf, Principles of electron tunneling spec-troscopy, (Oxford University Press, New York, 1985).
[37]M. K. Crawford, G. Burns, G. V. Chandrashekhar, F. H. Dacol, W. E. Farneth, E. M. McCarron, III and R. J. Smalley, Solid State Commun. 73, 507 (1990).
[38]J. Zhao, P. C. Dai, S. L. Li, P. G. Freeman, Y. Onose, and Y. Tokura, Phys. Rev. Lett. 99, 017001 (2007).
[39]E. W. Hudson, S. H. Pan, A. K. Gupta, K.-W. Ng, and J. C. Davis, Science 285, 88 (1999).
[40]B. Renker, F. Gompf, D. Ewert, P. Adelmann, H. Schmidt, E. Gering, and H. Mutka, Z. Phys. B: Condens. Matter 77, 65 (1989).
[41]G. M. Zhao, Phys. Rev. B 64, 024503 (2001).
[42]G. M. Zhao, Phys. Rev. B 82, 012506 (2010).
[43]G. M. Zhao and J. Wang, J. Phys.: Condens. Matter 22, 352202 (2010).
[44]W. Meevasana, T. P. Devereaux, N. Nagaosa, Z.-X. Shen, and J. Zaanen, Phys. Rev. B 74, 174524 (2006).
[45]T. Bauer and C. Falter, Phys. Rev. B 80, 094525 (2009).
[46]P. Zhang, S. G. Louie, and M. L. Cohen, Phys. Rev. Lett. 98, 067005 (2007).
[47]A. Abanov and A. V. Chubukov, Phys. Rev. Lett. 83, 1652 (1999).
[48]J. F. Zasadzinski, L. Coffey, P. Romano, and Z. Yusof, Phys. Rev. B 68, 180504(R) (2003).
[49]J.-H. Lee et al., Nature 442, 546 (2006).
[50]J. Brinckmann and P.A. Lee, Phys. Rev. Lett. 82, 2915 (1999).
[51]J. X. Li, C. Y. Mou, and T. K. Lee, Phys. Rev. B 62, 640 (2000).
[52]J. X. Li, and C. D. Gong, Phys. Rev. B 66, 014506 (2002).
[53]H. He et al., Phys. Rev. Lett. 86, 1610 (2001).
[54]S. Pailhes, C. Ulrich, B. Fauque, V. Hinkov, Y. Sidis, A. Ivanov, C. T. Lin, B. Keimer, and P. Bourges, Phys. Rev. Lett. 96, 257001 (2006).
[55]J. C. Campuzano et al., Phys. Rev. Lett. 83, 3709 (1999).
[56]W. S. Lee, K. Tanaka, I. M. Vishik, D. H. Lu, R. G. Moore, H. Eisaki, A. Iyo, T. P. Devereaux, and Z. X. Shen, Phys. Rev. Lett. 103, 067003 (2009).
[57]A. S. Alexandrov and C. Sricheewin, Europhys. Lett. 58, 576 (2002).

[58]A. S. Alexandrov, EPL 95, 27004 (2011).
[59]A. S. Alexandrov, J. H. Samson, and G. Sica, condmat/arXiv:1205.3436.
[60]G. M. Zhao, Phys. Rev. B 76, 020501R (2007).
[61]A. S. Alexandrov and V. V. Kabanov, Phys. Rev. Lett. 106, 136403 (2011).
[62]T. Aimi and M. J. Imada, Phys. Soc. Jpn. 76, 113708 (2007).
[63]T. M. Hardy, J. P. Hague, J. H. Samson, and A. S. Alexandrov, Phys. Rev. B 79, 212501 (2009).
[64]Y. DeWilde, N. Miyakawa,P. Guptasarma, M. Iavarone,1 L. Ozyuzer, J. F. Zasadzinski, P. Romano, D. G. Hinks, C. Kendziora, G. W. Crabtree, and K. E. Gray, Phys. Rev. Lett. 80, 153 (1998)
[65]H. Ding, J. R. Engelbrecht, Z. Wang, J. C. Campuzano, S.-C. Wang, H.-B. Yang, R. Rogan, T. Takahashi, K. Kadowaki, and D. G. Hinks, Phys. Rev. Lett. 87, 27001 (2001).
[66]M. Kugler, G. Levy de Castro, E. Giannini, A. Piriou, A. A. Manuel, C. Hess, and . Fischer, J. of Phys. Chem. Solids 67, 353 (2006).
[67]A.S. Alexandrov, Physica C 305, 46 (1998).
[68]Yucel Yildirim and Wei Ku, Phys. Rev. X 1, 011011 (2011).
[69]C. C. Tsuei and J. R. Kirtley, Rev. Mod. Phys. 72, 969 (2000).
[70]Q. Li, Y. N. Tsay, M. Suenaga, R. A. Klemm, G. D. Gu, and N. Koshizuka, Phys. Rev. Lett. 83, 4160 (1999).
[71]K. C. Hewitt, T. P. Devereaux, X. K. Chen, X.-Z. Wang, J. G. Naeini, A. E. Curzon, J. C. Irwin, and A. Martin, Phys. Rev. Lett. 78, 4891 (1997).
[72]I. Vobornik, R. Gatt, T. Schmauder, B. Frazer, R. J. Kelley, C. Kendziora, M. Grioni, M. Onellion, and G. Margaritondo, Physica C 317-318, 589 (1999).
[73]H. Ding, J. C. Campuzano, A. F. Bellman, T. Yokoya, M. R. Norman, M. Randeria, T. Takahashi, H. Katayama-Yoshida, T. Mochiku, K. Kadowaki, and G. Jennings, Phys. Rev. Lett. 74, 2784 (1995).
[74]W. S. Lee, I. M. Vishik, K. Tanaka, D. H. Lu, T. Sasagawa, N. Nagaosa, T. P. Devereaux, Z. Hussain, and Z.-X. Shen, Nature 450, 81 (2007).
[75]I. M. Vishik, W. S. Lee, F. Schmitt, B. Moritz, T. Sasagawa, S. Uchida, K. Fujita, S. Ishida, C. Zhang, T. P. Devereaux, and Z. X. Shen, Phys. Rev. Lett. 104, 207002 (2010).
[76]J. D. Koralek, J. F. Douglas, N. C. Plumb, Z. Sun, A.V. Fedorov, M. M. Murnane, H. C. Kapteyn, S. T. Cundiff, Y. Aiura, K. Oka, H. Eisaki, and D. S. Dessau, Phys.Rev. Lett. 96, 017005 (2006).
[77]D. L. Feng, A. Damascelli, K. M. Shen, N. Motoyama, D. H. Lu, H. Eisaki, K. Shimizu, J.-I. Shimoyama, K. Kishio, N. Kaneko, M. Greven, G. D. Gu, X. J. Zhou, C. Kim, F. Ronning, N. P. Armitage, and Z.-X Shen, Phys. Rev. Lett. 88, 107001 (2002).
[78]T. Kondo, R. Khasanov, T. Takeuchi, J. Schmalian, and A. Kaminski, Nature 457, 296 (2009).
[79]J. Betouras and R. Joynt, Physica C 250, 256 (1995).
[80]J. Mannhart and H. Hilgenkamp, Physica C 317-318, 383 (1999).
[81]F. Giustino, F. Cohen, and S. G. Louie, Nature 452, 975 (2008).
[82]R. Heid, K. P. Bohnen, R. Zeyher, and D. Manske, Phys. Rev. Lett. 100, 137001 (2008).
[83]S. Johnston, W. S. Lee, Y. Chen, E. A. Nowadnick, B. Moritz, Z.-X. Shen, and T. P.

Devereaux, Advances in Condensed Matter Physics 2010, Article ID 968304 (2010).
[84]D.M. Newns and C. C. Tsuei, Nature Physics **3**, 184 (2007).
[85]J. H. Hsiao, G. J. Martyna, and D. M. Newns, Phys. Rev. Lett. **114, 107001**(2015).

Author Index

Lightning Source UK Ltd.
Milton Keynes UK
UKOW06n1519260816

281590UK00001B/59/P

9 781119 234487